$9.65

ELEMENTARY
ALGEBRA

ELEMENTARY ALGEBRA

Roy Dubisch
University of Washington

Vernon R. Hood
Portland Community College

A Staples Press Book

W. A. Benjamin, Inc.

Menlo Park, California ● Reading, Massachusetts ●
London ● Amsterdam ● Don Mills, Ontario ● Sydney

Copyright © 1977 by W. A. Benjamin, Inc.
Philippines copyright 1977 by W. A. Benjamin, Inc.

Library of Congress Catalog Number 76-3846

ISBN-0-8053-2338-4
ABCDEFGHIJKL-HA-7987

W. A. Benjamin, Inc.
2727 Sand Hill Road
Menlo Park, California 94025

CONTENTS

PREFACE

Because mathematics at any level is best learned by doing, this text places its main emphasis on problem solving — both through text examples and through numerous exercises. In particular, word problems are introduced early and emphasized throughout.

But learning by doing without understanding the why of the doing is learning that is easily forgotten. On the other hand, once the *basic* principles of algebra are understood and combined with practice, the manipulative skills of algebra will be more easily retained. Thus, this text utilizes the basic principles of algebra to provide a firm foundation for the development and retention of manipulative skills.

Much of your ability to understand the basic principles of algebra depends on a knowledge of the basic principles of arithmetic. Hence, the early part of the text reviews these basic principles, and chapter and cumulative reviews provide an opportunity for you to keep in mind what you have learned earlier. Answers are given for all odd-numbered exercises.

A knowledge of algebra is useful in almost any trade or profession. We hope that this text will help you gain this knowledge. Keep in mind, however, that only through active involvement can you gain proficiency in mathematics. No one can do it for you!

Roy Dubisch
Vernon R. Hood

1

WHAT IS ALGEBRA?

Have you ever listened to crickets on a summer evening? Did you know that a change in temperature causes a change in their chirping? In fact, you can determine the temperature if you know the number of chirps a cricket makes in one minute. To do this you divide the number of chirps per minute by 4 and add 40. The resulting number is equal to the temperature in degrees Fahrenheit. Thus, if on a summer evening you counted 144 chirps in one minute, the temperature could be found in the following manner: $144 \div 4 = 36$ and $36 + 40 = 76$. Hence the temperature would be $76°$ Fahrenheit.

Mathematicians would say that the chirping of the crickets is a *function* of the temperature and can be expressed by the formula

$$t = \frac{n}{4} + 40$$

where n represents the number of chirps the crickets make in one minute and t represents the temperature in degrees Fahrenheit.

There are many interesting relationships in the world about us that can be expressed algebraically. Algebra provides a compact way of describing complicated situations and a method for analyzing difficult problems.

This chapter is meant to give you an overview of the subject of algebra. Many of the ideas mentioned will be unfamiliar to you and will be discussed in detail in later chapters. That is, you are not expected to be familiar with all of these ideas at this time — the important thing is to get some feeling of what algebra is all about.

1.1 Algebra Is a Language

Perhaps the first thing that most people think of when they hear the word "algebra" is that letters are involved (especially the letter x!). They think of formulas such as

$$F = \frac{9}{5}C + 32$$

and of equations such as

$$2w + 2(4w + 3) = 106$$

Algebra does indeed use symbols — including letters — but not, as many persons seem to think, as a device for making simple matters complicated! Once you understand the language of algebra, you will see that it imparts information in a compact way that is easier to understand than ordinary English.

Thus the formula $F = \frac{9}{5}C + 32$, which relates temperature in degrees Fahrenheit (F) to temperature in degrees Celsius (C), is a compact way of saying, "To find the temperature in degrees Fahrenheit given the temperature in degrees Celsius, multiply the temperature in degrees Celsius by $\frac{9}{5}$ and then add 32."

Similarly, the equation $2w + 2(4w + 3) = 106$ can be regarded as a compact way of stating that we are looking for the width of a rectangle whose perimeter is 106 feet and whose length is 3 feet more than four times the width.

It takes time and effort to learn the language of algebra — as it takes time and effort to learn any language. But your time and effort will be amply rewarded by the ability to use this language in many situations. Algebra can be of use not only in almost every trade and profession, but also in everyday life where, for example, formulas involving interest, car performance, and so on frequently need to be applied.

1.2 Algebra Is a Tool

Consider again the formula $F = \frac{9}{5}C + 32$. This formula enables us to find F very quickly if we are given C. But what if we are given F and wish to find C? Clearly, it would be useful to have a formula $C = \ldots$. From the physical meaning of the two thermometer scales (freezing point of water is 0°C and 32°F; boiling point of water is 100°C and 212°F), it is possible to reason out such a formula directly and conclude that

$$C = \frac{5}{9}(F - 32)$$

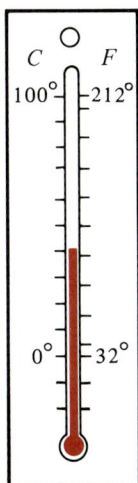

It is one of the nice features of algebra, however, that once its basic principles are understood we can, so to speak, just "turn the crank" to get results. As you will see, learning how to use the "algebra machine" will enable you to go from $F = \frac{9}{5}C + 32$ to $C = \frac{5}{9}(F - 32)$ without any difficulty.

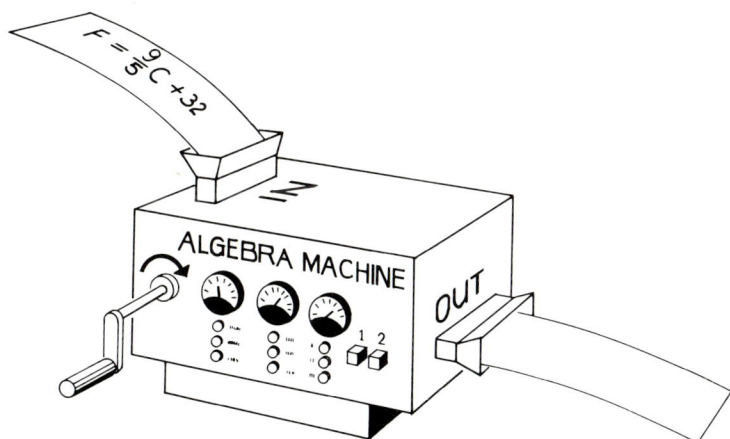

Similarly, it is certainly possible — but with a certain amount of mental strain — for an intelligent person not knowing algebra to conclude that the width of the rectangle described previously is 10 feet. But, again, the "algebra machine" will produce this answer very easily.

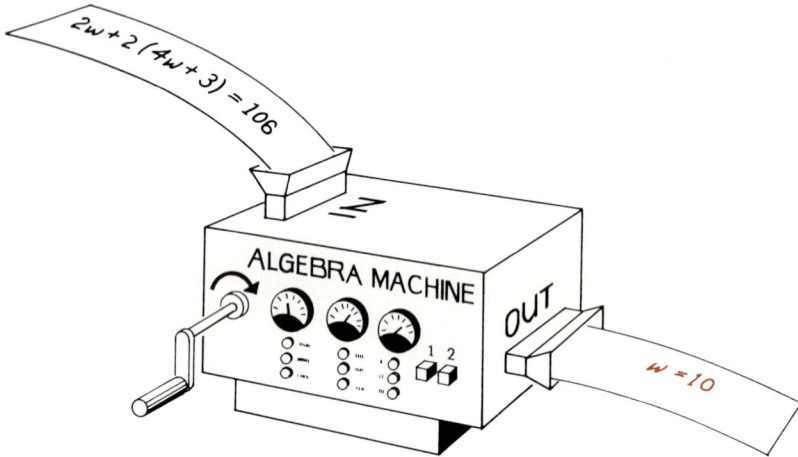

1.3 Algebra Is a Logical System

In this sense, algebra is basically the same as arithmetic, the only difference being in the greater use of symbols and the emphasis on general principles rather than on special cases and computation. All first graders know that $2 + 3 = 3 + 2$. But later they learn that this equality is a special case of the commutative property of addition:

$$a + b = b + a$$

for *all* numbers a and b.

Now from this property and other properties which we will consider in the next chapter, we can *prove* other properties, such as the fact that

$$(x + 1)(y + 1) = xy + x + y + 1$$

for all numbers x and y.

This illustrates the logical and formal side of algebra, which is basic for an understanding of the *why* of much of mathematics as opposed to the *how*.

1.4 Algebra Is a Search for Patterns and Generalizations

Algebra — like all of mathematics — can be regarded as a search for patterns that enable us to organize information and simplify procedures. For example, suppose that we are asked to find the sum

$$101 + 102 + 103 + 104 + 105 + 106$$

We could, of course, simply add the six numbers directly. Or, we could first regroup the summands to get

$$(101 + 106) + (102 + 105) + (103 + 104)$$

which gives us

$$207 + 207 + 207 = 3 \times 207 = 621$$

Now no great saving of time is made by using this regrouping procedure for this particular problem. But suppose that we were asked to find the sum of all the consecutive whole numbers from 100 through 199. We can write this sum as

$$S = 100 + 101 + 102 + \cdots + 197 + 198 + 199$$

using ellipsis dots to indicate the missing numbers. Here direct computation of the sum would be a rather painful process! Using the same pattern as before, however, we note that

$$100 + 199 = 299, \ 101 + 198 = 299,$$
$$102 + 197 = 299, \ldots, \ 149 + 150 = 299$$

and quickly get

$$S = 50 \times 299 = 14{,}950$$

Finally (and, again, the details are not important for this overview), we can extend the pattern to the sum

$$S = n + (n + 1) + \cdots + (n + k)$$

where n and k are any whole numbers and get

$$S = \frac{k + 1}{2} \ (2n + k)$$

With this formula in hand we need do no further thinking to handle sums of this type. Our two examples are simply special cases where, in the first one, $n = 101$ and $k = 5$ and, in the second, $n = 100$ and $k = 99$. We have, for these sums,

$$\frac{6}{2}(202 + 5) = 3 \times 207 = 621$$

and

$$\frac{100}{2}(200 + 99) = 50 \times 299 = 14,950$$

respectively.

This concludes our overview of algebra: as a language, as a tool, as a logical system, and as a search for patterns and generalizations. The remaining chapters will enlarge upon these topics.

EXERCISES

In exercises 1–3 use the relationship between the number of chirps a cricket makes and the temperature, $t = \frac{n}{4} + 40$, to answer each question.

1. If crickets are chirping at the rate of 128 chirps per minute, what is the temperature?

2. Crickets: 135 chirps per minute. Temperature?

3. Complete the following table:

Chirps per minute	28	44	60	88	100	128	144	160
Temperature in degrees F	?	51	?	?	?	72	?	?

The following graph shows the relationship between the number of chirps per minute of a cricket and the temperature in degrees Fahrenheit. Use the graph to answer questions 4–6.

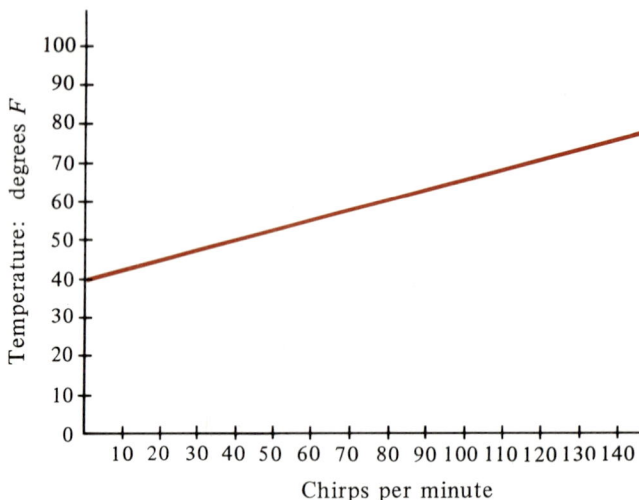

Graph showing relation between the temperature and the chirps of a cricket.

4. If you heard crickets chirping at the rate of 110 chirps per minute, what would be the approximate temperature?
5. If a cricket chirped 24 times in 15 seconds, what should be the temperature?
6. If the temperature were 70° Fahrenheit, how many chirps per minute would you expect crickets to make?

In exercises 7–9 use the relationship between the Fahrenheit and Celsius temperature scales, $F = \frac{9}{5} C + 32$, to answer each question.

7. If a Paris paper reported that the temperature today would be 30° Celsius, would you expect a warm or cold day? What would be the temperature in Fahrenheit?
8. Suppose that, in Italy, a doctor took your temperature and reported it as 39.5° Celsius. If a normal temperature is 98.6° Fahrenheit, did you have a fever? If so, how much above normal was your temperature?
9. Complete the following table:

Celsius	0	25	40	50	60	75	100
Fahrenheit	32	?	?	?	?	?	212

The following graph shows the relationship between the Celsius and the Fahrenheit temperature scales. Use the graph to answer questions 10–12.

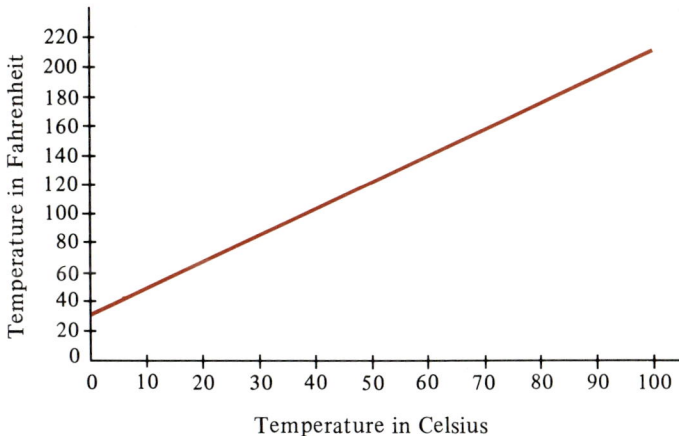

Graph showing the relationship between Celsius and Fahrenheit temperature scales.

10. A hot water tank thermostat is set at 140° Fahrenheit. What would the water temperature be if measured with a Celsius thermometer?
11. If eggs hatch at 104° Fahrenheit, what would the approximate Celsius temperature be?

12.   Alcohol boils at 173° Fahrenheit. What, approximately, would this
      temperature be on the Celsius scale?

In exercises 13–16 use the idea from Section 1.4 to find the given sums.

13. $1 + 2 + 3 + 4 + 5$
14. $8 + 9 + 10 + 11 + 12$
15. $11 + 12 + 13 + 14 + \cdots + 50$
16. $101 + 102 + 103 + 104 + \cdots + 500$

Exercises 17–30 consist of a series of well-known puzzles involving reasoning
and sometimes a little arithmetic. How many can you solve?

17.   Mac and Jack have an 8-quart container of root beer. They wish to divide the
      beverage equally but only have an empty 5-quart jar and an empty 3-quart
      container. How can they do it?
18.   A frog decided to jump out of a well that is 12 feet deep. Every 10 minutes
      he jumps up 3 feet on the side of the well and slides back 1 foot. How long
      will it take him to get out of the well?
19.   Sue and Bill are 40 miles apart and are riding bicycles toward each other. Sue
      is traveling at 12 miles per hour and Bill is traveling at 8 miles per hour. At
      the start a fly on Bill's nose leaves, travels 24 miles per hour, lands on Sue's
      nose, and promptly returns to Bill's nose, flying at the same speed. If the fly
      continues its trips back and forth between the noses until Sue and Bill meet,
      how far does the fly fly?
20.   José travels a distance of 90 miles and averages 45 miles per hour. On the
      return trip he averages 30 miles per hour. Find his average speed for the
      180-mile trip.
21.   Where on earth can you walk 5 miles due north, then 5 miles due west, and
      then 5 miles due south and find yourself where you started?
22.   A bookworm starts at page 1 of Volume I and bores its way in a straight line
      to the last page of Volume II. If the books are in order next to each other on
      the bookshelf and each cover is $\frac{1}{4}$ inch thick and each book without the cover
      is 2 inches thick, how far does the bookworm travel?
23.   A freight train 1 mile long travels through a tunnel that is 1 mile long. If the
      train travels at a speed of 20 miles per hour, how long does it take to pass
      completely through the tunnel?
24.   Two ducks before a duck, two ducks behind a duck, and a duck in the middle.
      How many ducks?
25.   Sally went bird hunting and killed 20 birds. She fired 8 times and killed an
      odd number of birds each time. How many did she kill with each shot?

26. A man is $\frac{3}{8}$ of the way across a trestle when he hears a train coming. If the train is traveling at 40 miles per hour and he can just make it to either end, how fast does he have to run?

27. Two coins have a total value of 55¢, and yet one of them is not a nickel. What are the two coins?

28. Jane sold two horses for $75 each. She made a 30% profit on one and took a 30% loss on the other. Did Jane have a profitable or unprofitable day? If there was a profit or a loss, approximately how much was it?

29. You are hired to do a job that will take 3 weeks (15 working days). You are given your choice of wages: (a) $10 per day, or (b) a penny the first day and each day afterward your wage is doubled. Which method of payment would you choose?

30. You have 12 pieces of chain, each piece consisting of 3 links. You take them to a blacksmith to join them into one circular or endless chain. If it costs 10¢ to cut a link and 10¢ to weld a link, what would you expect the blacksmith to charge for the job?

2

NUMBER SYSTEMS

2.1 The Natural Numbers and Zero

Mathematics begins for the small child with comparing *sets* of objects. From this activity comes the concept of the *counting numbers*: 1, 2, 3, ... Thus, the concept of 2, for example, arises from the fact that all of the sets shown in Figure 2.1 have something in common: their *members* can be *matched* as shown by the arrows. Other names for the set of counting numbers are *natural numbers* and *positive integers*

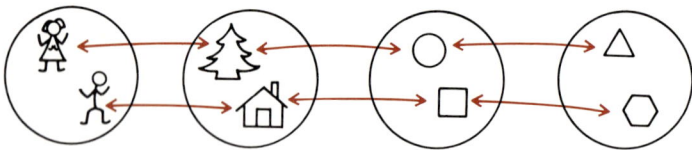

Figure 2.1

From the idea of the *empty set* arises the need for the number 0. The natural numbers and 0 together comprise the set of *whole numbers*, or *nonnegative integers*. Whole numbers may also be associated with certain points on a *number line*, as shown in Figure 2.2.

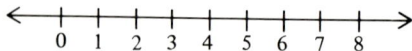

Figure 2.2

After the concept of a whole number is clearly in mind, the child must learn the meaning of *place value* in writing numerals where, for example, each 2 in 222 means something different. The first 2 on the left means 2 hundreds, the second 2, 2 tens, and the third 2, 2 ones. Thus $222 = 200 + 20 + 2$.

Now many hours must be spent in developing the concepts of addition, subtraction, multiplication, and division of whole numbers and in learning procedures for performing these operations. Later comes work with both *common* fractions, such as $\frac{2}{3}$, and *decimal* fractions, such as 0.23.

Some of this work will be briefly reviewed in this chapter, but we will also be looking at algebraic ways of expressing our conclusions. For this purpose we need to use letters, as discussed in Chapter 1. When we do so, we call the letters *variables*. Thus if a and b represent two numbers, then $a + b$ represents the sum of those two numbers and $a \times b$ the product of these two numbers. We also often write $a \times b$ as $a \cdot b$ or, even more simply, as ab. Similarly, we can write $2 \cdot a \cdot b$ or $2ab$ for $2 \times a \times b$. Note, however, that although we can write $2 \cdot 3$ for 2×3, we cannot write 2×3 as 23.

If a is replaced by 10 and b by 5, then

$$a + b = 10 + 5 = 15$$

and

$$2ab = 2 \times 10 \times 5 = 100$$

The following exercises are intended to help you review some of the basic operations of addition, multiplication, subtraction, and division, using the language of algebra.

EXERCISES 2.1

1. Add:
   2349
   +1415

2. Add:
   1415
   +2349

3. Compare your answers to exercises 1 and 2. What can you say about $a + b$ and $b + a$ if a and b are any whole numbers?

4. Add: $(345 + 219) + 64$. (This means to first add 345 and 219, and then to this sum add 64.)

5. Add: $345 + (219 + 64)$. (This means first add 219 and 64, and then add this sum to 345.)

6. Compare your answers to exercises 4 and 5. What can you say about $(a + b) + c$ and $a + (b + c)$ where a, b, and c are any whole numbers?

7. Multiply:
    569
    X124

8. Multiply:
    124
    X569

9. Compare your answers to exercises 7 and 8. What can you say about $a \times b$ and $b \times a$ if a and b are any whole numbers?

10. Multiply: $(69 \times 15) \times 23$.

11. Multiply: $69 \times (15 \times 23)$.

12. Compare your answers to exercises 10 and 11. What can you say about $(a \times b) \times c$ and $a \times (b \times c)$ if a, b, and c are any whole numbers?

13. Compute: $29 \times (15 + 87)$.

14. Compute: $(29 \times 15) + (29 \times 87)$.

15. Compare your answers to exercises 13 and 14. What can you say about $a \times (b + c)$ and $(a \times b) + (a \times c)$ if a, b, and c are any whole numbers?

16. When does $a - b = b - a$?

17. When does $(a - b) - c = a - (b - c)$?

18. Compute: $26 \times (169 - 123)$.

19. Compute: $(26 \times 169) - (26 \times 123)$.

20. Compare your answers to exercises 18 and 19. What can you say about $a \times (b - c)$ and $(a \times b) - (a \times c)$ if a, b, and c are any whole numbers such that b is greater than or equal to c?

21. When does $a \div b = b \div a$?

22. When does $(a \div b) \div c = a \div (b \div c)$?

23. What can you say about $a + 0$ for any whole number a?

24. What can you say about $a \times 0$ for any whole number a?

25. What can you say about $a \times 1$ for any whole number a?

26. Name the properties given in your answers to exercises 3, 6, 9, 12, 15, 23, 24, and 25.

2.2 Properties of Addition and Multiplication

Doing exercises 2.1 should have suggested to you the following properties of addition and multiplication of whole numbers, which we give here with their names.

For all whole numbers a, b, and c, it is true that:

1. $a + b = b + a$ (**Commutative property of addition**)
    Example: $2 + 3 = 5$ and also $3 + 2 = 5$

2. $a \times b = b \times a$ (**Commutative property of multiplication**)
    Example: $2 \times 3 = 6$ and also $3 \times 2 = 6$

3. $a + (b + c) = (a + b) + c$ (**Associative property of addition**)
   Example: $2 + (3 + 4) = 2 + 7 = 9$ and also $(2 + 3) + 4 = 5 + 4 = 9$

4. $a \times (b \times c) = (a \times b) \times c$ (**Associative property of multiplication**)
   Example: $2 \times (3 \times 4) = 2 \times 12 = 24$ and also $(2 \times 3) \times 4 = 6 + 4 = 24$

5. $a \times (b + c) = (a \times b) + (a \times c)$ (**Distributive property of multiplication over addition** — commonly referred to simply as the **distributive property**)
   Example: $2 \times (3 + 4) = 2 \times 7 = 14$ and also $(2 \times 3) + (2 \times 4) = 6 + 8 = 14$

6. $a + 0 = 0 + a = a$ (**Addition property of 0**)
   Example: $2 + 0 = 0 + 2 = 2$

7. $a \times 0 = 0 \times a = 0$ (**Multiplication property of 0**)
   Example: $2 \times 0 = 0 \times 2 = 0$

8. $a \times 1 = 1 \times a = a$ (**Multiplication property of 1**)
   Example: $2 \times 1 = 1 \times 2 = 2$

EXERCISES 2.2

In exercises 1–22 determine if the given statement is true or false. For those that are true, identify the property of whole numbers illustrated.

Examples:

a. $2 + 3 = 3 + 2$ True. Commutative property of addition.

b. $5 \times 4 = 5 + 4$ False.

c. $3 \cdot (2 + 5) = (3 \cdot 2) + (3 \cdot 5)$ True. Distributive property.

1. $6 + 9 = 6 + 7$
2. $25 + 9 = 9 + 25$
3. $(2 + 3) + 8 = 2 + (3 + 8)$
4. $(2 + 5) + 7 = 7 + (2 + 5)$
5. $4 + (3 + 5) = (4 + 3) + 5$
6. $8 \times 12 = 12 \times 8$
7. $9 \cdot 7 = 7 \cdot 9$
8. $(5 \times 6) \times 3 = 5 \times (6 \times 3)$
9. $(3 \cdot 4) \cdot 2 = 2 \cdot (3 \cdot 4)$
10. $7 \times (4 \times 2) = (7 \times 4) \times 2$
11. $2 \times (3 + 4) = (2 \times 3) + (2 \times 4)$
12. $(5 \cdot 8) + (5 \cdot 7) = 5 \cdot (8 + 7)$
13. $6 + (2 \times 3) = (6 + 2) \times (6 + 3)$
14. $(7 + 3) \times (7 + 1) = 7 \times (3 + 1)$
15. $(7 - 6) \times (5 + 3) = 5 + 3$
16. $(2 + 9) + (3 + 1) = (3 + 1) + (2 + 9)$
17. $(5 \times 2) \times (1 \times 1) = 5 \times 2$

18. $(5 + 2) \cdot (4 - 3) =$
    $(4 - 3) \cdot (5 + 2)$
19. $(5 - 5) \cdot (3 + 4) = 7 - 7$

20. $(7 - 5) \times (7 - 7) = 7 \times (7 - 5)$
21. $5 \cdot (7 + 3) = 5 \cdot (3 + 7)$
22. $4 \times (3 + 2) = (3 + 2) \times 4$

In exercises 23–32 complete the given statement to make it a true statement.

Example: $(3 + ?) + 5 = 5 + (3 + 2)$
$(3 + 2) + 5 = 5 + (3 + 2)$

23. $? \times 8 = 8 \times 9$
24. $4 \cdot ? = 17 \cdot 4$
25. $(? \times 6) \times 7 = 5 \times (6 \times 7)$
26. $(23 + 27) \times 10 = (23 \times ?) + (27 \times ?)$
27. $15 \cdot (? + ?) = (15 \cdot 30) + (15 \cdot 20)$
28. $176 \times (? \times 47) = (176 \times 321) \times 47$
29. $137 \cdot (? + 485) = (137 \cdot 8) + (? \times 485)$
30. $? \times (3 \times 4) = 0$
31. $(5 - 4) \times (? + 7) = 10$
32. $(? + ?) \cdot 5 = (3 \cdot ?) + (7 \cdot ?)$

In exercises 33–42 use the commutative and associative properties of addition and multiplication of whole numbers to compute the sums and products quickly.

Example: $(67 \times 50) \times 2 = 67 \times (50 \times 2) = 67 \times 100 = 6700$

33. $7 + (3 + 6)$
34. $8 + (17 + 2)$
35. $(35 \times 2) \times 5$
36. $25 \cdot (4 \cdot 77)$
37. $(13 + 9) + 1$

38. $17 + (29 + 3)$
39. $97 + (3 + 28)$
40. $4 \cdot (250 \cdot 39)$
41. $67 + (10 + 3)$
42. $(13 + 14) + (2 \cdot 3)$

In exercises 43–51 use the distributive property to write the given expression in a different way.

Examples:

a. $(3 + 5) \cdot 2 = (3 \cdot 2) + (5 \cdot 2)$

b. $7(x + y) = 7x + 7y$

c. $(3 \cdot 8) + (3 \cdot 5) = 3 \cdot (8 + 5)$

d. $13a + 13b = 13(a + b)$

43. $4 \cdot (2 + 3)$	46. $5(x + y)$	49. $(13 \cdot 3) + (9 \cdot 3)$
44. $7 \cdot (6 + 5)$	47. $3(a + b)$	50. $4x + 4y$
45. $(8 + 4) \cdot 9$	48. $(4 \cdot 2) + (4 \cdot 7)$	51. $(a \cdot 3) + (b \cdot 3)$

2.3 Fractions

Success in working with algebraic fractions such as $\frac{x}{y}$, $\frac{a + b}{a - b}$, etc., depends strongly on an understanding of arithmetic calculations with fractions. In this section we will review these procedures.

Addition: If two fractions have the same denominator, there is no problem. For example:

$$\frac{3}{5} + \frac{4}{5} = \frac{3 + 4}{5} = \frac{7}{5}$$

and

$$\frac{5}{11} + \frac{2}{11} = \frac{7}{11}$$

Expressing this fact in general terms, we have

$$\frac{a}{b} + \frac{c}{b} = \frac{a + c}{b}$$

When the denominators are different we can use the very important *basic principle* for fractions that, for any fraction $\frac{a}{b}$ and any nonzero number m,

$$\frac{a}{b} = \frac{a \times m}{b \times m}$$

Thus, to add $\frac{2}{3}$ and $\frac{5}{10}$ we can write

$$\frac{2}{3} + \frac{5}{10} = \frac{2 \times 10}{3 \times 10} + \frac{5 \times 3}{10 \times 3} = \frac{20}{30} + \frac{15}{30} = \frac{35}{30}$$

We can also use this principle to express fractions in *lowest terms,* i.e., as fractions whose numerators and denominators have 1 as the only *common factor.* Thus the answer to the addition problem above can be rewritten as

$$\frac{35}{30} = \frac{7 \times 5}{6 \times 5} = \frac{7}{6}$$

Here are two other examples of addition of fractions:

$$\frac{4}{11} + \frac{3}{5} = \frac{4 \times 5}{11 \times 5} + \frac{3 \times 11}{5 \times 11} = \frac{20}{55} + \frac{33}{55} = \frac{53}{55}$$

$$\frac{4}{9} + \frac{5}{6} = \frac{4 \times 2}{9 \times 2} + \frac{5 \times 3}{6 \times 3} = \frac{8}{18} + \frac{15}{18} = \frac{23}{18}$$

Notice that in the second example we could have proceeded as in the first two examples to write

$$\frac{4}{9} + \frac{5}{6} = \frac{4 \times 6}{9 \times 6} + \frac{5 \times 9}{6 \times 9} = \frac{24}{54} + \frac{45}{54} = \frac{69}{54}$$

and then written

$$\frac{69}{54} = \frac{23 \times 3}{18 \times 3} = \frac{23}{18}$$

by using the basic principle. We simplified the computation, however, by using the *least common denominator* (*LCD*), 18, of the two denominators, 9 and 6. The LCD is the smallest number divisible by both of the denominators: 18, 36, 54, 72, ... are all divisible by both 9 and 6 but 18 is the *smallest* number divisible by both 9 and 6.

If the LCD is not obvious, it can be found by writing the denominators as a product of *prime* numbers. (A natural number is a prime number if it is not equal to 1 and has only 1 and itself as divisors. Thus the first five prime numbers are 2, 3, 5, 7, and 11.) So, for the problem

$$\frac{4}{9} + \frac{5}{6}$$

we write

$$9 = 3 \times 3 \quad \text{and} \quad 6 = 2 \times 3$$

We call 3 a *factor* of 9, and 2 and 3 factors of 6. Then 3×3 is the *factorization* of 9 into a product of primes. Similarly, 2×3 is the factorization of 6 into a product of primes. Then the LCD is

$$2 \times 3 \times 3 = 18$$

where two factors of 3 are needed because two factors of 3 appear in the factorization of 9.

As another example of the use of the LCD, consider the addition

$$\frac{5}{18} + \frac{7}{12} + \frac{1}{2}$$

Here we write

$$18 = 2 \times 3 \times 3, \qquad 12 = 2 \times 2 \times 3, \qquad 2 = 2$$

and have for the LCD,

$$(2 \times 2) \times (3 \times 3) = 36$$

Then

$$\frac{5}{18} + \frac{7}{12} + \frac{1}{2} = \frac{5 \times 2}{18 \times 2} + \frac{7 \times 3}{12 \times 3} + \frac{1 \times 18}{2 \times 18} = \frac{10}{36} + \frac{21}{36} + \frac{18}{36} = \frac{49}{36}$$

Subtraction: Again, as for addition, when the denominators are the same there is no problem. For example:

$$\frac{5}{7} - \frac{2}{7} = \frac{5 - 2}{7} = \frac{3}{7}$$

and

$$\frac{7}{12} - \frac{5}{12} = \frac{7 - 5}{12} = \frac{2}{12} = \frac{1 \times 2}{6 \times 2} = \frac{1}{6}$$

Expressing this fact in general terms, we have

$$\frac{a}{b} - \frac{c}{b} = \frac{a - c}{b}$$

When the denominators are different we can again use the LCD as shown in the next example:

$$\frac{11}{15} - \frac{7}{12} = \frac{11 \times 4}{15 \times 4} - \frac{7 \times 5}{12 \times 5} = \frac{44}{60} - \frac{35}{60} = \frac{9}{60} = \frac{3 \times 3}{20 \times 3} = \frac{3}{20}$$

Here

$$15 = 3 \times 5 \quad \text{and} \quad 12 = 2 \times 2 \times 3$$

and so the LCD is

$$2 \times 2 \times 3 \times 5 = 60$$

Multiplication: For any two fractions $\frac{a}{b}$ and $\frac{c}{d}$, we have

$$\frac{a}{b} \times \frac{c}{d} = \frac{a \times c}{b \times d}$$

Thus

$$\frac{2}{3} \times \frac{4}{5} = \frac{2 \times 4}{3 \times 5} = \frac{8}{15}$$

The basic principle can sometimes be used to simplify multiplication of fractions. For example, consider the product

$$\frac{13}{6} \times \frac{15}{26}$$

Instead of multiplying directly to get

$$\frac{13}{36} \times \frac{15}{26} = \frac{13 \times 15}{36 \times 26} = \frac{195}{936} = \frac{5 \times 39}{24 \times 39} = \frac{5}{24}$$

we can write

$$\frac{13}{36} \times \frac{15}{26} = \frac{13 \times 15}{36 \times 26} = \frac{13 \times 3 \times 5}{(12 \times 3) \times (13 \times 2)} = \frac{5 \times (13 \times 3)}{(12 \times 2) \times (13 \times 3)}$$

$$= \frac{5}{12 \times 2} = \frac{5}{24}$$

Division: The basic principle can be used to explain the usual "invert and multiply" rule for dividing fractions by using the fact that

$$a \div b = \frac{a}{b}$$

For example:

$$\frac{2}{3} \div \frac{5}{7} = \frac{\frac{2}{3}}{\frac{5}{7}} = \frac{\frac{2}{3} \times (3 \times 7)}{\frac{5}{7} \times (3 \times 7)} = \frac{2 \times 7}{5 \times 3} = \frac{2 \times 7}{3 \times 5} = \frac{2}{3} \times \frac{7}{5}$$

Thus

$$\frac{2}{3} \div \frac{5}{7} = \frac{2}{3} \times \frac{7}{5} = \frac{14}{15}$$

Similarly

$$\frac{21}{8} \div \frac{15}{7} = \frac{21}{8} \times \frac{7}{15} = \frac{21 \times 7}{8 \times 15} = \frac{(3 \times 7) \times 7}{(2 \times 4) \times (3 \times 5)} = \frac{7 \times 7}{2 \times 4 \times 5} = \frac{49}{40}$$

In general, if $\frac{a}{b}$ and $\frac{c}{d}$ are any two fractions with $\frac{c}{d}$ not equal to 0, we have

$$\frac{a}{b} \div \frac{c}{d} = \frac{a}{b} \times \frac{d}{c} = \frac{a \times d}{b \times c}$$

Note that every whole number can be written as a fraction: $\frac{a}{1} = a$ for all whole numbers a. Thus the set of whole numbers can be regarded as a *subset* of the set of fractions.

Note also that whereas the numerator of a fraction can be 0 $\left(\frac{0}{1} = \frac{0}{2} = \frac{0}{3} = \ldots = 0 \right)$, the denominator cannot: $\frac{1}{0}, \frac{2}{0}, \frac{3}{0}, \ldots$ *are not* fractions. Whenever we write a symbol for a fraction such as $\frac{a}{b}$ it is to be understood that b is not equal to zero (in symbols, $b \neq 0$ where the symbol "$\neq$" means "not equal to").

EXERCISES 2.3

In exercises 1-20 perform the indicated operation and express your answer in lowest terms.

1. $\dfrac{5}{3} + \dfrac{8}{11}$

2. $\dfrac{1}{2} + \dfrac{5}{6}$

3. $\dfrac{7}{12} + \dfrac{2}{3}$

4. $\dfrac{2}{3} + \dfrac{1}{4}$

5. $\dfrac{6}{5} + \dfrac{4}{3}$

6. $\dfrac{7}{8} - \dfrac{1}{4}$

7. $\dfrac{3}{4} - \dfrac{2}{3}$

8. $\dfrac{5}{7} - \dfrac{2}{3}$

9. $\dfrac{11}{12} - \dfrac{2}{5}$

10. $\dfrac{5}{6} - \dfrac{3}{8}$

11. $\dfrac{5}{6} \times \dfrac{2}{3}$

12. $\dfrac{8}{9} \cdot \dfrac{5}{6}$

13. $\dfrac{21}{8} \times \dfrac{15}{7}$

14. $\dfrac{3}{10} \cdot \dfrac{2}{3}$

15. $\dfrac{8}{5} \times \dfrac{3}{2}$

16. $\dfrac{18}{7} \div \dfrac{4}{3}$

17. $\dfrac{7}{9} \div \dfrac{5}{18}$

18. $\dfrac{6}{7} \div \dfrac{2}{3}$

19. $\dfrac{5}{3} \div \dfrac{3}{8}$

20. $\dfrac{4}{15} \div \dfrac{2}{9}$

21 - 22. Copy the squares and insert numbers in the vacant spaces so that magic squares are formed. (Magic squares are those in which each row, column, and diagonal all have the same sum.)

21.

$\dfrac{4}{3}$	$\dfrac{1}{6}$	$\dfrac{1}{4}$	?
?	?	$\dfrac{5}{6}$	$\dfrac{2}{3}$
$\dfrac{3}{4}$	$\dfrac{7}{12}$	$\dfrac{1}{2}$	?
$\dfrac{1}{3}$	$\dfrac{7}{6}$	$\dfrac{5}{4}$	?

22.

$\dfrac{8}{5}$	$\dfrac{1}{5}$	?	?
$\dfrac{1}{2}$	?	?	$\dfrac{4}{5}$
?	$\dfrac{7}{10}$	$\dfrac{3}{5}$	$\dfrac{6}{5}$
$\dfrac{2}{5}$	$\dfrac{7}{5}$	$\dfrac{3}{2}$	$\dfrac{1}{10}$

23 – 27. Complete the following table. (For each column determine the value of $\frac{a}{b}$ and $\frac{c}{d}$, using the given information, and then compute the numbers for the blank squares.)

	23.	24.	25.	26.	27.
$\frac{a}{b}+\frac{c}{d}+\frac{3}{2}$	?	?	?	$\frac{83}{40}$	$\frac{29}{16}$
$\frac{a}{b} \div \frac{c}{d}$	?	?	$\frac{14}{5}$	?	?
$\frac{7}{5} - \frac{c}{d}$	$\frac{11}{15}$	?	$\frac{3}{20}$	$\frac{41}{40}$	$\frac{87}{80}$
$\frac{c}{d} \times \frac{a}{b}$	?	$\frac{5}{8}$	?	?	?
$3 \times \frac{a}{b}$	$\frac{3}{2}$	$\frac{9}{4}$	?	?	?

28. Show that $\frac{1}{0}$ is not a fraction by using the fact that $\frac{1}{0} = 1 \div 0$. (Hint: If $1 \div 0 = a$, then it must be true that $1 = a \times 0$.)

2.4 Rational Numbers

Fractions, as well as whole numbers, can be pictured on a number line as suggested by Figure 2.3. Here $\frac{1}{2}$ is associated with the point halfway between the points corresponding to 0 and 1; $\frac{9}{4}$ (= $2\frac{1}{4}$) is associated with the point $\frac{1}{4}$ of the way between 2 and 3; etc.

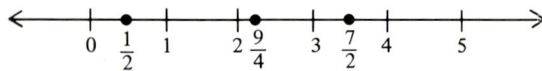

Figure 2.3

What about the points to the left of the point corresponding to 0? They will correspond to *negative numbers*, as indicated in Figure 2.4. We read $^-1$ as "negative one," $^-(\frac{1}{2})$ as "negative one-half," and so on, and call them negative numbers. Numbers such as $1, \frac{1}{2}$, and so on are called *positive numbers*; 0 is neither positive nor negative. The part of a number line to the right of 0 is called the *positive half* of the number line and the part to the left is called the *negative half*

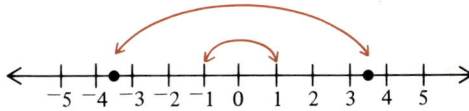

⁻1 and 1 are opposites   ⁻3.5 and 3.5 are opposites

Figure 2.4

As shown in Figure 2.4, we can think of the point corresponding to ⁻1 on a number line as being "opposite" to the point corresponding to 1 and, conversely, the point corresponding to 1 as being opposite to ⁻1. For this reason, ⁻1 is sometimes called the *opposite* of 1 and 1 is called the opposite of ⁻1. 0 is its own opposite, i.e., ⁻0 = 0.

There are many physical interpretations of positive and negative numbers: temperatures above zero (positive) and below zero (negative); debts (negative) and assets (positive); heights above sea level (positive) and depths below sea level (negative); and the now common rocket countdown before launch (negative) and after launch (positive).

In general, ⁻*a* and *a* are opposites of each other. Thinking in terms of "opposite" makes it easy to see that, for example,

$$\text{⁻(⁻5) (the opposite of ⁻5) is } 5$$

and, in general, that

$$\text{⁻(⁻}a\text{)} = a$$

for any number *a*. For example,

$$\text{if } a = 5$$
$$\text{then ⁻(⁻}a\text{)} = \text{⁻(⁻5)} = 5 = a$$

and

$$\text{if } a = \text{⁻5}$$
$$\text{then ⁻(⁻}a\text{)} = \text{⁻[⁻(⁻5)]} = \text{⁻5} = a$$

Note that when we read "⁻5" as "negative five" no confusion can arise: ⁻5 is indeed a negative number. But when we read "⁻*a*" as "negative *a*" it does sound as if we are again talking about a negative number. However, as we have noted, if *a* = ⁻5, then ⁻*a* = ⁻(⁻5) = 5, a positive number. For this reason, it might be better always to refer to ⁻*a* as the opposite of *a* or as *the* negative *of a*. Nevertheless, it is customary to read "⁻*a*" simply as "negative *a*."

The set of numbers consisting of the positive fractions, the negative fractions, and zero is called the set of *rational numbers*. Positive fractions are, then, also called *positive rational numbers* and negative fractions are also called *negative rational numbers*. Similarly, the set of *integers* consists of the set of *positive integers* together with their opposites (the set of *negative integers*) and zero.

Diagramatically, we have

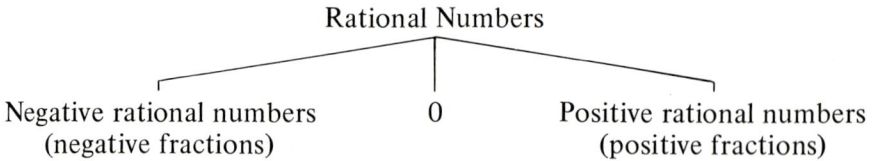

and

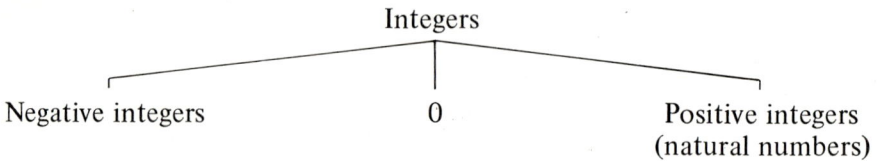

EXERCISES 2.4

In exercises 1–15 determine the opposite of the given number. Then locate both the number and its opposite on a number line.

Example: ⁻2.5   The opposite of ⁻2.5 is 2.5

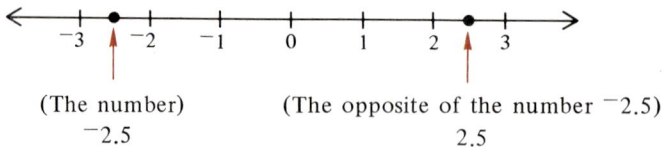

(The number)   (The opposite of the number ⁻2.5)
    ⁻2.5   2.5

1. 3   4. 5   7. ⁻$\left(\dfrac{3}{4}\right)$

2. ⁻7   5. 0   8. ⁻$\left(\dfrac{5}{8}\right)$

3. ⁻5   6. $\dfrac{1}{2}$ 9. $\dfrac{8}{5}$

10. $3\frac{2}{3}$

12. $^-\left(\frac{13}{3}\right)$

14. $^-2.7$

11. $^-6\frac{1}{5}$

13. 1.5

15. $^-(^-8)$

In exercises 16–30 first determine the number indicated and then give the opposite of the number.

Example: $^-(^-9) = 9$ and the opposite of 9 is $^-9$.

16. 5×3

21. $\frac{2}{3} \times 0$

26. $^-(^-(^-1))$

17. $^-(2 + 3)$

22. $^-\left(^-\left(\frac{2}{3} + \frac{1}{2}\right)\right)$

27. $^-\left(^-\left(^-\left(^-3\frac{1}{2}\right)\right)\right)$

18. $^-(^-5 + 0)$

23. $^-(^-100)$

28. $^-(^-(^-1.5))$

19. $^-(^-(4 \times 6))$

24. $^-\left(^-\left(^-\frac{1}{2}\right)\right)$

29. $^-\left(^-\left(^-\frac{37}{5}\right)\right)$

20. $^-\left(\frac{1}{2} \times \frac{3}{5}\right)$

25. $^-(^-(^-13))$

30. $^-(^-(^-(^-(^-2))))$

31. What kind of number is $-x$ if
    (a) x is a positive rational number?
    (b) x is a negative rational number?
    (c) x is the rational number 0?

32. What kind of number is x if
    (a) $-x$ is a positive rational number?
    (b) $-x$ is a negative rational number?
    (c) $-x$ is the rational number 0?

33. Is every rational number the opposite of some rational number?

34. Is the set of opposites of all rational numbers the same as the set of rational numbers?

35. Is the set of negative rational numbers a subset of the set of opposites of rational numbers?

36. Is the set of all opposites of rational numbers a subset of the set of rational numbers?

37. Is every opposite of a rational number a negative rational number?

38. What is the opposite of the rational number zero?

39. Does every rational number have an opposite?

40. How many opposites does each rational number have?

2.5 Addition of Rational Numbers

From our physical interpretations of positive and negative numbers it is easy to see how we should add rational numbers. For example, a debt of $2 added to a debt of $3 gives a total indebtedness of $5. So

$$^-2 + ^-3 = ^-5$$

Similarly, a debt of $2, with $5 in your pocket, gives you a net worth of $3. So

$$^-2 + 5 = 3$$

Finally, a debt of $5, with $2 in your pocket, gives you a net indebtedness of $3. So

$$^-5 + 2 = ^-3$$

We can show addition of rational numbers on a number line as illustrated in Figure 2.5. Our first "jump" or move takes us to the

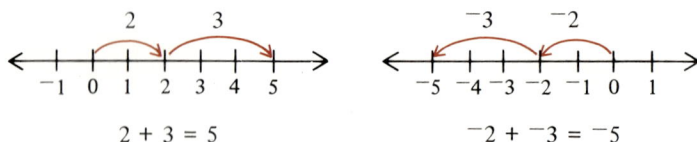

2 + 3 = 5 $^-2 + ^-3 = ^-5$

Figure 2.5

point corresponding to the first of the two numbers we are adding. Now we move the proper number of spaces to the *right* if the number we are adding is *positive* and to the *left* if the number we are adding is *negative*. Figure 2.6 shows the other two examples of addition that we have considered.

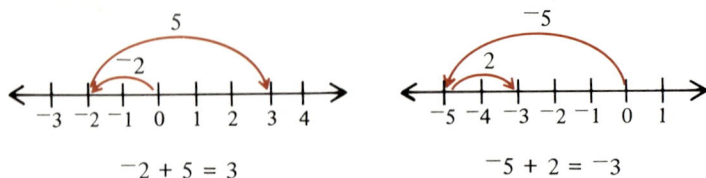

$^-2 + 5 = 3$ $^-5 + 2 = ^-3$

Figure 2.6

From this number line interpretation of addition we can see that $a + (^-a) = 0$ for all rational numbers a. For this reason, ^-a, the opposite or negative of a, is also called the *additive inverse* of a.

Up to this point we have been using the raised dash, as in ⁻1, to indicate negative numbers, in contrast to the dash as used in 3 – 2 to indicate subtraction. However, since it is awkward to write or print the raised dash, it is customary to write

$$^-1 \text{ as } -1, \qquad ^-\!\!\left(\frac{1}{2}\right) \text{ as } -\frac{1}{2}, \qquad ^-a \text{ as } -a$$

etc. (but still reading –1, for example, as "negative one" and not as "minus one").

Using this notation, we now ask what interpretation can be given to a symbol such as $\frac{-2}{3}$ and, in general, to $\frac{-a}{b}$. To arrive at a reasonable interpretation, recall that $\frac{2}{3}$ is associated with the point obtained by dividing the interval from 0 to 1 into 3 parts and then taking 2 of them, as shown in Figure 2.7. It is reasonable, then, to conclude that the

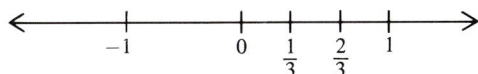

Figure 2.7

point corresponding to $\frac{-2}{3}$ can be obtained by dividing the interval from 0 to –1 into 3 parts and taking 2 of them, as shown in Figure 2.8.

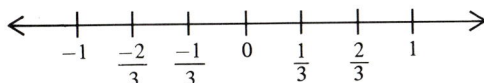

Figure 2.8

But this shows that the point corresponding to $\frac{-2}{3}$ is the point corresponding to the opposite of $\frac{2}{3}$, that is, to $-\frac{2}{3}$. Hence we have

$$\frac{-2}{3} = -\frac{2}{3}$$

and, in general, make the definition

$$\frac{-a}{b} = -\frac{a}{b}$$

where a is any integer and b is any positive integer. (Later we will consider the case where b is a negative integer.)

We can use this definition to reduce problems involving the addition of rational numbers to problems involving only the addition of integers, as shown in the following examples.

Example 1: $-\dfrac{2}{5}+\dfrac{3}{5}=\dfrac{-2}{5}+\dfrac{3}{5}=\dfrac{-2+3}{5}=\dfrac{1}{5}$

Example 2: $\dfrac{2}{7}+\left(-\dfrac{4}{7}\right)=\dfrac{2}{7}+\dfrac{-4}{7}=\dfrac{2+(-4)}{7}=\dfrac{-2}{7}=-\dfrac{2}{7}$

Example 3: $\left(-\dfrac{5}{3}\right)+\left(-\dfrac{3}{4}\right)=\left(-\dfrac{5\times 4}{3\times 4}\right)+\left(-\dfrac{3\times 3}{4\times 3}\right)$

$$=\left(-\dfrac{20}{12}\right)+\left(-\dfrac{9}{12}\right)=\dfrac{-20}{12}+\dfrac{-9}{12}$$

$$=\dfrac{(-20)+(-9)}{12}=\dfrac{-29}{12}=-\dfrac{29}{12}$$

Example 4: $-\dfrac{5}{8}+1=\dfrac{-5}{8}+\dfrac{8}{8}=\dfrac{-5+8}{8}=\dfrac{3}{8}$

EXERCISES 2.5

In exercises 1–39, find the sum. When the answer is not an integer, write it in the form $\dfrac{a}{b}$ or $-\dfrac{a}{b}$, where $\dfrac{a}{b}$ is a positive fraction in lowest terms.

1. $1+4$

2. $1+(-4)$

3. $-1+4$

4. $-1+(-4)$

5. $8+(-7)$

6. $-5+3$

7. $-16+(-18)$

8. $7+(-12)$

9. $-9+5$

10. $\dfrac{2}{3}+\dfrac{3}{4}$

11. $\dfrac{2}{3}+\left(-\dfrac{3}{4}\right)$

12. $-\dfrac{2}{3}+\dfrac{3}{4}$

13. $-\dfrac{2}{3}+\left(-\dfrac{3}{4}\right)$

14. $-1\dfrac{1}{4}+3\dfrac{3}{4}$

15. $1\dfrac{1}{4}+\left(-3\dfrac{3}{4}\right)$

16. $1\dfrac{4}{5}+\left(-2\dfrac{3}{5}\right)$

17. $-1\dfrac{4}{5}+\left(-2\dfrac{3}{5}\right)$

18. $-\dfrac{2}{3}+2$

19. $\dfrac{2}{3}+(-2)$

20. $-2+\dfrac{2}{3}$

21. $[5+(-3)]+4$

22. $5+(-3+4)$

23. $(-6+3)+(-3)$

24. $(7+5)+(-3)$

25. $3+(-7+5)$

26. $[-7+(-5)]+3$

27. $-7+[5+(-3)]$

28. $(-3 + 0) + 3$

32. $0 + (-5)$

36. $\frac{1}{3} + \left(-\frac{1}{3}\right)$

29. $\left(\frac{2}{3} + \frac{5}{3}\right) + \left(-\frac{10}{3}\right)$

33. $0 + \left(-\frac{1}{2}\right)$

37. $-\frac{2}{5} + \frac{2}{5}$

30. $\left(-\frac{2}{3}\right) + \left(-\frac{1}{2}\right) + \frac{5}{2}$

34. $5 + (-5)$

38. $\frac{13}{7} + \left(-\frac{13}{7}\right)$

31. $-2 + 0$

35. $-15 + 15$

39. $-5 + \frac{1}{5}$

In exercises 40-48 determine the additive inverse of the given number.

40. -4

43. $3 + 5$

46. $\frac{1}{2} + \frac{3}{2}$

41. 5

44. $5 + (-3)$

47. $\frac{2}{3} + \left(-\frac{2}{3}\right)$

42. $-\frac{2}{3}$

45. 0

48. $\frac{267}{315}$

49. From a mild $48°$ on December 18 the temperature dropped 54 degrees in 7 days.
    (a) What was the temperature on Christmas Day?
    (b) If the temperature was 10 degrees warmer on New Year's Eve than it was on December 18, what was the temperature on New Year's Eve?

50. In the game of Boffo a player can either win points or lose them. If he loses more points than he has won, his score goes "in the hole." (a) Sue's score is 45 points. If Jack wins 135 points from her, what will Sue's score be? Express this as a positive or a negative number. (b) On the next play Sue wins 65 points. What is her score now, expressed as a positive or a negative number?

2.6 Subtraction of Rational Numbers

Compare the following subtractions and additions:

$$5 - 4 = 1 \qquad\qquad 5 + (-4) = 1$$
$$8 - 3 = 5 \qquad\qquad 8 + (-3) = 5$$
$$\frac{7}{8} - \frac{3}{4} = \frac{1}{8} \qquad\qquad \frac{7}{8} + \left(-\frac{3}{4}\right) = \frac{1}{8}$$

Clearly, we have, in these cases,

$$a - b = a + (-b) = a + (\text{opposite of } b)$$

But how about 4 – 5, –3 – (–2), –3 – 2, etc.? Is it true that, for *all* rational numbers a and b, we have $a - b = a + (-b)$?

It is indeed useful to agree that, for all rational numbers a and b,

(1) $a - b = a + (-b) = a + $ (opposite of b)

Basically, the reason behind this definition is the following relation between addition and subtraction of positive rational numbers:

(2) $a - b = c$ if and only if $c + b = a$

Thus

$$5 - 4 = 1 \quad \text{because} \quad 1 + 4 = 5$$

$$8 - 3 = 5 \quad \text{because} \quad 5 + 3 = 8$$

$$\frac{7}{8} - \frac{3}{4} = \frac{1}{8} \quad \text{because} \quad \frac{1}{8} + \frac{3}{4} = \frac{7}{8}$$

(Indeed, this is the basis for checking a subtraction problem by using addition.)

If we agree that (2) should hold for all rational numbers, we have

$$a - b = a + (-b) \text{ if and only if } \left[a + (-b)\right] + b = a$$

Then, to show that $[a + (-b)] + b = a$, we note that, under the assumption that the associative property of addition holds for rational numbers,

$$[a + (-b)] + b = a + [(-b) + b] = a + 0 = a$$

Using (1), then, we can transform every subtraction problem into an addition problem as illustrated by the following examples.

Example 1: $5 - (-1) = 5 + 1 = 6$

Example 2: $-5 - (-1) = -5 + 1 = -4$

Example 3: $-\dfrac{7}{2} - \dfrac{3}{4} = -\dfrac{7}{2} + \left(-\dfrac{3}{4}\right) = -\dfrac{14}{4} + \left(-\dfrac{3}{4}\right) = \dfrac{-14}{4} + \dfrac{-3}{4}$

$$= \frac{(-14) + (-3)}{4} = \frac{-17}{4} = -\frac{17}{4}$$

Example 4: $-\dfrac{7}{8} - \left(-\dfrac{1}{8}\right) = -\dfrac{7}{8} + \dfrac{1}{8} = -\dfrac{6}{8} = -\dfrac{3}{4}$

Note that if we let $a = 0$ in (1), we have

$$0 - b = 0 + (-b) = -b$$

which gives us a relation between subtraction and taking the opposite of a number. It is still important, however, to keep in mind the difference between the two uses of $-$. For example,

$$-3 - (-2) = -1$$

should be read as "negative three minus negative two is equal to negative one" and *not* as "minus three minus minus two is equal to minus one."

Finally, we note that because one operation "undoes" the other, addition and subtraction of rational numbers are *inverse* operations; that is,

$$(a + b) - b = a \text{ and } (a - b) + b = a$$

for all rational numbers a and b.

EXERCISES 2.6

In exercises 1–20 write the additive inverse (opposite) of the given number in the form a or $-a$, where a is a positive number.

1. 3

2. -3

3. 0

4. $\frac{1}{2}$

5. $-\frac{1}{2}$

6. $\frac{-3}{2}$

7. $\frac{4}{5}$

8. $-(-2)$

9. $-\left(-\frac{3}{4}\right)$

10. $-\left(-\frac{-2}{3}\right)$

11. $-\left(-\frac{-8}{7}\right)$

12. $-(-1)$

13. $-\left(\frac{-2}{3}\right)$

14. $-(-5)$

15. $-\frac{5}{3}$

16. 5

17. -5

18. $-\frac{1}{2}$

19. $\frac{3}{4}$

20. $-\left(\frac{-1}{3}\right)$

21. What kind of number is represented by $-x$ if
    (a) x is a positive integer?
    (b) x is a negative integer?
    (c) x is 0?

22. What rational number is its own additive inverse?

23. Is the number $-a$ a positive or a negative number? Explain your conclusion.

In exercises 24–44 do the given subtraction problem by first writing a related addition problem.

Example: $5 - 7 = 5 + (-7) = -2$

24. $7 - 12$

25. $15 - 9$

26. $13 - (-2)$

27. $23 - (-7)$

28. $-13 - 2$

29. $-16 - 14$

30. $-2 - (-2)$

31. $-34 - (-54)$

32. $-\dfrac{1}{3} - \left(-\dfrac{1}{3}\right)$

33. $-\dfrac{2}{5} - \left(-\dfrac{3}{4}\right)$

34. $-\dfrac{5}{6} - \dfrac{3}{4}$

35. $-\dfrac{3}{10} - \dfrac{4}{5}$

36. $3 - \dfrac{1}{8}$

37. $-2\dfrac{3}{4} - 5$

38. $3\dfrac{1}{5} - 4\dfrac{3}{5}$

39. $-\left(-\dfrac{1}{2}\right) - \dfrac{1}{4}$

40. $-\left(-\dfrac{5}{6}\right) - \dfrac{5}{12}$

41. $-\left(-\dfrac{2}{3}\right) - \left(-\dfrac{3}{4}\right)$

42. $-\left(-\dfrac{3}{4}\right) - \left(-\dfrac{2}{3}\right)$

43. $\left(\dfrac{3}{8} - \dfrac{4}{8}\right) - \dfrac{1}{8}$

44. $\dfrac{3}{8} - \left(\dfrac{4}{8} - \dfrac{1}{8}\right)$

2.7 Multiplication of Rational Numbers

Just as $3 \times 2 = 2 + 2 + 2 = 6$, we can think of $3 \times (-2)$ as $(-2) + (-2) + (-2) = -6$. Then, just as $3 \times 2 = 2 \times 3$, it should seem reasonable to agree that because $3 \times (-2) = -6$, then $(-2) \times 3 = -6$. Generalizing these conclusions, we have the following rule:

$$a \times (-b) = (-b) \times a = -(a \times b)$$

for all positive rational numbers a and b. Thus

$$\left(-\dfrac{1}{2}\right) \times \dfrac{3}{4} = -\left(\dfrac{1}{2} \times \dfrac{3}{4}\right) = -\dfrac{3}{8}$$

and

$$\dfrac{5}{12} \times \left(-\dfrac{3}{4}\right) = -\left(\dfrac{5}{12} \times \dfrac{3}{4}\right) = -\dfrac{15}{48} = -\dfrac{5}{16}$$

It is not so easy to decide what a reasonable answer should be to $(-2) \times (-3)$. An argument to show that a reasonable answer is 6 goes as follows:

We know that

(1) $-3 + 3 = 0$

Now if we assume that the multiplication property of 0 holds for rational numbers, we can conclude from (1) that multiplying both sides of the equality by -2 gives us

(2) $(-2) \times (-3 + 3) = (-2) \times 0 = 0$

Still further, if we assume the distributive property, we know that

(3) $(-2) \times (-3 + 3) = (-2) \times (-3) + (-2) \times 3$

Putting (2) and (3) together gives us

(4) $(-2) \times (-3) + (-2) \times 3 = 0$

We know that $(-2) \times 3 = -6$, so (4) becomes

(5) $(-2) \times (-3) + (-6) = 0$

That is, $(-2) \times (-3)$ is the opposite of -6. But we know that 6 is the opposite of -6 and hence we must have $(-2) \times (-3) = 6$. Thus

$$(-2) \times (-3) = 2 \times 3 = 6$$

A similar argument can be used to show that for any two rational numbers $\frac{a}{b}$ and $\frac{c}{d}$ we have

$$\left(-\frac{a}{b}\right) \times \left(-\frac{c}{d}\right) = \frac{a}{b} \times \frac{c}{d}$$

Thus,

$$\left(-\frac{2}{3}\right) \times \left(-\frac{5}{7}\right) = \frac{2}{3} \times \frac{5}{7} = \frac{10}{21}$$

and

$$\left(-\frac{5}{3}\right) \times \left(-\frac{4}{15}\right) = \frac{5}{3} \times \frac{4}{15} = \frac{20}{45} = \frac{4}{9}$$

We have previously seen (Section 2.5) that $-\frac{a}{b} = \frac{-a}{b}$. Now if we assume that the basic principle (Section 2.3),

$$\frac{a}{b} = \frac{a \times m}{b \times m}$$

holds for all rational numbers, we can show that

$$\frac{a}{-b} = \frac{-a}{b} \text{ and } \frac{-a}{-b} = \frac{a}{b}$$

For, by this basic principle, we have

$$\frac{a}{-b} = \frac{a \times (-1)}{(-b) \times (-1)} = \frac{-a}{b}$$

and

$$\frac{-a}{-b} = \frac{(-a) \times (-1)}{(-b) \times (-1)} = \frac{a}{b}$$

Thus

$$-\frac{2}{3} = \frac{-2}{3} = \frac{2}{-3}$$

and

$$\frac{-2}{-3} = \frac{2}{3}$$

Note that, for any rational number $\frac{a}{b}$ with $a \neq 0$, we have

$$\frac{a}{b} \times \frac{b}{a} = 1$$

We say that $\frac{b}{a}$ is the *multiplicative inverse (reciprocal)* of $\frac{a}{b}$. So

$$\frac{2}{3} \times \frac{3}{2} = \frac{6}{6} = 1$$

and

$$\frac{-4}{3} \times \frac{3}{-4} = \frac{-12}{-12} = 1$$

Thus

$\frac{3}{2}$ is the multiplicative inverse of $\frac{2}{3}$

$\frac{2}{3}$ is the multiplicative inverse of $\frac{3}{2}$

$-\frac{3}{4}$ is the multiplicative inverse of $-\frac{4}{3}$

and

$-\frac{4}{3}$ is the multiplicative inverse of $-\frac{3}{4}$

EXERCISES 2.7

In exercises 1–30 find the product. Write any fractions in your answers in lowest terms.

Examples: $3 \times 4 = 12$; $-3 \times 4 = -12$; $3 \times (-4) = -12$; $-3 \times (-4) = 12$

1. 4×5

2. $-6 \times (-2)$

3. $7 \times (-3)$

4. 2×9

5. $(-5) \times (-4)$

6. $-15 \times (-2)$
7. 5×0
8. 0×6
9. -29×0

10. $0 \times (-17)$

11. $1 \times (-1)$

12. $-1 \times (-1)$

13. $-\dfrac{1}{4} \times \dfrac{1}{2}$

14. $\dfrac{-2}{3} \times \left(-\dfrac{1}{2}\right)$

15. $\dfrac{5}{6} \times \left(-\dfrac{6}{7}\right)$

16. $\dfrac{2}{5} \times \dfrac{55}{12}$

17. $-\dfrac{6}{5} \times \left(-\dfrac{4}{5}\right)$

18. $-\dfrac{7}{8} \times \dfrac{12}{21}$

19. $-\dfrac{7}{8} \times \left(-\dfrac{12}{21}\right)$

20. $-1 \times \dfrac{5}{6}$

21. $3 \times (-5) \times 2$

22. $-4 \times 6 \times 3$

23. $(-5) \times (-6) \times (-10)$

24. $(-2) \times (-11) \times (-1)$

25. $4 \times \left(-\dfrac{1}{2}\right) \times (-2)$

26. $7 \times \left(-\dfrac{1}{3}\right) \times (-3)$

27. $\left(-\dfrac{1}{2}\right) \times \left(\dfrac{2}{3}\right) \times 0 \times (-12)$

28. $[2 + (-3)] \times (-5)$

29. $\left(-\dfrac{3}{4} + \dfrac{5}{8}\right) \times (-1)$

30. $\left[-\dfrac{2}{3} + \left(-\dfrac{1}{6}\right)\right] \times \left[\dfrac{3}{4} + \left(-\dfrac{6}{8}\right)\right]$

In exercises 31–45 evaluate the given expression if the replacements for the variables are as follows:

$$a = -1, \; b = 2, \; c = -2, \; y = -3, \text{ and } x = 3$$

Examples:

a. $5ab = 5 \times (-1) \times 2 = -10$

b. $(a + b)(x + y) = (-1 + 2)[3 + (-3)] = 1 \cdot 0 = 0$

31. $4xy$	36. $-2xy$	41. $a(b + c)$
32. $\frac{1}{2}ab$	37. $-3abc$	42. $3x(a + c)$
33. $5bc$	38. $-ac$	43. $(a + x)(y + c)$
34. abc	39. $-(ac)$	44. $(a - x)(y - c)$
35. xy	40. $(a + b)(c + y)$	45. $-2abcxy$

In exercises 46–50 write the given rational number in two other ways.

Example: $\dfrac{-12}{5} = \dfrac{12}{-5} = -\dfrac{12}{5}$

46. $-\dfrac{15}{7}$	48. $\dfrac{-3}{-16}$	50. $-\dfrac{-3}{-7}$
47. $\dfrac{13}{-2}$	49. $\dfrac{-5}{3}$	

In exercises 51–56 determine whether the given statement is true or false. If the statement is false, change the second number to make the statement true.

Examples:

a. $\dfrac{-8}{2} = -\dfrac{8}{2}$ True.

b. $\dfrac{-5}{-3} = -\dfrac{5}{3}$ False. Change $-\dfrac{5}{3}$ to $\dfrac{5}{3}$. Now $\dfrac{-5}{-3} = \dfrac{5}{3}$.   True.

51. $-\dfrac{15}{5} = \dfrac{-15}{-5}$	53. $\dfrac{-12}{-3} = \dfrac{12}{3}$	55. $\dfrac{18}{-6} = \dfrac{-18}{6}$
52. $-\dfrac{6}{3} = -\dfrac{6}{-3}$	54. $\dfrac{3}{4} = \dfrac{-3}{4}$	56. $\dfrac{-7}{5} = \dfrac{7}{-5}$

In exercises 57–64 write the multiplicative inverse of the given number if such an inverse exists.

57. 5	60. $-\dfrac{7}{8}$	63. $-\dfrac{8}{5}$
58. $\dfrac{1}{3}$	61. 0	64. $\dfrac{13}{27}$
59. -27	62. $\dfrac{5}{3}$	

2.8 Division of Rational Numbers

The same argument used in Section 2.3 enables us to conclude that, for all rational numbers $\frac{a}{b}$ and $\frac{c}{d}$ with $\frac{c}{d} \neq 0$, we have

$$\frac{a}{b} \div \frac{c}{d} = \frac{a}{b} \times \frac{d}{c}$$

For example:

$$\left(-\frac{2}{3}\right) \div \left(-\frac{5}{4}\right) = \frac{-2}{3} \div \frac{-5}{4} = \frac{-2}{3} \times \frac{4}{-5} = \frac{-8}{-15} = \frac{8}{15}$$

and

$$\left(-\frac{4}{7}\right) \div \frac{2}{3} = \frac{-4}{7} \div \frac{2}{3} = \frac{-4}{7} \times \frac{3}{2} = \frac{-12}{14} = -\frac{6}{7}$$

Note that we require that $\frac{c}{d} \neq 0$ here; division by 0 is impossible. To see why this is so, let us go back to the basic principle for division:

$$a \div b = c \text{ if and only if } a = b \times c$$

For example:

$$6 \div 2 = c \text{ if and only if } 6 = 2 \times c$$

and hence $c = 3$.

Now suppose we want to divide 6 by 0. Then if

$$6 \div 0 = c$$

we would have

$$6 = 0 \times c$$

But, by the multiplication property of 0, we know that $0 \times c = 0$ for all rational numbers c. Hence there is no number c such that $6 \div 0 = c$. The same argument applies, of course, for any other nonzero number a: $a \div 0 = c$ is impossible if $a \neq 0$.

What is the situation if $a = 0$? That is, what about $0 \div 0$? By the same argument as above we conclude that if

$$0 \div 0 = c$$

we would have to have

$$0 = 0 \times c$$

Now there certainly exists a number c such that $0 = 0 \times c$. We have, for example, $0 = 0 \times 1$. But we also have $0 = 0 \times 2, 0 \times \frac{1}{2} = 0$ and, indeed, $0 = 0 \times c$ for *any* rational number c. Faced with the fact that there is no *unique* answer to $0 \div 0$, we agree that $a \div 0$ does not exist for any rational number a.

EXERCISES 2.8

In exercises 1–15 write the indicated quotient in the form $\frac{a}{b}$ or $-\frac{a}{b}$ where a and b are positive integers and $\frac{a}{b}$ is in lowest terms.

Example: $\dfrac{3}{4} \div \dfrac{-7}{8} = \dfrac{3}{4} \times \dfrac{8}{-7} = \dfrac{24}{-28} = \dfrac{6 \times 4}{(-7) \times 4} = \dfrac{6}{-7} = -\dfrac{6}{7}$

1. $\dfrac{1}{3} \div \dfrac{3}{4}$

2. $\dfrac{-7}{8} \div \dfrac{-2}{3}$

3. $-\dfrac{5}{8} \div \dfrac{5}{6}$

4. $\dfrac{2}{5} \div -\dfrac{3}{10}$

5. $\dfrac{11}{12} \div \dfrac{-4}{7}$

6. $\dfrac{5}{8} \div \dfrac{-5}{8}$

7. $\dfrac{-4}{5} \div \dfrac{7}{10}$

8. $\dfrac{5}{8} \div -4$

9. $-18 \div \dfrac{9}{10}$

10. $-1\dfrac{1}{2} \div -3$

11. $-1\dfrac{5}{6} \div \dfrac{5}{12}$

12. $-2\dfrac{7}{16} \div 13$

13. $\dfrac{5}{6} \div 1\dfrac{1}{9}$

14. $-2\dfrac{1}{4} \div -3\dfrac{3}{8}$

15. $1\dfrac{2}{5} \div -2\dfrac{2}{3}$

2.9 The Field of Rational Numbers

With the definitions we have made concerning addition and multiplication of rational numbers, the properties we listed in Section 2.2 for whole numbers continue to hold for all rational numbers.

That is, for all *rational* numbers a, b, and c:

1. $a + b = b + a$ and $a \times b = b \times a$ (Commutative properties of addition and multiplication)

2. $a + (b + c) = (a + b) + c$ and $a \times (b \times c) = (a \times b) \times c$ (Associative properties of addition and multiplication)

3. $a \times (b + c) = (a \times b) + (a \times c)$ (Distributive property)

4. $a + 0 = 0 + a = a$ (Addition property of 0)

5. $a \times 1 = 1 \times a = a$ (Multiplication property of 1)

Indeed, these last two properties can be strengthened to read:

4′. There is a *unique* number, 0, such that $a + 0 = 0 + a = a$ for all rational numbers a.

5′. There is a *unique* number, 1, such that $a \times 1 = 1 \times a = a$ for all rational numbers a.

The number 0 is called the *additive identity* and the number 1 is called the *multiplicative identity*.

In addition to these properties, however, we now have:

6. For every rational number a, there exists a unique rational number b such that $a + b = b + a = 0$. **(Existence of additive inverses)**
Examples: If $a = \frac{1}{2}$, then $b = -\frac{1}{2}$ since $\frac{1}{2} + (-\frac{1}{2}) = 0$.

If $a = -\frac{5}{3}$, then $b = \frac{5}{3}$ since $-\frac{5}{3} + \frac{5}{3} = 0$.

If $a = 0$, then $b = 0$ since $0 + 0 = 0$.

7. For every rational number $a \neq 0$, there exists a unique rational number c such that $a \times c = c \times a = 1$. **(Existence of multiplicative inverses)**
Examples: If $a = \frac{1}{2}$, then $c = 2$ since $\frac{1}{2} \times 2 = 1$.

If $a = -2$, then $c = -\frac{1}{2}$ since $(-2) \times (-\frac{1}{2}) = 1$.

If $a = 1$, then $c = 1$ since $1 \times 1 = 1$.

If $a = \frac{3}{2}$, then $c = \frac{2}{3}$ since $\frac{2}{3} \times \frac{3}{2} = 1$.

There are two other properties which we have tacitly assumed. These are called the *closure properties.*

8. If a and b are rational numbers, then $a + b$ and $a \times b$ are rational numbers.

A set of numbers having properties 1–8 is called a *field,* and these properties are called the *field properties.* The set of real numbers. discussed in Section 2.10, is also a field, as is the set of complex numbers considered in more advanced mathematics courses.

By use of the field properties it is possible to justify other properties and procedures. For example, from $(-4) + (-3) = -7$ we can conclude that $-7 + 3 = -4$. For, if $(-4) + (-3) = -7$, then

$$-7 + 3 = [(-4) + (-3)] + 3 \qquad \text{(Substitute } (-4) + (-3) \text{ for } -7)$$

$$= (-4) + [(-3) + 3] \qquad \text{(Associative property of addition)}$$

$$= -4 + 0 \qquad \text{(Additive inverse property)}$$

$$= -4 \qquad \text{(Addition property of 0)}$$

Note that we have not listed the multiplication property of 0 as one of the basic field properties. This is because the multiplication property of 0 can be *proved* from the field properties as follows:

$$1 + 0 = 1 \qquad \text{(Addition property of 0)}$$

$$(1 + 0) \times a = 1 \times a \qquad \text{(Multiplying both } 1 + 0 \text{ and 1 by } a\text{)}$$

$$(1 \times a) + (0 \times a) = 1 \times a \qquad \text{(Distributive property)}$$

$$a + (0 \times a) = a \qquad \text{(Multiplication property of 1)}$$

$$0 \times a = 0 \qquad \text{(0 is the } \textit{unique} \text{ additive identity)}$$

EXERCISES 2.9

In exercises 1–15 state the field property of the rational number system that is illustrated by the equality.

1. $5 \times 6 = 6 \times 5$

2. $(3 + 4) + 7 = 3 + (4 + 7)$

3. $12 + 15 = 15 + 12$

4. $5 \times 0 = 0$

5. $1 \times \dfrac{1}{2} = \dfrac{1}{2}$

6. $4 \times (6 + 5) = (4 \times 6) + (4 \times 5)$

7. $\dfrac{3}{4} + \left(\dfrac{-3}{4}\right) = 0$

8. $\dfrac{7}{8} \times \dfrac{8}{7} = 1$

9. $\dfrac{1}{2} + \left(\dfrac{2}{3} + \dfrac{4}{5}\right) = \dfrac{1}{2} + \left(\dfrac{4}{5} + \dfrac{2}{3}\right)$

10. $\dfrac{7}{8} \times \left(\dfrac{3}{4} + \dfrac{5}{6}\right) = \left(\dfrac{3}{4} + \dfrac{5}{6}\right) \times \dfrac{7}{8}$

11. $\left(\dfrac{1}{3} \times \dfrac{5}{6}\right) + \left(\dfrac{1}{3} \times \dfrac{3}{4}\right) = \dfrac{1}{3} \times \left(\dfrac{5}{6} + \dfrac{3}{4}\right)$

12. $xy = yx$

13. $(x + y) + z = z + (x + y)$

14. $ab + ac = a(b + c)$

15. $x \cdot \dfrac{1}{x} = 1$

In exercises 16–35 determine if the given statement is true for every rational number a, b, x, or y. If it is true, state the field property of rational numbers which it illustrates; if it is false, give a reason why it is false.

Examples:

a. $3(x + y) = 3x + 3y$
    True. Distributive property.

b. $3 + (x \cdot y) = (3 + x)(3 + y)$
False. If $x = 5$ and $y = 2$, then

$$3 + (5 \cdot 2) = (3 + 5)(3 + 2)$$

$$3 + 10 = 8 \cdot 5$$

$$13 = 40 \text{ is a false statement.}$$

16. $x(2 + 5) = (2 + 5) \cdot x$

17. $(y + 1) \cdot 2 = (1 + y) \cdot 2$

18. $5(x + 1) = 5x + 1$

19. $x \cdot (y \cdot a) = (x \cdot y) \cdot a$

20. $\frac{1}{2} \times (a + b) = \frac{1}{2}a + \frac{1}{2}b$

21. $13 + a = a + 13$

22. $2a(b + x) = 2ab + 2ax$

23. $2x + 3x = 5x$

24. $\frac{1}{4}a + \frac{1}{2}a = \frac{3}{4}a$

25. $(2x)(ab) = (2x)(ba)$

26. $a + 0 = 0 + a$

27. $\frac{a}{b} \cdot \frac{b}{a} = 1$

28. $(2x + a) + 3y = 2x + (a + 3y)$

29. $(7y + xy) = (7 + x)y$

30. $-x + x = 0$

31. $13x + b = b + 13x$

32. $2a(x + y) = 2ax + y$

33. $ab + xy = ba + yx$

34. $3x + \frac{1}{2}x = 3\frac{1}{2}x$

35. $2a + 2b = 2(a + b)$

In exercises 36–45 use the commutative and associative properties for multiplication of rational numbers to write the given product in a simpler form.

Example: $(2x)(3y) = (2 \cdot 3)(x \cdot y) = 6xy$

36. $(6c)(3d)$

37. $(2y)(7x)$

38. $(5pq)(3r)$

39. $(15b)(15ac)$

40. $\left(\frac{1}{2}x\right)\left(\frac{1}{6}y\right)$

41. $\left(\frac{3}{4}p\right)\left(\frac{1}{2}q\right)$

42. $\left(\frac{2}{3}a\right)\left(\frac{1}{4}\right)$

43. $\left(\frac{3}{8}\right)\left(\frac{1}{3}xy\right)$

44. $(2x)\left(\frac{1}{4}y\right)(3bc)$

45. $(4a)(b)(c)\left(\frac{1}{4}\right)$

In exercises 46–55 multiply by using the distributive property.

Examples:

a. $6(p + q) = 6p + 6q$

b. $2a(6 + 5b) = (2a \cdot 6) + (2a \cdot 5b) = 12a + 10ab$

46. $y(x + 4)$ 50. $a(b + c)$ 54. $\frac{2}{3}(3a + 6b + 9c)$

47. $9(4 + m)$ 51. $\frac{1}{6}(2x + 3)$ 55. $ab(x + 2y)$

48. $(x + 4) \cdot y$ 52. $\frac{1}{4}(12w + 32y)$

49. $(4 + m) \cdot 9$ 53. $2(3x + 4y + 2z)$

In exercises 56–70 use the distributive property to rewrite the given sum as a product.

Examples:

a. $6a + 5a = (6 + 5)a = 11a$

b. $6a + 8b = (2 \cdot 3a) + (2 \cdot 4b) = 2(3a + 4b)$

56. $2x + 3x$ 64. $6y + 6y + 6y$

57. $5y + 6y$ 65. $x + x$ (Hint: $x = 1 \cdot x$)

58. $\frac{1}{2}a + \frac{3}{2}a$ 66. $3rs + 3rt$

59. $\frac{3}{4}y + \frac{3}{4}y$ 67. $3p + p$

60. $2x + 6y$ 68. $\frac{1}{4}x + x + \frac{3}{4}x$

61. $15a + 20b$ 69. $abc + abd + acd$

62. $\frac{3}{5}x + \frac{2}{5}y$ 70. $y + y + y + y$

63. $6x + 9y + 12z$

2.10 Decimal Fractions

We can easily write some rational numbers as *decimal fractions.* Thus

$$\frac{2}{5} = \frac{2 \cdot 2}{5 \cdot 2} = \frac{4}{10} = 0.4 \qquad \text{or} \qquad 5 \overline{\smash{)}2.0} \atop \text{...}$$

$$\begin{array}{r} 0.4 \\ 5\overline{)2.0} \\ \underline{2.0} \\ 0 \end{array}$$

$$\frac{-3}{50} = -\frac{3 \cdot 2}{50 \cdot 2} = -\frac{6}{100} = -0.06 \qquad \text{or} \qquad 50\overline{)\begin{array}{r} -0.06 \\ -3.00 \\ \underline{3.00} \\ 0 \end{array}}$$

Because 3 is not a factor of 10, the fraction $\frac{2}{3}$ cannot be written as a fraction with a denominator of 10, 100, 1000, etc. Thus to write $\frac{2}{3}$ as a decimal we must calculate $2 \div 3$. Doing this we obtain

$$3\overline{)\begin{array}{r} 0.66 \ldots \\ 2.000 \ldots \\ \underline{18} \\ 20 \\ \underline{18} \\ 2 \\ \ldots \end{array}}$$

where 6's repeat endlessly. How about $\frac{8}{7}$?

We have

$$7\overline{)\begin{array}{r} 1.142857 \ldots \\ 8.000000 \ldots \\ \underline{7} \\ 10 \\ \underline{7} \\ 30 \\ \underline{28} \\ 20 \\ \underline{14} \\ 60 \\ \underline{56} \\ 40 \\ \underline{35} \\ 50 \\ \underline{49} \\ 10 \\ \ldots \end{array}}$$

and, at this point, it is clear that we begin to repeat, since a remainder of 1 has occurred for the second time. Thus

$$\frac{8}{7} = 1.142857 \ 142857 \ 142857 \ldots$$

where the segment of digits, 142857, repeats endlessly.

It is no accident that we obtain a sequence of repeating digits. The possible remainders when dividing by 7 are 0 (but only if the division terminates), 1, 2, 3, 4, 5, and 6. (Why not 7, 8, . . . ?) Thus at some point a remainder will be repeated and hence a segment of digits begin to repeat. A similar argument applies to any fraction $\frac{a}{b}$. If the division of a by b does not terminate, the possible remainders are $1, 2, \ldots$, $b-1$ and so, at some point, a remainder must be repeated.

We call a decimal such as 0.4 or 0.06 a *terminating* decimal and one such as 0.666 ... or 1.142857 142857 ... a *repeating* decimal. Every fraction can be written either as a terminating decimal or a repeating decimal. (Of course, a terminating decimal can be considered as a repeating decimal where the repeating digit is simply 0: 0.4 = 0.40000. . . and 0.06 = 0.0600000. . . .)

Conversely, it is certainly obvious that every terminating decimal can be written as an ordinary (*common*) fraction: $0.62 = \frac{62}{100}$, $0.037 = \frac{37}{1000}$, etc. We will also show, in Chapter 3, that every repeating decimal is equal to a common fraction.

If we regard whole numbers as decimals in the sense that 0 = 0.0, 1 = 1.0, etc., we can say, then, that the set of rational numbers is the set consisting of 0, the positive and negative terminating decimals, and the positive and negative repeating decimals. In diagramatic form we have

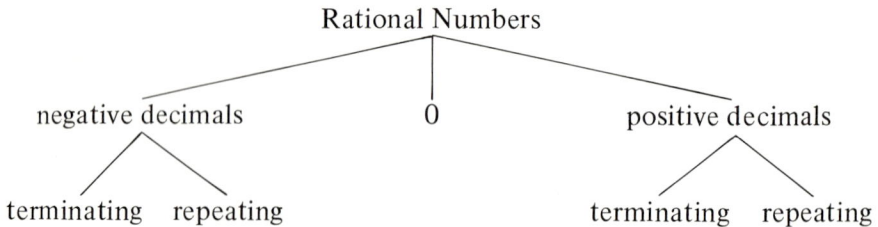

Rational Numbers

negative decimals   0   positive decimals

terminating   repeating   terminating   repeating

It is easy to construct nonrepeating decimals. For example, consider

$$4.101001000100001\ldots$$

where after each 1 we have first one 0, then two 0's, then three, etc. This is a nonrepeating and nonterminating decimal and hence it is not a rational number. It is an example of an *irrational* number. Other irrational numbers, such as $\sqrt{2}, \sqrt{7}$, and $\sqrt[3]{5}$, will be considered in Chapter 5. The set of irrational numbers together with the set of rational numbers form the set of *real* numbers.

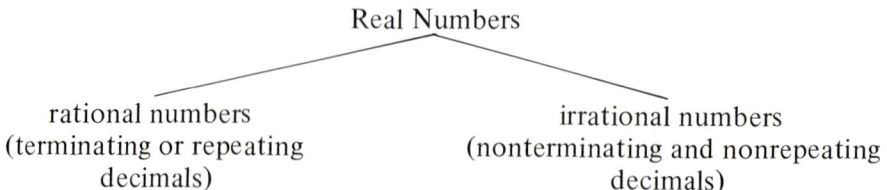

Real Numbers

rational numbers   irrational numbers
(terminating or repeating   (nonterminating and nonrepeating
decimals)   decimals)

There is a *one-to-one correspondence* between the set of real numbers and the points on a number line; i.e., to every point on a given number line there corresponds a real number, and to every real number there corresponds a point on the number line.

A bar over the repeating cycle is commonly used to indicate a repeating decimal. Thus we write

$$0.666\ldots = 0.\overline{6}, \quad 0.142857142857\ldots = 0.\overline{142857}$$

and

$$-1.325757\ldots = -1.32\overline{57}$$

(Sometimes $0.\overline{6}$ is written as $0.\dot{6}$, $0.\overline{142857}$ as $0.\dot{1}42857\dot{7}$, and $-1.32\overline{57}$, as $-1.32\dot{5}\dot{7}$, etc.)

EXERCISES 2.10

In exercises 1-10 show that the given common fraction can be written as a terminating decimal.

Example: $\dfrac{9}{20} = \dfrac{9}{2 \cdot 2 \cdot 5} \cdot \dfrac{5}{5} = \dfrac{45}{10 \cdot 10} = \dfrac{45}{100} = 0.45$

1. $\dfrac{7}{20}$ 5. $\dfrac{57}{200}$ 9. $\dfrac{3}{32}$

2. $\dfrac{3}{8}$ 6. $\dfrac{23}{125}$ 10. $\dfrac{121}{80}$

3. $\dfrac{27}{25}$ 7. $\dfrac{17}{40}$

4. $\dfrac{3}{4}$ 8. $\dfrac{5}{16}$

In exercises 11-20 show that the decimal representation of the given rational number does not terminate but does repeat.

Example: $\dfrac{4}{15} = \dfrac{4}{3 \cdot 5}$ 3 is not a factor of 10; thus $\frac{4}{15}$ is not a terminating decimal.

$$\begin{array}{r} 0.266\ldots \\ 15\overline{)4.000\ldots} \\ \underline{30} \\ 100 \\ \underline{90} \\ 100 \\ \cdots \end{array} \qquad \dfrac{4}{15} = 0.2\overline{6}$$

11. $\dfrac{5}{9}$ 15. $\dfrac{5}{6}$ 19. $\dfrac{4}{7}$

12. $\dfrac{9}{11}$ 16. $\dfrac{5}{12}$ 20. $\dfrac{5}{37}$

13. $\dfrac{1}{7}$ 17. $\dfrac{8}{9}$

14. $\dfrac{2}{3}$ 18. $\dfrac{17}{48}$

Write the fractions in exercises 21, 22, and 23 as decimal fractions. Study your answers to exercises 21, 22, and 23 and try to find a way of determining the decimal representation of such fractions without actually doing the division; then write the fractions in exercises 24–30 as decimal fractions.

21. $\dfrac{35}{99}$ 25. $\dfrac{13}{99}$ 29. $\dfrac{5}{999}$

22. $\dfrac{47}{99}$ 26. $\dfrac{28}{99}$ 30. $\dfrac{5}{99}$

23. $\dfrac{213}{999}$ 27. $\dfrac{500}{999}$

24. $\dfrac{7}{9}$ 28. $\dfrac{50}{999}$

31. Express $\frac{1}{3}$ and $\frac{2}{3}$ as repeating decimals and then find their sum in terms of a repeating decimal. Since $\frac{1}{3} + \frac{2}{3} = 1$, what conclusion can you draw?

32. Could there be as many as 20 digits in the repeating cycle for the decimal representation of $\frac{3}{17}$? Explain.

33. What would be the greatest possible number of digits in the repeating cycle for the decimal representation of $\frac{26}{37}$? How many digits are there in the repeating cycle?

34. If a fraction in lowest terms has a denominator of 99, how many digits will be in the repeating cycle of its decimal representation? Answer the same question for a fraction in lowest terms whose denominator is 999.

35. (a) Find the prime factors of each of the denominators in exercises 1–10.

    (b) Find the prime factors of each of the denominators of the fractions in exercises 11–20.

    (c) On the basis of the results in (a) and (b), develop a rule which would enable you to determine before you do the division when the decimal representation for a rational number will be a terminating decimal.

2.11 Inequalities

We use the symbol " $>$ " to mean "greater than." Thus $3 > 2$ is read as "three is greater than two." The symbol " $<$ " means "less than," so we read $2 < 3$ as "two is less than three." On the number line shown in Figure 2.9 we see that the point corresponding to 3 is to the

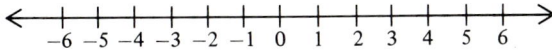

Figure 2.9

right of the point corresponding to 2. We extend the idea of greater than to negative numbers with the definition that $a > b$ whenever the point corresponding to a on a number line is to the right of the point corresponding to b. Thus, as shown on the number line in Figure 2.10,

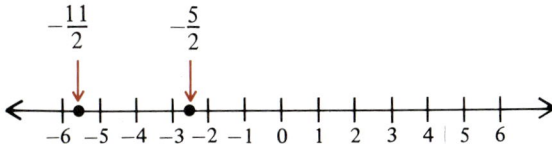

Figure 2.10

$-3 > -4$, $-\frac{5}{2} > -\frac{11}{2}$, and $0 > -1$. Note that every positive number is greater than any negative number. We can also write, of course, that $-4 < -3$, $-\frac{11}{2} < -\frac{5}{2}$, and $-1 < 0$ and note that every negative number is less than 0 and less than any positive number.

It is easy to see that $\frac{3}{4} > \frac{1}{2}$. But what about $\frac{13}{14}$ and $\frac{15}{17}$? Which of the following statements is true?

$$\frac{13}{14} = \frac{15}{17}, \quad \frac{13}{14} > \frac{15}{17}, \quad \text{or} \quad \frac{13}{14} < \frac{15}{17} \text{ ?}$$

A quick way to test for equality or inequality of fractions is to use the following theorem:

If b and d are positive numbers (i.e., $b > 0$ and $d > 0$), then

$$\frac{a}{b} = \frac{c}{d} \qquad \text{if and only if } ad = bc$$

$$\frac{a}{b} > \frac{c}{d} \qquad \text{if and only if } ad > bc$$

$$\frac{a}{b} < \frac{c}{d} \qquad \text{if and only if } ad < bc$$

For example:

$$\frac{13}{14} > \frac{15}{17} \text{ because } 13 \times 17 \,(= 221) > 14 \times 15 \,(= 210)$$

$$\frac{13}{14} = \frac{26}{28} \text{ because } 13 \times 28 = 14 \times 26$$

and

$$\frac{-12}{11} < \frac{-13}{12} \text{ because } (-12) \times 12 \,(= -144) < (-13) \times 11 \,(= -143)$$

Note also that

$$-\frac{12}{11} < -\frac{13}{12}$$

because

$$-\frac{12}{11} = \frac{-12}{11} \quad \text{and} \quad -\frac{13}{12} = \frac{-13}{12}$$

To apply the test, however, fractions must be written in the form $\frac{a}{b}$ where $b > 0$.

There are four basic properties of inequalities. They are, for any rational numbers a, b, and c:

1. One and only one of the following statements is true: $a = b, a > b,$ or $a < b.$ **(Trichotomy property)**

2. If $a > b$, then $a + c > b + c.$ **(Addition property)**
   Example: $3 > 2$ and $3 + 1 > 2 + 1.$

3. If $a > b$ and $c > 0$, then $a \times c > b \times c.$
   Example: $3 > 2, 4 > 0,$ and $3 \times 4 > 2 \times 4.$

4. If $a > b$ and $c < 0$, then $a \times c < b \times c.$
   Example: $3 > 2, -4 < 0,$ and $3 \times (-4) < 2 \times (-4).$

(Multiplication properties)

Note that in (4) we begin with $a > b$ and get $a \times b < b \times c.$ That is, $> \rightarrow <$. When we go from $>$ to $<$, or vice versa, we often say that we have *reversed the sense* of the inequality.

We can also write properties (2) – (4) in terms of $<$. That is,

2'. If $a < b$, then $a + c < b + c.$
   Example: $2 < 3$ and $2 + 1 < 3 + 1.$

3'. If $a < b$ and $c > 0$, then $a \times c < b \times c.$
   Example: $2 < 3, 4 > 0,$ and $2 \times 4 < 3 \times 4.$

4′. If $a < b$ and $c < 0$, then $a \times c > b \times c$.
    Example: $2 < 3$, $-4 < 0$, and $2 \times (-4) > 3 \times (-4)$.

Sometimes the symbols " $\geqslant$ " and " $\leqslant$ " are used, as in $x \geqslant 5$. This means that $x = 5$ or $x > 5$. Similarly, $x \leqslant \frac{1}{2}$ means that $x = \frac{1}{2}$ or $x < \frac{1}{2}$.

EXERCISES 2.11

1. From the number line shown below, determine whether $a < b$ or $a > b$ and whether $c > d$ or $c < d$.

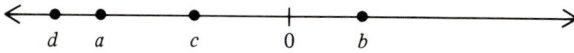

In exercises 2–15 replace ⟨?⟩ with $<, >$, or $=$ so as to form a correct statement.

2. 5 ⟨?⟩ -2

3. -2 ⟨?⟩ 5

4. $\frac{1}{3}$ ⟨?⟩ $\frac{1}{2}$

5. $\frac{1}{2}$ ⟨?⟩ $\frac{1}{3}$

6. $\frac{-4}{6}$ ⟨?⟩ $\frac{14}{-21}$

7. $1\frac{3}{5}$ ⟨?⟩ $-3\frac{1}{5}$

8. $\frac{-9}{16}$ ⟨?⟩ $\frac{-7}{12}$

9. $\frac{5}{16}$ ⟨?⟩ $\frac{-2}{5}$

10. -7 ⟨?⟩ -5

11. $\frac{-5}{6}$ ⟨?⟩ $-\frac{7}{6}$

12. $\frac{8}{12}$ ⟨?⟩ $\frac{10}{15}$

13. $\frac{17}{20}$ ⟨?⟩ $\frac{15}{17}$

14. $\frac{-28}{3}$ ⟨?⟩ $\frac{30}{-4}$

15. $\frac{121}{125}$ ⟨?⟩ $\frac{-364}{488}$

In exercises 16–25 indicate whether each statement is true or false.

16. $2 + 7 > 8 - 17$

17. $\frac{1}{2} + \frac{1}{3} \geqslant \frac{1}{3} + \frac{1}{2}$

18. $-7 + 5 < -2$

19. $-2 \geqslant -7 + 5$

20. $\frac{7}{8} + \frac{-5}{6} > \frac{-3}{4} + \frac{2}{3}$

21. $\frac{1}{2} - \frac{2}{5} < \frac{3}{10}$

22. $\frac{5}{8} + \frac{3}{5} \geqslant \frac{1}{8} + \frac{11}{10}$

23. $2 < 6 < 8$

24. $-10 < -\frac{1}{2} < 0$

25. $-2 < 0 < -3$

In exercises 26–35 use the given inequality and the number in parentheses to write two new inequalities by (a) adding the indicated number to both sides of the inequality and (b) multiplying both sides of the inequality by the indicated number.

Example: $-5 < 7, (3)$

a. by the addition property: $-5 + 3 < 7 + 3 \rightarrow -2 < 10$

b. by the multiplication property: $-5(3) < 7(3) \rightarrow -15 < 21$

26. $3 > 2, (5)$

27. $-6 < -5, (2)$

28. $-2 < 3, (-5)$

29. $3 > -4, (-3)$

30. $\dfrac{1}{2} < \dfrac{3}{4}, \left(\dfrac{1}{4}\right)$

31. $-\dfrac{1}{2} > \dfrac{-5}{8}, \left(-\dfrac{1}{2}\right)$

32. $3x < 5x, (2)$

33. $2x > -3x, (-1)$

34. $5x + 3 > 2x + 7, (-3)$

35. $2x - 5 < 3x + 2, (-2)$

In exercises 36–40 (a) replace ⑦ with $<, >,$ or $=$ so as to form a true statement and (b) indicate what property of inequalities is illustrated.

Examples:

a. If $3a < 5a$, then $6a$ ⑦ $10a$.

Solution: If $3a < 5a$, then $6a < 10a$ by the multiplication property of inequalities using a positive number (2) as a multiplier.

b. If $2x > -3y$, then $-6x$ ⑦ $9y$.

Solution: If $2x > -3y$, then $-6x < 9y$ by the multiplication property of inequalities using a negative number (-3) as a multiplier.

36. If $a < b$, then $-3a$ ⑦ $-3b$.

37. If $7 > c$, then 9 ⑦ $c + 2$.

38. If $r > s$, then $-\dfrac{1}{2}r$ ⑦ $-\dfrac{1}{2}s$.

39. If $m < n$, then $m - 2$ ⑦ $n - 2$.

40. If $4y < 3x$, then $-8y + 2$ ⑦ $-6x + 2$.

41. Arrange the following rational numbers in order from smallest to largest:
$\dfrac{5}{6}$, 0.41, $-\dfrac{37}{41}$, $0, \dfrac{4}{5}$, 2.3, $-3, \dfrac{49}{60}$, $-\dfrac{20}{7}, \dfrac{7}{3}, \dfrac{2}{5}$, $-\dfrac{5}{2}$.

42. Arrange the following rational numbers in order from largest to smallest:
$\dfrac{3}{8}$, 0.52, $\dfrac{23}{14}$, $-5\dfrac{1}{2}$, -2.6, $-\dfrac{40}{7}, \dfrac{7}{3}$, $-\dfrac{3}{8}$.

43. Is there a smallest *integer* n such that $n < 435$ and $n > -157$? A largest integer? If such numbers exist, what are they?

44.   Is there a smallest *rational number* x such that $x < 5\frac{1}{2}$ and $x > -\frac{3}{4}$? A largest rational number? If such numbers exist, what are they?

45.   Is there a smallest integer n such that $\frac{1}{2} n > -24$? A largest integer? If such numbers exist, what are they?

46.   Is there a smallest rational number x such that $\frac{1}{2} x > -24$? A largest rational number? If such numbers exist, what are they?

47.   Is there a smallest positive integer? A largest?

48.   Is there a smallest negative integer? A largest?

49.   Is there a smallest positive rational number? A largest?

50.   Is there a smallest negative rational number? A largest?

2.12 Absolute Value

We can think of the correspondence between points on a number line and the real numbers as measuring the distance from the point corresponding to 0 (the 0-point) to a given point. Thus the distance from the 0-point to the 2-point is 2 units. But, also the –2-point is at a distance of 2 units from the 0-point, as shown in Figure 2.11. In

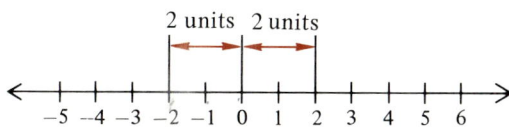

Figure 2.11

general, for any real number a, the –a-point is the same distance from the 0-point as is the a-point. In each case we consider the distance as being a positive number except that, of course, we consider 0 as the distance between the 0-point and itself.

Looking at distances on a number line leads to the concept of the *absolute value* of a real number r as the distance of the r-point from the 0-point. We write $|r|$ for the absolute value of r and make the formal definition:

1. $|0| = 0$

2. $|r| = r$ if r is a positive number

3. $|r| = -r$ is r is a negative number

Thus, $\left|\frac{1}{2}\right| = \frac{1}{2}$ and $\left|-\frac{1}{2}\right| = -\left(-\frac{1}{2}\right) = \frac{1}{2}$.

EXERCISES 2.12

In exercises 1–15 write each expression in the form a or $-a$ where a is a positive rational number.

Examples: $|-5| = 5$ and $-\left|\frac{3}{2}\right| = -\frac{3}{2}$.

1. $|5|$
2. $|-7|$
3. $-\left|\frac{1}{2}\right|$
4. $\left|-\frac{3}{4}\right|$
5. $|-(-5)|$

6. $|0|$
7. $|-13 + 1|$
8. $-|-8 + (-6)|$
9. $\left|\frac{1}{2} \times 0\right|$
10. $\left|-\frac{1}{2} + \frac{3}{4}\right|$

11. $-|5 \times (-3)|$
12. $|(-5) \times (-3)|$
13. $|-(-(-5))|$
14. $-|-(-(-(-9)))|$
15. $|-[5 + (-3)]|$

16. If x is a negative rational number, i.e., $x < 0$, what kind of number is
    (a) $|x|$?   (b) $-|x|$?
17. If x is a positive rational number, i.e., $x > 0$, what kind of number is
    (a) $|x|$?   (b) $-|x|$?
18. If x is replaced by zero, what is (a) $|x|$? (b) $-|x|$?

In exercises 19–30 perform the indicated operations.

Example: $|7| + |-3| = 7 + 3 = 10$

19. $|-3| + |-1|$
20. $|5| + |0|$
21. $-(|6| + |-4|)$
22. $|-17| + |5|$
23. $\left|-\frac{1}{2}\right| - \left|\frac{3}{4}\right|$
24. $-\left|\frac{2}{3}\right| + \left|\frac{-2}{3}\right|$

25. $7 - (|-4| + |-3|)$
26. $-|-2|(|3| - |-4|)$
27. $-(|2| + 3 \cdot |5|)$
28. $6(|-3| + |-2|) + 3(-|-5|)$
29. $(|-10| + |-2|) + (|14| + |-3|)$
30. $\left(\left|\frac{1}{2}\right| - \left|\frac{3}{4}\right|\right) - \left(-\left|\frac{7}{8} - \frac{5}{8}\right|\right)$

In exercises 31–40 indicate whether the given statement is true or false.

31. $|-5| < 2$
32. $|-4| < |-5|$
33. $|13| = |-13|$
34. $\left|\frac{3}{7}\right| > \left|\frac{5}{11}\right|$
35. $|-11| = -(-11)$

36. $\left|\frac{7}{13}\right| > \left|-\frac{13}{25}\right|$
37. $|2 - 5| = |2| - |5|$
38. $|4 - 9| = -(4 - 9)$
39. $|-3| \cdot |5| = -|3| \cdot (-|-5|)$
40. $\left|\frac{21}{23} - \frac{4}{17}\right| < 0$

In exercises 41–50 replace ⑦ with <, =, or > so as to form a true statement.

41. $\left| 5 + 8 \right|$ ⑦ $\left| 5 \right| + \left| 8 \right|$
42. $\left| -5 + (-8) \right|$ ⑦ $\left| -5 \right| + \left| -8 \right|$
43. $\left| -5 + 8 \right|$ ⑦ $\left| -5 \right| + \left| 8 \right|$
44. $\left| 5 + (-8) \right|$ ⑦ $\left| 5 \right| + \left| -8 \right|$
45. $\left| \left(-\frac{3}{4} \right) + \left(-\frac{5}{4} \right) \right|$ ⑦ $\left| -\frac{3}{4} \right| + \left| \frac{-5}{4} \right|$
46. $\left| \frac{3}{4} + \left(\frac{-5}{4} \right) \right|$ ⑦ $\left| \frac{3}{4} \right| + \left| \frac{-5}{4} \right|$
47. $\left| 5 \cdot 7 \right|$ ⑦ $\left| 5 \right| \cdot \left| 7 \right|$
48. $\left| 5 \cdot (-7) \right|$ ⑦ $\left| 5 \right| \cdot \left| -7 \right|$
49. $\left| (-5)(-7) \right|$ ⑦ $\left| -5 \right| \cdot \left| -7 \right|$
50. $\left| (-5)(7) \right|$ ⑦ $\left| -5 \right| \cdot \left| 7 \right|$

In exercises 51 and 52 replace ⑦ with ⩽, =, or ⩾ so that each statement is true for all real numbers x and y. Use the results of exercises 41–50 to help you arrive at your answer.

51. $\left| x + y \right|$ ⑦ $\left| x \right| + \left| y \right|$
52. $\left| x \cdot y \right|$ ⑦ $\left| x \right| \cdot \left| y \right|$

REVIEW EXERCISES

Section 2.2

In exercises 1–10 determine if the given statement is true or false. For those that are true, identify the property of whole numbers illustrated by the statement.

1. $13 + 2 = 2 + 13$ T
2. $7 + 5 = 5 + 7$ T
3. $7 \times 6 = 6 \times 7$ T
4. $17 \cdot 13 = 13 \cdot 17$ T
5. $(5 + 3) + 4 = 5 + (3 + 4)$ T
6. $2 \times (3 \times 4) = (2 \times 3) \times 4$ T
7. $5 \times (2 + 4) = (5 \times 2) + (5 \times 4)$ F
8. $(7 + 6) \times 2 = (7 \times 2) + (6 \times 2)$ F
9. $(8 \cdot 5) + (8 \cdot 3) = 8 \cdot (5 + 3)$ T
10. $(3 + 5) \times (2 + 4) = (5 + 3) \times (4 + 2)$

In exercises 11–20 use the distributive property to write the given expression in a different way.

Examples: $7x + 7y = 7(x + y)$ and $5(a + b) = 5a + 5b$

11. $9x + 9y$ $9(x+y)$
12. $3a + 3b$ $3(a+b)$
13. $2a + 3a$ $5a$
14. $7y + 6y$ $13y$
15. $8(y + 3)$ $11y$

16. $12(x + y)$
17. $4(c + d)$
18. $(x + y) \cdot 7$
19. $(2 + 3) \cdot x$
20. $(5 + 6) \cdot y$

Section 2.3

In exercises 21–30 perform the indicated operation and write your answer in lowest terms.

21. $\dfrac{4}{7} - \dfrac{1}{2}$ $\dfrac{8}{14} - \dfrac{7}{14} = \dfrac{1}{14}$

22. $\dfrac{4}{5} + \dfrac{1}{2}$

23. $\dfrac{6}{11} + \dfrac{4}{5}$

24. $\dfrac{7}{11} - \dfrac{2}{9}$

25. $\dfrac{3}{8} \times \dfrac{2}{5}$

26. $\dfrac{6}{8} \times \dfrac{4}{5}$

27. $\dfrac{5}{7} \times \dfrac{21}{25}$

28. $\dfrac{3}{8} \div \dfrac{4}{9}$ $\dfrac{27}{32}$

29. $\dfrac{3}{16} \div \dfrac{6}{15}$

30. $\dfrac{5}{12} \div \dfrac{1}{6}$

Section 2.4

In exercises 31–35 give the simplest form of the opposite (additive inverse) of the given number.

31. -7 7

32. $-(-5)$ -5
33. $-(-(-3))$ $+3$

34. $-\dfrac{2}{3}$ $\dfrac{2}{3}$

35. $-\left(-\dfrac{3}{4}\right)$ $-\dfrac{3}{4}$

Section 2.5

In exercises 36–45 find the sums.

36. $5 + 7$ 12

37. $-5 + 7$ 2

38. $-5 + (-7)$ -12

39. $5 + (-7) = -12$

40. $-\dfrac{2}{3} + \dfrac{5}{6}$ $\dfrac{4}{6} + \dfrac{1}{6}$

41. $-\dfrac{1}{2} + \left(-\dfrac{2}{3}\right)$ —

42. $\dfrac{3}{4} + \dfrac{5}{8}$

43. $-2\dfrac{1}{3} + 5\dfrac{1}{4}$

44. $\dfrac{5}{7} + \left(-\dfrac{2}{9}\right)$

45. $-\dfrac{3}{5} + \left(-\dfrac{5}{11}\right)$

Section 2.6

In exercises 46–55 do the subtraction problem by first writing a related addition problem.

46. $\dfrac{8}{15} - \dfrac{3}{5}$

47. $\dfrac{3}{7} - \left(\dfrac{-2}{9}\right)$

48. $-\dfrac{7}{8} - \dfrac{2}{3}$

49. $-\dfrac{4}{5} - \dfrac{3}{8}$

50. $-\dfrac{3}{5} - \left(-\dfrac{2}{3}\right)$

51. $\dfrac{3}{4} - \dfrac{4}{7}$

52. $-\dfrac{7}{5} - \dfrac{6}{7}$

53. $\dfrac{5}{6} - \left(-\dfrac{9}{8}\right)$

54. $\dfrac{2}{3} - \dfrac{1}{8}$

55. $-\dfrac{3}{8} - \left(-\dfrac{2}{7}\right)$

Section 2.7

In exercises 56–65 find the product. Any fraction obtained should be reduced to lowest terms.

56. 5×7

57. -5×7

58. 5×-7

59. -5×-7

60. $\dfrac{-3}{4} \times \dfrac{5}{8}$

61. $-\dfrac{3}{4} \times -\dfrac{8}{15}$

62. $\dfrac{9}{16} \times -\dfrac{5}{6}$

63. $-\dfrac{4}{5} \times \dfrac{15}{16}$

64. $-1\dfrac{3}{4} \times -3\dfrac{1}{7}$

65. $5\dfrac{1}{3} \times -1\dfrac{1}{8}$

In exercises 66-70 give two other ways of writing the given fraction.

Example: $-\dfrac{7}{8} = \dfrac{-7}{8} = \dfrac{7}{-8}$

66. $-\dfrac{3}{4}$

69. $\dfrac{-2}{-3}$

67. $\dfrac{-5}{6}$

70. $-\dfrac{4}{3}$

68. $\dfrac{7}{-5}$

Section 2.8

In exercises 71-80 do the indicated division and write your answer in lowest terms.

71. $\dfrac{3}{7} \div \left(-\dfrac{4}{5}\right)$

76. $\dfrac{7}{8} \div \dfrac{8}{7}$

72. $-\dfrac{2}{5} \div \dfrac{3}{4}$

77. $-\dfrac{11}{7} \div \left(-\dfrac{7}{11}\right)$

73. $\dfrac{6}{11} \div \left(-\dfrac{6}{11}\right)$

78. $\left(\dfrac{5}{3} \div -\dfrac{1}{2}\right) \div \dfrac{2}{5}$

74. $-\dfrac{3}{2} \div \left(-\dfrac{3}{2}\right)$

79. $\dfrac{5}{3} \div \left(-\dfrac{1}{2} \div \dfrac{2}{5}\right)$

75. $-\dfrac{4}{3} \div \left(-\dfrac{3}{4}\right)$

80. $\left(-\dfrac{3}{7} \div \dfrac{7}{3}\right) \div \left(-\dfrac{14}{3} \div \dfrac{7}{9}\right)$

Section 2.9

In exercises 81-90 state the field property of the rational number system which is illustrated by the equality.

81. $\left(\dfrac{5}{8} + \dfrac{7}{6}\right) + \dfrac{8}{3} = \dfrac{5}{8} + \left(\dfrac{7}{6} + \dfrac{8}{3}\right)$

86. $\left(\dfrac{2}{5} + \dfrac{5}{7}\right) \times \dfrac{2}{3} = \left(\dfrac{5}{7} + \dfrac{2}{5}\right) \times \dfrac{2}{3}$

82. $\dfrac{7}{9} \times \dfrac{8}{7} = \dfrac{8}{7} \times \dfrac{7}{9}$

87. $ab = ba$

83. $\dfrac{4}{5} \times \left(\dfrac{6}{7} \times \dfrac{5}{3}\right) = \left(\dfrac{4}{5} \times \dfrac{6}{7}\right) \times \dfrac{5}{3}$

88. $a(x + y) = ax + ay$

84. $\dfrac{2}{3} \times \left(\dfrac{7}{9} + \dfrac{9}{11}\right) = \left(\dfrac{2}{3} \times \dfrac{7}{9}\right) + \left(\dfrac{2}{3} \times \dfrac{9}{11}\right)$

89. $(x + a) + b = (a + x) + b$

85. $\left(\dfrac{3}{4} \times \dfrac{2}{3}\right) + \left(\dfrac{3}{4} \times \dfrac{5}{7}\right) = \dfrac{3}{4} \times \left(\dfrac{2}{3} + \dfrac{5}{7}\right)$

90. $\dfrac{1}{a} \cdot a = 1$

In exercises 91–105 use the distributive property to rewrite the given expression in a different way.

Examples: $2(a + 3b) = 2a + 6b$ and $3x + 15y = 3(x + 5y)$.

91. $9(a + b)$

92. $y(a + b)$

93. $\frac{1}{2}(2x + 4y)$

94. $(3x + 6y) \cdot \frac{1}{3}$

95. $5c + 10d$

96. $6x + 2y$

97. $\frac{1}{2}a + 2b$

98. $\frac{3}{4}(8x + 12y)$

99. $\frac{2}{3}a\,(6b + 9c)$

100. $4ab + 12ab$

101. $\frac{2}{3}\left(\frac{6}{5}x + \frac{9}{8}y\right)$

102. $\frac{1}{2}\left(\frac{2}{3}a + \frac{4}{5}b\right)$

103. $ab(x + y)$

104. $abc + abd$

105. $6ac + 9ad$

Section 2.10

In exercises 106–115 change the given common fraction to a decimal fraction by first determining by inspection whether the decimal representation will be a terminating or a repeating decimal.

Examples: $\frac{2}{5}$, terminating decimal, 0.4

$\frac{3}{7}$, repeating decimal, $0.\overline{428571}$

106. $\frac{1}{8}$

107. $\frac{5}{12}$

108. $\frac{3}{4}$

109. $\frac{4}{7}$

110. $\frac{7}{9}$

111. $\frac{3}{16}$

112. $\frac{21}{28}$

113. $\frac{8}{5}$

114. $\frac{10}{7}$

115. $\frac{80}{128}$

Section 2.11

In exercises 116–125 replace ⑦ by $<, >$, or $=$ so as to form a true statement.

116. $\dfrac{7}{8}$ ⑦ $\dfrac{5}{6}$

117. $\dfrac{2}{3}$ ⑦ $\dfrac{4}{6}$

118. $\dfrac{-5}{8}$ ⑦ $\dfrac{-3}{4}$

119. $-\dfrac{1}{2}$ ⑦ $\dfrac{-5}{8}$

120. $\dfrac{-3}{4}$ ⑦ $\dfrac{-4}{3}$

121. $\dfrac{11}{12}$ ⑦ $\dfrac{27}{32}$

122. $\dfrac{7}{8}$ ⑦ $\dfrac{-9}{10}$

123. $\dfrac{-5}{14}$ ⑦ $\dfrac{1}{4}$

124. $\dfrac{1}{-2}$ ⑦ $\dfrac{1}{2}$

125. $\dfrac{3}{4}$ ⑦ $\dfrac{4}{-3}$

In exercises 126–135 use the given inequality and the number in parentheses to write two new true inequalities by (a) adding the indicated number to both sides of the inequality and (b) multiplying both sides of the inequality by the indicated number.

Example: $5 > -3; (-2)$

a.   by addition: $5 + (-2) > -3 + (-2) \rightarrow 3 > -5$

b.   by multiplication: $5(-2) < -3(-2) \rightarrow -10 < 6$

126. $6 < 9; (3)$

127. $6 < 9; (-3)$

128. $-3 < -2; (5)$

129. $-3 < -2; (-5)$

130. $-\dfrac{1}{2} > \dfrac{-2}{3}; (3)$

131. $\dfrac{3}{4} > \dfrac{2}{3}; \left(\dfrac{1}{12}\right)$

132. $\dfrac{-3}{4} < \dfrac{-2}{3}; \left(-\dfrac{1}{12}\right)$

133. $x > y; (2)$

134. $x > y; (-2)$

135. $x + 2 < a + 4; (2)$

Section 2.12

In exercises 136–145 write the given expression as a rational number without the absolute value sign.

136. $|7|$

137. $|-5|$

138. $-\left|\dfrac{1}{4}\right|$

139. $-\left|-\dfrac{1}{4}\right|$

140. $\left| -\left(-\dfrac{3}{4} \right) \right|$

141. $\left| -9 + 2 \right|$

142. $-\left| -\dfrac{3}{4} + \dfrac{1}{2} \right|$

143. $\left| -\left(-\dfrac{1}{5} \right) \right|$

144. $\left| -5 + 5 \right|$

145. $-\left| \dfrac{4}{7} \times \dfrac{7}{4} \right|$

3

SOLUTION OF LINEAR EQUATIONS AND INEQUALITIES

3.1 Mathematical Sentences

The statements

$$2 + 3 = 5, \quad 3 < 7, \quad \text{and } 3 + 4 \neq 8$$

are all examples of *mathematical sentences.* Each of these examples is a true statement. But a mathematical sentence may also be false, as are the sentences

$$2 + 3 = 6, \quad 3 > 7, \quad \text{and } 3 + 4 \neq 7$$

Another mathematical sentence — one that was mentioned in Chapter 1 — is

$$2w + 2(4w + 3) = 106$$

Because it contains the variable w, we cannot say whether it is true or false. If we replace w by 5 we get

$$2 \times 5 + 2(4 \times 5 + 3) = 106$$

which is false. On the other hand, if we replace w by 10 we get

$$2 \times 10 + 2(4 \times 10 + 3) = 106$$

which is true.

Mathematical sentences containing one or more variables, such as $2w + 2(4w + 3) = 106$, are sometimes called *open sentences* since the question of whether or not such a sentence is true or false remains "open" until the variable is replaced by a number.

Other examples of open sentences are:

1. $2x - 5 = 5x + \dfrac{1}{2}$

2. $x + y = 5$

3. $n + 5 < 8$

4. $2x + y < 6$

(1) and (2) are examples of *equations* and (3) and (4) are examples of *inequalities*; (1) and (3) are examples of open sentences with one variable and (2) and (4) are examples of open sentences with two variables.

   The process of finding all values of the variables that will produce a true sentence is commonly referred to as *solving* an equation or inequality.

   Consider the equation $n^2 = 25$ $(n^2 = n \times n)$. Now $5^2 = 25$ and $(-5)^2 = 25$, and 5 and -5 are the only two numbers whose square is 25. Thus 5 and -5 are the only *solutions* of the equation $n^2 = 25$. The *solution set* (also called *truth set*) of the equation consists of the numbers 5 and -5 and we write this as $\{5, -5\}$. Similarly, the only solution of the equation $2w + 2(4w + 3) = 106$ is 10, so that its solution set is $\{10\}$.

   What about the equation

$$x = x?$$

Since $x = x$ for every real number x, the solution set is the set of all real numbers.

   What about the equation

$$x + 3 = x?$$

Certainly for no real number x is it true that $x + 3 = x$. Thus the solution set of this equation is the *empty* set (also called the *null* set), which is written as $\{\ \}$ or $\emptyset$.

   The solution set of an open sentence depends upon the *domain* of the variables: the set of all numbers from which a replacement for the variables may be selected. Thus if the domain of n is the set of positive rational numbers, the equation $n^2 = 25$ has the solution set $\{5\}$; but if the domain of n is the set of all rational numbers, then the solution set is $\{5, -5\}$.

   In this chapter we will henceforth assume, unless stated otherwise, that the domain of each variable is the set of real numbers.

   We will consider first equations in one variable and, later in the chapter, inequalities in one variable. In the next chapter we will consider equations with more than one variable.

EXERCISES 3.1

In exercises 1–14 state whether the given sentence is true or false. For each false sentence change the $=$, $<$, or $>$ to make the sentence true.

Example: $2 + 3 = 7$, false; $2 + 3 < 7$, true.

1. $7 + 8 > 17$ F (<)

2. $25 - 6 = 19$ T

3. $9 + 8 > 8 + 9$ F (=)
   17 17

4. $(4 \times 5) + (3 \times 5) = 7 \times 5$ T
   20   15   35

5. $-3 \times -6 > 15 + 3$ F (=)

6. $(6 \times 100) + (3 \times 10) + 7 < 637$ F (=)

7. $3421 + 315 + 72 < 5975 + 2553$

8. $2\frac{1}{5} - \frac{4}{5} = 1\frac{2}{5}$ T

9. $\frac{31}{40} > 0.771$

10. $\frac{2}{3} \div \frac{3}{4} < \frac{3}{4} \div \frac{7}{8}$ F (>)

11. $\frac{1}{8} \div \frac{2}{7} > \left(\frac{1}{2} \times 3\right) \div 2$

12. 25% of $600 = 50\%$ of 300 T

13. $6.07 + 7.006 = 13.13$ F (<)

14. $84 \times 0.35 = 0.28 \times 100$

In exercises 15–29 find the solution set of the given open sentence if the domain of each variable is $\{1, 2, 3, 4, 5, 6, 7, 8, 9\}$.

Examples:

a. $4y > 16$;   solution set $= \{5, 6, 7, 8, 9\}$

For example: $4 \cdot 5 = 20 > 16$, whereas
$4 \cdot 1 = 4 < 16$

b. $n + 17 = 29$;   solution set $= \{\ \}$.

15. $y - 2 = 5$ {7}

16. $x + 5 < 10$ 4321

17. $c - 8 < 4$

18. $6n < 48$

19. $12n > 100$

20. $4a + 9 > 33$

21. $7x - 10 \neq 25$

22. $\frac{x}{4} = 2$

23. $\frac{a}{3} > 2$

24. $t + 10 \neq 17$

25. $x + 2 < 10$

26. $x^2 = 9$ 3

27. $x^2 < 9$ 2

28. $\frac{a}{2} - 5 < 12$

29. $13 > 2x + 1$

In exercises 30–44 find the solution set of the given open sentence if the domain of each variable is $\{-5, -4, -3, -2, -1, 0, 1, 2, 3, 4, 5\}$.

30. $n + 8 < 10$ $\{1,-1\}$

31. $y + 4 > 8$ $\{5,-5\}$

32. $3x = 21$ $7\to\emptyset$

33. $2y < -3$ -2

34. $\dfrac{x}{2} < 0$ $\{-1,\cdots-5\}$

35. $\dfrac{b}{3} > -1$ $0\cdots-5$

36. $p + 6 < 16$ $\{-5\cdots 5\}$

37. $x + 1 < 0$ $\{-2\cdots-5\}$

38. $6y - 4 = 26$ 5

39. $x^2 = -4$ -2

40. $x^2 > 0$ $\{1\cdots 5\}$

41. $\dfrac{s}{2} + 3 < 0$ $-5\}$

42. $5 + \dfrac{x}{3} > 8$

43. $4x - 8 = 0$ 2

44. $5x + 7 < 1$ $\{-2\cdots 5\}$

3.2 Equivalent Equations

The equation

(1) $$x = 2$$

is much simpler than the equation

(2) $$\tfrac{1}{2}x - 1.2 = 0.3 - \tfrac{1}{4}x$$

The solution set of (1) is obviously $\{2\}$, but it is certainly not obvious that $\{2\}$ is also the solution set of (2). It is nevertheless true, as we shall show later, that (1) and (2) have the same solution set.

Equations that have the same solution set are said to be *equivalent equations,* and the process of solving an equation generally consists of finding an equivalent equation whose solution set is immediately apparent.

It is not true, however, that we can always find such a nice, simple, equivalent equation as (1). There is, for example, no such equation equivalent to

$$x^2 - 5x + 9 = 0$$

In this chapter, however, we will be concerned only with *linear equations.* These are equations that are equivalent to equations of the form $x = k$ where k is a real number. (Or $y = k$, $w = k$, etc., according to the letter we use for the variable.)

The procedure for solving such equations is based upon the following properties of equality:

If a, b, and c are any real numbers and $a = b$, then

A.	$a + c = b + c$	(**Addition property of equality**)
B.	$a - c = b - c$	(**Subtraction property of equality**)
C.	$a \times c = b \times c$	(**Multiplication property of equality**)
D.	$a \div c = b \div c$ if $c \neq 0$	(**Division property of equality**)

Let us now see how these properties and the properties of addition can be applied to solve various linear equations.

Example 1: Find the solution set of the equation $x + 7 = 12$.

a.	$(x + 7) + (-7) = 12 + (-7)$	(Addition property of equality)
b.	$x + [7 + (-7)] = 5$	(Associative property of addition)
c.	$x + 0 = 5$	(Additive inverse property)
d.	$x = 5$	(Addition property of 0)

This process can be reversed. Starting with (d) we can get our original equation simply by adding 7 (the opposite of -7) to both sides of our equality. Hence

$$\text{if } x + 7 = 12, \text{ then } x = 5$$

and

$$\text{if } x = 5, \text{ then } x + 7 = 12$$

Thus

$$x + 7 = 12 \quad \text{and} \quad x = 5$$

are equivalent equations and the solution set of $x = 5$, namely, $\{5\}$, is also the solution set of the equation $x + 7 = 12$.

Of course, even a small child can see that if $\square + 7 = 12$, the number to be put in the box to make a true sentence must be 5. We have given all the details and pointed out how the various properties of numbers are used in order to make the logic clear. In the remaining examples we omit some of these details, but it would be well if you would supply them mentally there and in the exercises that follow.

Example 2: Solve the equation $\frac{4}{5}x = 20$.

$$\frac{5}{4} \cdot \frac{4}{5}x = \frac{5}{4} \cdot 20 \qquad \text{Why?}$$

$$\left(\frac{5}{4} \cdot \frac{4}{5}\right)x = \frac{5}{4} \cdot 20 \qquad \text{Why?}$$

$$x = 25 \qquad \text{Why?}$$

(Note that the solution set of the equation is, of course, {25}. However, when the instructions are simply "solve the equation," an answer such as $x = 25$ is sufficient.)

Example 3: Solve the equation $\frac{3}{4}x + 11 = 32$.

$$\frac{3}{4}x = 32 - 11 \qquad \text{Why?}$$

$$\frac{4}{3} \cdot \frac{3}{4}x = \frac{4}{3} \cdot 21 \qquad \text{Why?}$$

$$x = 28 \qquad \text{Why?}$$

Our next example shows what went on inside one of the "algebra machines" of Chapter 1. (See page 4.)

Example 4. Solve the equation $2w + 2(4w + 3) = 106$.

$$2w + (8w + 6) = 106 \qquad \text{Why?}$$
$$(2w + 8w) + 6 = 106 \qquad \text{Why?}$$
$$10w + 6 = 106 \qquad \text{Why?}$$
$$(10w + 6) - 6 = 106 - 6 \qquad \text{Why?}$$
$$10w = 100 \qquad \text{Why?}$$
$$(10w) \div 10 = 100 \div 10 \qquad \text{Why?}$$
$$w = 10 \qquad \text{Why?}$$

Note that the properties of equality (B) and (D) involving subtraction and division are not really necessary (although they are convenient) since

$$a - c = a + (-c) \qquad \text{and} \qquad a \div c = a \times \frac{1}{c}$$

Our next example involves fractions where it is a good procedure, as a first step, to multiply both sides of the equality by the LCD (see Section 2.3) of the fractions in the equation.

Example 5. Find the solution set of the equation $\frac{1}{2}(x - 7) = \frac{1}{3}(2x + 5)$.

$$6 \cdot \frac{1}{2}(x - 7) = 6 \cdot \frac{1}{3}(2x + 5) \qquad \text{Why?}$$

$$3(x - 7) = 2(2x + 5) \qquad \text{Why?}$$

$$3x - 21 = 4x + 10 \qquad \text{Why?}$$

$$(3x - 21) + 21 = (4x + 10) + 21 \qquad \text{Why?}$$

$$3x = 4x + 31 \qquad \text{Why?}$$

$$3x - 4x = (4x + 31) - 4x \qquad \text{Why?}$$

$$-x = 31 \qquad \text{Why?}$$

$$(-1)(-x) = (-1)(31) \qquad \text{Why?}$$

$$x = -31 \qquad \text{Why?}$$

and so the solution set is $\{-31\}$.

For our next example we return to the equation given at the beginning of this section.

Example 6: Solve the equation $\frac{1}{2}x - 1.2 = 0.3 - \frac{1}{4}x$.

$$4\left(\frac{1}{2}x - 1.2\right) = 4\left(0.3 - \frac{1}{4}x\right) \qquad \text{Why?}$$

$$2x - 4.8 = 1.2 - x \qquad \text{Why?}$$

$$(2x - 4.8) + 4.8 = (1.2 - x) + 4.8 \qquad \text{Why?}$$

$$2x = 6 - x \qquad \text{Why?}$$

$$2x + x = (6 - x) + x \qquad \text{Why?}$$

$$3x = 6 \qquad \text{Why?}$$

$$x = 2 \qquad \text{Why?}$$

It is, of course, always a good idea to check solutions as a safeguard against errors in computation. Thus, to check our solution to Example 6, we note that if $x = 2$,

$$\frac{1}{2}x - 1.2 = \frac{1}{2} \times 2 - 1.2 = 1 - 1.2 = -0.2$$

and also

$$0.3 - \frac{1}{4}x = 0.3 - \frac{1}{4} \times 2 = 0.3 - \frac{1}{2} = 0.3 - 0.5 = -0.2$$

You may have noted that for all of the equations considered so far the arithmetical computations are rather simple. This is because the important things to learn here are the algebraic procedures. It is a fact, however, that equations occurring in science and industry are all too likely to involve considerably more difficult arithmetic – difficult, that is, without the use of a calculator or computer. To remind you that it is frequently necessary to carry out more extended arithmetical calculations in connection with various jobs, a small number of exercises which involve such computations (indicated by *) will be included from time to time. You are not expected to do these problems unless you have access to a calculator!

Our final example, then, concerns such a "real life" problem.

Example 7: The approximate speed, in feet per second, of sound in air at a temperature of t degrees Celsius is given by the formula $V = 1087.5 + 1.97t$. Find the temperature at which the speed of sound will be 1200 feet per second.

Solution: We have

$$1200 = 1087.5 + 1.97t$$

$$1.97t = 1200 - 1087.5 = 112.5$$

$$t = \frac{112.5}{1.97} = 57.106598 = 57.1°C \text{ (to the nearest tenth of a degree)}$$

Note that if you use a calculator in this problem, six decimal places are obtained "free of charge." In problems like this you will be asked to carry out your computations to the limit of your calculator so that your ability to use a calculator can be checked. In actual practice, however, answers are *rounded off* to correspond to the limits of accuracy of the measurements involved (all measurements are only approximations!). In the example above, because the numbers 1087.5 and 1.97 are only approximations and it is difficult to measure temperature correctly to more than a tenth of a degree, rounding off 57.106598 to 57.1 is reasonable.

As another example of rounding off, suppose the voltage, E, applied to a circuit with a resistence, R, of 0.12 ohms (measured correct to the nearest hundreth of an ohm) is 6.1 volts (measured correct to the nearest tenth of a volt). Then the current, I, in amperes, obtained from the equation $I = \frac{E}{R}$ is

$$I = \frac{6.1}{0.12} \text{ amperes}$$

Doing this calculation on an eight-place calculator gives

$$I = 50.833333 \text{ amperes}$$

But such an answer is really ridiculous: The voltage and the resistance were each measured to only two *significant figures* and so all we really know is that the current is 51 amperes correct to the nearest ampere. However, since, the topics of rounding off (50.833333 to 51 in this case) and significant figures are rather complex ones and since the purpose of the exercises is simply to provide practice in the use of the calculator, we will not consider these aspects of the calculator problems any further.

When you use a calculator, do not be concerned with a difference between your answer and the answer given in the text (or in a classmate's answer) if it only involves the last one or two digits. Such a difference can easily arise if the computation is done in different ways. For example, consider the calculation of

$$\frac{952}{126} \times \frac{15}{123}$$

First multiplying 952 by 15, then dividing by 126, and finally dividing by 123 gives 0.9214091. On the other hand, first dividing 952 by 126, then dividing by 123, and finally multiplying by 15 gives 0.9214092. (The difference in the answers is caused by the fact that at each stage the calculator is rounding off rather than working with the exact numbers.)

EXERCISES 3.2

Solve the following equations.

1. $3x + 7 = 28$

2. $2y + 5 = 19$

3. $3a - 9 = 21$

4. $6b - 8 = 4$

5. $29 = 4y + 5$

6. $56 = 9t - 25$

7. $\frac{1}{4}x + 5 = 10$

8. $\frac{1}{10}y + 1 = 6$

9. $\frac{1}{3}n + 2 = 9$

10. $\frac{1}{8}x - 9 = 0$

11. $5x + 2x = 14$

12. $5a + a = 24$

13. $8x - 5x + x = 12$

14. $8y + y - y = 8$

15. $x + x + 2x = 8$

16. $3(y + 5) = 60$

17. $4(2x - 3) = 20$

18. $3(x + 4) = 15$

19. $-3(2x - 3) = 15$

20. $\frac{1}{2}(6r + 8) = 19$

21. $\frac{1}{3}(x + 6) = 8$

22. $2x + 9 = x + 10$

23. $3x + 1 = 25 - x$

24. $18 + 14x = 7x - 24$

25. $1 - 3(2x - 4) = 4x - 9$

26. $9y + 2(y + 1) = 24$

27. $4(b + 1) + 3(b + 2) = 94$

28. $4(a + 7) = 2(a + 15)$

29. $\frac{1}{3}(6x - 9) = \frac{1}{2}(8x + 2)$

30. $\frac{8x}{5} - 2 = \frac{4x}{3} - 4$

31. $\frac{3x}{4} - \frac{x}{5} = 11$

32. $\frac{5x}{7} - 30 = \frac{5x}{3}$

33. $\frac{x}{3} - 5 + \frac{x}{2} = \frac{x}{6} - 1$

34. $x + \frac{x}{2} - 10 = \frac{x}{5}$

35. $\frac{13}{2} + \frac{a}{3} = 5 + \frac{a}{2}$

36. $y + (5y + 3) - 7 = 0$

37. $12.5y - (2 - y) = 19 - (2y - 10)$

38. $3b - (8 - b) = b - 2(b - 5)$

39. $3(x - 2) = 2(x - 3)$

40. $2y - 5(y + 1) = 7$

41. $3(t - 14) - 7(t - 18) = 0$

42. $2(x - 1) - 3(2 - x) = 4(3 - x)$

43. $2m - 4(5 - m) = 5(m - 3) - 2$

44. $\frac{5}{2}(4x + 2) - 7 = \frac{3}{4}(4x - 7) + 51 + 3x$

45. $\frac{5}{3}(12 - 3y) - 10y = \frac{7}{6}(8 - 24y) + \frac{35}{3}$

46. $18 - (2x - 18) = 5x + 2(3x - 8)$

47. $37 - (x + 12) = 20 + 4x$

48. $2(y + 6) + 32 = 5(y + 2)$

49. $28 - 8\left(\frac{1}{2}x - \frac{3}{4}\right) = x + 1$

50. $\frac{x + 2}{3} = 14$

51. $\frac{y + 7}{4} + 1 = 5$

52. $\frac{b + 6}{2} + 3 = 24$

53. $11 = \frac{19x + 2}{5} + 3$

54. $10 = 1 + \frac{3y + 3}{4}$

In exercises 55-65 each equation has been taken from a scientific, engineering, trade, or business application in which the equation has been determined and is now ready to solve. Use a calculator to solve these equations, giving your answer as it appears on your calculator.

*55. $P.\ E. = (911)(3.14)(62.5)(22.3)$

*56. $K.\ E. = \dfrac{(3.14)\left(\frac{1}{16}\right)^2 (50)(62.5)(50)^2}{64.4}$

*57. $200 - 6.22A - 60 - 12.44A = 10.49A$

*58. $\dfrac{111y}{16} = \dfrac{216 + 741}{128}$

*59. $4.316 = 4.5 - (2 \times 0.43301) + 2M + 0.250$

*60. $150 + 9.3a + 26 = 200 - 6.2a$

*61. $(600)(6) = (100)(12) + \dfrac{600V}{64.4} + \dfrac{200V}{64.4} + (34.64)(12)$

*62. $500T = 25(500) + \dfrac{(5000)(44)^2}{64.4} - (1498.5)(500)$

*63. $\dfrac{500(44)^2}{64.4} - 45(1.962) - 1500S = 0$

*64. $3.817 = 4.5 - 2(D \times 0.57735)$

*65. $10V - (17.32)(0.3)V - 500 = 0$

3.3   Common and Decimal Fractions Again

In Section 2.10 we showed that every common fraction can be written as a terminating or repeating decimal and that every terminating decimal can be written as a common fraction. Now we show that every repeating decimal can also be written as a common fraction.

Example 1: Find a common fraction equal to $0.\overline{3}$. We write $0.\overline{3}$ as $0.333 \ldots$ and let

(1) $x = 0.333 \cdots$

Multiplying both sides of (1) by 10 we have

$$10x = 3.333 \cdots$$
$$10x = 3 + 0.333 \cdots$$
$$10x = 3 + x$$
$$9x = 3$$

and so

$$x = \frac{3}{9} = \frac{1}{3}$$

Example 2: Find a common fraction equal to $0.\overline{24}$. We write

(2) $x = 0.242424 \ldots$

and now multiply both sides of (2) by 100 because there are two digits in the repeating cycle. We have

$$100x = 24.2424 \cdots$$
$$100x = 24 + 0.\overline{24}$$

$$100x = 24 + x$$

$$99x = 24$$

and so

$$x = \frac{24}{99} = \frac{8}{33}$$

Example 3: Find a common fraction equal to $2.3\overline{15}$. Now we let

(3) $x = 2.3\overline{15}$

and notice that, unlike the two previous examples, the decimal has a digit, 3, to the right of the decimal point that is not part of the repeating cycle. The first step, then, is to multiply both sides of equation (3) by 10 so that the only digits to the right of the decimal point are those in the repeating cycle. So we have

(4) $10x = 23.\overline{15}$

Now, because there are two digits in the repeating cycle, we multiply both sides of equation (4) by 100 and get

$$1000x = 2315.\overline{15}$$

Then

$$1000x - 10x = 2315.\overline{15} - 23.\overline{15}$$

$$990x = 2292$$

and so

$$x = \frac{2292}{990} = \frac{382}{165}$$

Note that in finding a common fraction equal to a repeating decimal we can check the answer by performing a division — and that division is extremely easy to do if a calculator is available!

From these results and those of Section 2.10, we can draw the following conclusions:

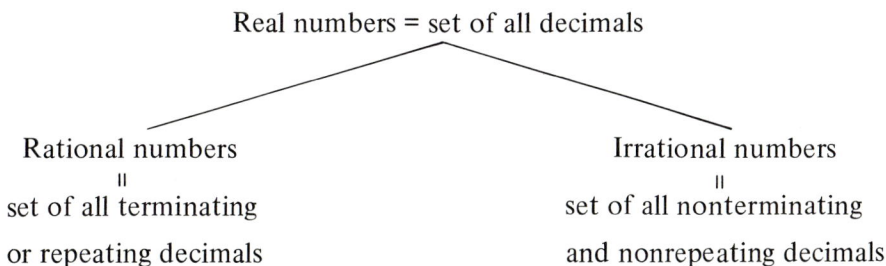

Real numbers = set of all decimals

Rational numbers
=
set of all terminating
or repeating decimals

Irrational numbers
=
set of all nonterminating
and nonrepeating decimals

EXERCISES 3.3

In exercises 1–20 find a common fraction in lowest terms which is equal to the given repeating decimal.

1. $0.\overline{4}$	6. $0.\overline{18}$	11. $3.\overline{5}$	16. $4.\overline{247}$
2. $0.\overline{5}$	7. $0.\overline{09}$	12. $6.\overline{01}$	17. $0.2\overline{7}$
3. $0.\overline{7}$	8. $0.\overline{80}$	13. $0.00\overline{9}$	18. $4.29\overline{35}$
4. $0.\overline{37}$	9. $0.\overline{405}$	14. $3.8\overline{45}$	19. $0.456\overline{654}$
5. $0.7\overline{2}$	10. $0.2\overline{45}$	15. $0.01\overline{79}$	20. $10.7\overline{891}$

21. Arrange the following rational numbers in order from largest to smallest: 0.70, $0.\overline{7}$, $0.\overline{707}$, 0.707, $0.\overline{70}$.

22. Arrange the following rational numbers in order from smallest to largest: $0.4\overline{52}$, $0.\overline{452}$, $0.5\overline{42}$, $0.\overline{542}$, $0.4\overline{25}$.

3.4   Word Problems: Introduction

Word problems vary from the quite practical: "How much of an alloy which contains 6 grams of gold in 10 grams must be melted with another alloy which contains 3 grams of gold in 10 grams to get an alloy containing 5 grams of gold in 10 grams?" to "just for fun" problems such as: "In 2 years Bruce will be 5 times as old as he was 2 years ago. Find his present age." Still others involve a knowledge of physics, chemistry, or engineering: "How far does a wooden ball of specific gravity 0.4 and radius 60 centimeters sink in water?"

The *basic* algebraic principles involved in solving such problems, however, are the same, and since practical-type problems that do not involve some knowledge of engineering or science are rather scarce, we will consider many "just for fun" problems. Keep in mind, however, that an understanding of the basic principles will enable you to handle practical-type problems when you encounter them in studying other subjects.

It should also be noted that, in the interest of simplicity, various physical situations are idealized; motorists travel long distances without a stop, people work long hours without a break, planes never have to stay in holding patterns before landing, etc.!

A restriction which we will place on the problems considered in this chapter is that, when they are put in algebraic form, the resulting equation or inequality will be linear. Thus, for example, the equation resulting from the sinking ball problem turns out to be

$$345{,}600 = h^2(6 - h)$$

which is not a linear equation.

3.5 Verbal and Mathematical Phrases

The biggest difficulty, of course, in solving many word problems is the "translation" of the verbal sentence or sentences into a mathematical sentence. The equation resulting from the problem about alloys is

$$\frac{6}{10}x + \frac{3}{10}(10 - x) = \frac{5}{10} \cdot 10$$

which is quite easy to solve by the technique that we have just discussed; the problem is to get the equation.

As a first step in this direction it is useful to practice translating verbal phrases into mathematical phrases — with the sentences to be tackled later.

Example 1: "Four more than 6 times a number x " is translated as "$6x + 4$."

Example 2: "Eight times a number which is 5 less than the number s" is translated as "$8(s - 5)$."

Example 3: "The time in hours it takes a plane to go 1000 kilometers at a speed of r kilometers an hour" is translated as "$1000 \div r$" or "$\frac{1000}{r}$."

Note that in Example 3 a metric unit of length, the *kilometer*, has been used. Since the metric system is in use throughout the world and is used increasingly in the United States, it is a good idea to practice using the vocabulary of the metric system. Note, however, that here and elsewhere in this text no knowledge of the metric system is assumed. Thus, in this example all you need to recognize is that, from the nature of the problem, a kilometer must be a unit of length! (The essential facts concerning the metric system are given in Table I at the back of the book.)

EXERCISES 3.5

In exercises 1-10 match each verbal phrase on the left with the proper mathematical phrase on the right.

Verbal Phrase		Mathematical Phrase
1. Five plus x. *B*	a.	$y - 4$
2. x added to y. *e*	b.	$5 + x$
3. r minus 2. *C*	c.	$r - 2$

4. *t* more than 3. G   d.   *r* − *s*
5. Four less than *y*. a   e.   *y* + *x*
6. The difference between *r* and *s*. D   f.   2*x*
7. Four increased by *t*. I   g.   3 + *t*
8. The product of *x* and 2. F h.   6*n* − 7
9. Three more than twice *x*. J i.   4 + *t*
10. Seven less than 6 times h   j.   2*x* + 3
    a number *n*.

In exercises 11–20 match each mathematical phrase on the left with the proper verbal phrase on the right.

Mathematical Phrase   ### Verbal Phrase

11. 4 + *x*   D   15 a.   One less than a certain number.

12. *m* + 3   C   19 b.   One-third of a number.
13. 2*x*   F   12 c.   Three added to a certain number.

14. 2*x* + 1   H   11 d.   Four plus a certain number.
15. *y* − 1   A   16 e.   The difference between two numbers.

16. *r* − *x*   E   13 f.   Twice a certain number.
17. 5*a* − 2   I   20 g.   Five less than $\frac{1}{3}$ of a number.
18. 2*a* + 3   J   14 h.   One more than twice a certain number.

19. $\frac{1}{3}k$ B   17 i.   Two less than 5 times a number.

20. $\frac{n}{3}$ − 5   G   18 j.   Three more than twice a number.

In exercises 21–45 give a mathematical phrase equivalent to the given verbal phrase. Use any letter you wish to hold the place of a number.

21. Three more than 5 times a number.   24. The product of a number and 5.

22. Six less than 3 times a number.   25. Five times the sum of two numbers.

23. A number decreased by 2.   26. Three times the product of two numbers.

27. Four added to the product of two numbers.

28. Fifteen more than twice a certain number.

29. The sum of two numbers divided by 2.

30. One-half the sum of 3 and a certain number.

31. Twelve more than twice the number of students in Jim's class.

32. Ten miles less then 3 times the number of miles Sarah walked.

33. The value in cents of a certain number of nickels.

34. The value in dollars of a certain number of quarters.

35. The cost of 60 pencils, all of the same kind.

36. The annual salary of a man who earns a certain number of dollars each week.

37. Half the age of a man 10 years from now.

38. Fifteen percent of a certain number.

39. Twice a number times the sum of two other numbers.

40. The difference of two numbers multiplied by a third number.

41. The number of liters in a 50-liter tank after h hours if the tank was full to start and a liter was drained off each hour.

42. The distance in centimeters a particle falls after t seconds if it falls 490 times as many centimeters as the square of the time in seconds.

43. The average rate of speed in miles per hour of a plane that travels 540 miles in h hours.

44. The time in hours it takes a plane to go 1200 kilometers at a speed of r kilometers per hour.

45. The distance in miles a plane travels in h hours at a speed of 330 miles per hour.

46. Write a mathematical phrase representing any even number.

47. Write a mathematical phrase representing any odd number.

48. Write a mathematical phrase representing the sum of two consecutive odd whole numbers.

49. Write a mathematical phrase representing the sum of two consecutive even whole numbers.

50. Write a mathematical phrase representing the sum of three consecutive whole numbers.

3.6 Verbal and Mathematical Sentences

Just as we can use verbal phrases to construct a verbal sentence, so we can use mathematical phrases to construct a mathematical sentence. For example, we can form the sentence (equation)

$$3x - 12 = 2x + 5$$

from the mathematical phrases

$$3x - 12 \text{ and } 2x + 5$$

Such an equation can arise, of course, from more than one verbal sentence. It could, for example, be a mathematical formulation of the problem: "Twelve less than 3 times a certain number is 5 more than twice the number. Find the number." Or, equally well, a mathematical formulation of the problem: "Pete has 3 times as much money as Joe. If Pete spends \$12 he will still have \$5 more than twice what Joe has. How many dollars does Joe have?"

EXERCISES 3.6

In exercises 1–10 match each statement on the left with a mathematical sentence on the right.

1. The area of a rectangle is equal to ℮ the product of its length and width.

   a. $T = \dfrac{m}{88}$

2. The area of a triangle is equal to D half the product of its altitude and base.

   b. $c = 150 + 10(n - 10)$

3. The interest for 1 year is equal ⊺ to the principal times the rate.

   c. $C = \dfrac{5}{9}(F - 32)$

4. The time in hours it takes a car to travel a certain number of kilometers at the rate of 88 kilometers per hour is equal to the distance divided by the rate.

   2 d. $A = \dfrac{1}{2}ab$

5. The cost in cents of sending a telegram to a certain city is \$1.50 for the first ten words and 10¢ for each additional word.

   ∣ e. $A = lw$

6. The cost in cents of sending a package via airmail to a certain zone is $1.00 for the first pound plus $0.56 for each additional pound.

    f. $A = \dfrac{h\,(b_1 + b_2)}{2}$

7. Celsius temperature is equal to $\frac{5}{9}$ the difference between the Fahrenheit temperature and 32.

    g. $C = 100 + 56(n - 1)$

8. The area of a trapezoid is equal to half the product of its altitude and the sum of its bases.

    h. $T = 110{,}980 + 0.70(I - 200{,}000)$

9. The cost in dollars of printing a certain number of pamphlets is $75 for the first 100 copies plus 25¢ for each additional pamphlet.

    i. $I = pr$

10. A certain income tax schedule states that the tax in dollars will be $110,980 plus 70% of the excess over $200,000 of taxable income.

    j. $C = \frac{1}{4}(n - 100) + 75$

In exercises 11–20 after each equation there is a partial description of a problem that the equation could represent. In each case copy the description replacing each blank. Then solve the equation.

Example: $10x + 5(x + 5) = 130$. A man has $\underline{\ \ ?\ \ }$ in change, consisting of nickels and $\underline{\ \ ?\ \ }$. There are $\underline{\ \ ?\ \ }$ more nickels than $\underline{\ \ ?\ \ }$. How many coins of each kind does he have?

Solution: A man has $\underline{\$1.30}$ in change, consisting of nickels and $\underline{\text{dimes}}$. There are $\underline{5}$ more nickels than $\underline{\text{dimes}}$.

$$10x + 5(x + 5) = 130$$
$$10x + 5x + 25 = 130$$
$$15x + 25 = 130$$
$$15x = 105$$
$$x = 7, \ x + 5 = 7 + 5 = 12$$

The man has 7 dimes and 12 nickels.

11. $x + (x + 35) = 115$. Frank spent __?__ for a ball and a knife. The __?__ cost __?__ more than the ball. Find the cost of each.

12. $x + (x + 12) = 98$. Two girls sold __?__ boxes of candy. One girl sold __?__ more boxes than the other. How many did each sell?

13. $2w + 2(w + 18) = 210$. It takes __?__ feet of fencing to enclose a rectangular garden. It is __?__ feet longer than it is __?__ . Find the length and the width.

14. $100x + (50)(2\frac{1}{2}x) = 900$. A man paid a debt of __?__ dollars with hundred dollar bills and __?__ , using __?__ times as many __?__ as __?__ . How many of each did he use?

15. $0.04x + 0.06(6000 - x) = 320$. A woman invested __?__ dollars, part of it at __?__ percent and the rest at __?__ percent. Her yearly income amounted to __?__ dollars. How much was invested at each rate?

16. $[10(5 - x) + x] - 9 = 10x + (5 - x)$. The sum of the digits of a two-digit number is 5. If __?__ is subtracted from the number, the digits will be interchanged. Find the number.

17. $10(x + 2) + x = 16x$. In a two-digit number the tens' digit is __?__ more than the units' digit. The number is __?__ times the __?__ digit. Find the number.

18. $0.04x + 0.24(290 - x) = 0.20 \times 290$. How much milk testing __?__ % butterfat should be used with cream testing __?__ % butterfat to make __?__ pounds that test __?__ %?

19. $0.80x + 1.80(10 - x) = 1.10 \times 10$. A dealer mixed candy selling at $ __?__ a kilogram and candy selling at $ __?__ a kilogram to make __?__ kilograms of candy to sell at $ __?__ a kilogram. How much of each did he use?

20. $40(t + 1) = 50t$. A car leaves a certain town at __?__ miles per hour. __?__ hour later another car leaves the same town in the same direction at __?__ miles per hour. How long will it take the second car to overtake the first car?

3.7 Solving Word Problems

There is no simple technique for translating word problems into mathematical sentences. Careful attention, however, to the following examples and the remarks concerning them should enable you to become at least reasonably proficient in handling word problems.

Example 1: In 2 years George will be 5 times as old as he was 2 years ago. Find his present age.

Solution: Let x = George's present age in years. Now it is often a good idea to form a semialgebraic equation (which we will call the *basic equation*) from the verbal statement. It reads:

(George's age in 2 years) = 5 (George's age 2 years ago)

Then, in terms of x, the left-hand side of the equation becomes $x + 2$ and the right-hand side becomes $5(x - 2)$. Hence, our final equation is $x + 2 = 5(x - 2)$. This gives us

$$x + 2 = 5x - 10$$
$$-4x = -12$$
$$x = 3$$

Check: George's present age is 3; in 2 years, George will be 5; 2 years ago he was 1, and $5 = 5 \times 1$.

The next problem involves an equation from Chapters 1 and 2.

Example 2: The perimeter of a rectangle is 106 feet and the length is 3 feet more than 4 times the width. Find the dimensions of the rectangle.

Solution: Here our basic equation is

2(width of the rectangle in feet) + 2(length of rectangle in feet) = 106

If we let w = width of rectangle in feet, the length will be $4w + 3$ feet and so we have

$$2w + 2(4w + 3) = 106$$

This gives us

$$2w + 8w + 6 = 106$$
$$10w = 100$$
$$w = 10$$

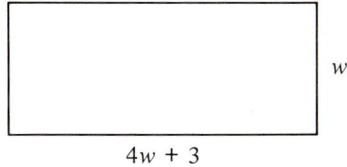

Thus the width is 10 feet and the length in feet is $4 \times 10 + 3 = 43$.

Check: $(2 \times 10) + (2 \times 43) = 20 + 86 = 106$.

Another device to assist in forming equations is to begin by guessing a solution — any guess, no matter how wild, will do. Thus in Example 1 we might guess that George is 10 years old. How can we check to find out whether or not our guess is correct? If George is 10 years old now, George's age in 2 years will be $10 + 2$ and his age 2 years ago was $10 - 2$. So we wish to find out whether $10 + 2 = 5(10 - 2)$. Since $10 + 2 = 12$ and $5(10 - 2) = 5 \times 8 = 40$, our guess is incorrect. But, replacing our guess of 10 by x, we have

$$x + 2 = 5(x - 2)$$

Similarly, in Example 2 we could guess 30 as the width of the rectangle in feet. Then its length would be (4 · 30 + 3) feet and so we would need to check to see if (2 · 30) + 2(4 · 30 + 3) = 106. Obviously, this is not a true statement, but from it we can get $2w + 2(4w + 3) = 106$ by replacing 30 by w.

After using this technique a few times you will soon come to realize that there is really no need for it! You can just as well "guess" x as 10 or w as 30. But, in the beginning, this device may help to make a problem sound less abstract.

Note that we were careful to say in each case what our variable represents. We wrote, for example, "x = George's present age in years" and not "x = George" or even "x = George's age."

All of these hints and cautions may not seem at all necessary for such simple problems as we have discussed. Careful attention to them, however, and practice in simple problems will stand you in good stead when we discuss more complicated problems.

Here are three more examples to help get you started on the solution of word problems.

Example 3: Find a number such that if 8 is subtracted from it and this difference multiplied by 5, the result is 35.

Solution:

1. Let x = the number.

2. Then $5(x - 8) = 35$

$$5x - 40 = 35$$

$$5x = 75$$

$$x = 15$$

3. The solution: The number is 15.

4. Check: $15 - 8 = 7$ and $5 \times 7 = 35$.

Example 4: The sum of two numbers is 18. The larger equals 5 times the smaller. Find the numbers.

Solution:

1. Let x = the smaller number.
   Then $5x$ = the larger number.

2.   The equation: $x + 5x = 18$

$$6x = 18$$

$$x = 3$$

3.   The solution: The smaller number is 3 and the larger number is $5 \times 3 = 15$.

4.   Check: $15 + 3 = 18$ and 15 is 5×3.

Alternate Solution:

1.   Let x = the smaller number.

2.   Then $18 - x$ represents the larger number.

3.   The equation: $18 - x = 5x$

$$18 = 6x$$

$$3 = x$$

4.   The solution: The smaller number is 3 and the larger number is $18 - 3 = 15$.

5.   Check: $3 + 15 = 18$, $18 - 3 = 15$, and $5 \times 3 = 15$.

Example 5: A girl had $1.30 in nickels and dimes. If she had 5 more nickels than dimes, how many of each did she have?

Solution:

1.   Let x = the number of dimes.

2.   Then $x + 5$ = the number of nickels.

3.   The basic equation is:

value of nickels in pennies + value of dimes in pennies =

value of 130 pennies

Then from the following table

	Number of coins	$\times$	Value of each coin	$=$	Total value
Dimes	x		10		$10x$
Nickels	$x + 5$		5		$5(x + 5)$
All coins					130

we get the equation

$$5(x + 5) + 10x = 130$$

so that

$$5x + 25 + 10x = 130$$
$$15x = 105$$
$$x = 7$$

4.   Solution: The girl had 7 dimes and $7 + 5 = 12$ nickels.

5.   Check: 7 dimes is 70¢; 12 nickels is 60¢; 60¢ + 70¢ = $1.30.

EXERCISES 3.7

Solve the following problems and check your answers.

1.   One number is 3 times a second number. The sum of the two numbers is 16. Find the numbers.

2.   One number is 6 more than the second number. The sum of the two numbers is 51. Find the numbers.

3.   The sum of two numbers is 16. If 1 were added to twice the smaller, the resulting number would be 3 more than the larger. Find the numbers.

4.   The sum of two numbers is 50 and their difference is 20. What are the two numbers?

5.   The value of a pile of 20 coins consisting of dimes and quarters is $3.95. How many coins of each kind are there?

6.   A piggy bank contained $1.80 in nickels and dimes. If there were 25 coins in all, how many coins of each kind were there?

7.   A boy had $2.90 in nickels, dimes, and quarters. If he had 2 more dimes than quarters and 4 more nickels than dimes, how many did he have of each?

8.   Jill had 18 coins, all dimes or quarters. If they amounted to $2.85, how many did she have of each?

9.   The length of a rectangle is 3 times the width. The perimeter of the rectangle is 56 centimeters. Find the length and width.

10.   A tennis court for singles is 24 feet longer than twice its width. The perimeter is 210 feet. Find its dimensions.

11.   The length of a rectangle is 5 meters more than twice the width. If the perimeter is 88 meters, what are the dimensions of the rectangle?

12.   The perimeter of a rectangle is 124 feet. If the length is 6 feet less than 3 times the width, what are the dimensions?

13. A farmer has a herd of 40 cattle. There are $\frac{2}{3}$ as many Jerseys as Holsteins. Find the number of each.

14. A tree 81 feet high is broken by the wind so that the upper part is $\frac{4}{5}$ of the lower. At what height did the tree break off?

15. Four men planned to form a partnership, each investing the same amount of money, and then buy a piece of property. If only three purchase it, each will have to invest $2000 more than he planned to invest. What was the cost of the property?

16. Two angles are called supplementary when the sum of their degree measures is $180°$. If one of two supplementary angles is $20°$ more than twice the other, what is the degree measure of each?

17. Jack is 4 years older than his sister. In 10 years the sum of their ages will be 42. How old is each now?

18. In 3 years Mary will be twice as old as Sue is now. Two years ago the sum of their ages was 20. How old is each now?

19. Bill is $\frac{1}{3}$ as old as his father, who is 2 years older than his mother. When Bill was born the sum of his parents' ages was 46. How old are his parents now?

20. A father, on being asked his age and that of his daughter, replied, "If you add 4 to my age and divide the sum by 5 you will have my daughter's age, but 6 years ago I was 10 times as old as she was." Find their ages.

21. The sum of two numbers is 15. Twice the first number is the same as 4 times the second number. Find the numbers.

22. Ramón and José together weigh 81 kilograms. The difference between twice José's weight and 3 times Ramón's weight is 27 kilograms. Find the weight of each.

23. A collection of 35 coins consists of 5 more dimes than nickels and 3 fewer nickels than pennies. Find the number of each kind of coin.

24. Side A of a triangle is 10 centimeters longer than side B. Side B is twice as long as side C. If the perimeter of the triangle is 100 centimeters, find the length of side A.

25. A man's age is 3 times that of his son. In 12 years he will be twice as old as his son is then. What are their present ages?

26. The perimeter of a piece of land is 630 feet. Find the dimensions if the length is 9 feet less than 3 times the width.

27. The sum of two numbers is 198. The difference between the larger and the smaller number is 118. What are the two numbers?

28. Joe has 5 times as much money as Frank. However, if Joe pays Frank the $5 he owes him, then Joe has twice as much as Frank. How much money did each have in the beginning?

29. A girl has a collection of 65 coins amounting to $9.15 and consisting of nickels, dimes, and quarters. If the number of quarters exceeds the number of nickels by 1, how many of each kind of coin does she have?

30. The difference between $\frac{2}{3}$ of a number and $\frac{1}{6}$ of the same number is 78. What is the number?

31. When Felicia was born, her mother was 25 years old. Now her mother's age is 3 more years than twice her age. What are their present ages?

32. Kaye had 5040 words to type in order to finish her term paper. Her friend Sally, who types at the rate of 35 words per minute, offered to use her typewriter and help. If Kaye types at 45 words per minute, how long should it take them to finish?

33. A man has a rectangular farm whose length is 48 feet less than 10 times its width. If its perimeter is 3204 feet, what are its dimensions?

34. The sum of three consecutive odd numbers is 381. Find the numbers (Note: If x = the first odd number, then $x + 2$ is the second one. What is the third?)

35. Fifty coins consisting of dimes and nickels are worth $3.50. How many of each are there?

36. Samuel is $\frac{2}{3}$ as old as Isaac. In 6 years he will be $\frac{3}{4}$ as old as Isaac. How old is each now?

37. The perimeter of a triangle is 192 centimeters. The length of the longest side is 4 times that of the shortest side, and the length of the third side is 6 centimeters less than the longest side. Find the length of each side.

38. The sum of two numbers is 25. If $\frac{2}{3}$ the smaller one is 2 more than $\frac{1}{4}$ the larger one, find the two numbers.

39. Mac's present age is just 3 times the age that his brother was 2 years ago. How old is his brother now if Mac is 18 years old?

40. Are there three consecutive whole numbers whose sum is 100?

3.8 More Difficult Word Problems

In the next four examples, we will concentrate on the idea of forming a basic equation. Do not, however, forget the technique of guessing a solution as an aid in obtaining the desired equation.

Example 1: Fernandez can paint a certain type of house in 3 days and Hernandez can paint it in 5 days. How long would it take them to do it together?

Solution: Whatever fraction of the job one does, the other must do the other fraction. Thus, if Fernandez does $\frac{2}{3}$ of the job, Hernandez must do the remaining $\frac{1}{3}$. Such thinking leads us to the basic equation:

$$\begin{array}{ccc} \text{(Fraction of work done} & & \text{(Fraction of work done} \\ \text{by Fernandez)} & + & \text{by Hernandez)} \end{array} = 1$$

	Number of days to do the job alone	Fractional part done in 1 day	Fractional part done in **x** days
Fernandez	3	$\frac{1}{3}$	$\frac{x}{3}$
Hernandez	5	$\frac{1}{5}$	$\frac{x}{5}$

So if x is the number of days that it takes the two men working together to finish the job, we have

$$\frac{x}{3} + \frac{x}{5} = 1$$

Then, multiplying both sides of the equation by 15 (Why?), we have

$$15\left(\frac{x}{3} + \frac{x}{5}\right) = 15 \cdot 1$$

Thus

$$5x + 3x = 15$$
$$8x = 15$$
$$x = \frac{15}{8} = 1\frac{7}{8}$$

It would take them $1\frac{7}{8}$ days to paint the house.

Check: $\left(\frac{15}{8} \div 3\right) + \left(\frac{15}{8} \div 5\right) = \frac{5}{8} + \frac{3}{8} = 1.$

Example 2: How much of alloy A which contains 6 grams of gold in 10 grams must be melted with another alloy B which contains 3 grams of gold in 10 grams in order to get an alloy containing 5 grams of gold in 10 grams?

Solution: Our basic equation is

(Amount of gold in alloy A) + (Amount of gold in alloy B) = (Amount of gold in final alloy)

Now we let x = number of grams of alloy A. Then $10 - x$ is the number of grams of alloy B and we have

	Grams of alloy	×	Grams of gold in *one* gram of alloy	=	Grams of gold in alloy
Alloy *A*	x		$\frac{6}{10}$		$\frac{6}{10}x$
Alloy *B*	$10 - x$		$\frac{3}{10}$		$\frac{3}{10}(10 - x)$
Final alloy	10		$\frac{5}{10}$		$\frac{5}{10} \cdot 10$

Thus we obtain the equation

$$\frac{6}{10}x + \frac{3}{10}(10 - x) = \frac{5}{10} \cdot 10$$

Multiplying both sides of the equation by 10 gives us

$$6x + 3(10 - x) = 50$$

So

$$6x + 30 - 3x = 50$$
$$3x = 20$$
$$x = \frac{20}{3} = 6\frac{2}{3}$$

Thus $6\frac{2}{3}$ grams of alloy *A* are needed and $10 - 6\frac{2}{3} = 3\frac{1}{3}$ grams of alloy *B*.

Check: Grams of gold in alloy *A*: $\frac{6}{10} \cdot \frac{20}{3} = 4$; grams of gold in alloy *B*: $\frac{3}{10} \cdot \frac{10}{3} = 1$; grams of gold in final alloy: $\frac{5}{10} \cdot 10 = 5$; $4 + 1 = 5$.

Example 3: A small plane leaves an airport and travels east at 500 kilometers per hour. One hour later a westbound jet leaves the same airport and travels at 900 kilometers per hour. In how many hours will the two planes be 3300 kilometers apart?

Solution: Our basic equation is

(distance small plane travels) + (distance jet travels) = 3300

Now we let h be the number of hours that the small plane flew. Then $h - 1$ is the number of hours that the jet flew and we have

	Speed of plane in kilometers per hour	X	Time in hours it travels	=	Distance in kilometers it travels
Small plane	500		h		$500h$
Jet	900		$h - 1$		$900(h - 1)$
Total distance in kilometers					3300

Thus we have the equation

$$500h + 900(h - 1) = 3300$$

Hence

$$500h + 900h - 900 = 3300$$
$$1400h = 4200$$
$$h = 3$$

and so the planes will be 3300 kilometers apart 3 hours after the first plane takes off.

Check: In 3 hours, the small plane will have traveled 3 X 500 kilometers = 1500 kilometers, the jet (2 X 900) kilometers = 1800 kilometers, and 1500 + 1800 = 3300.

Example 4: A man invested $8000, part at 7% per year and the remainder at 5% per year. If the yearly income on the two investments was $510, how much was invested at each rate?

Solution: Let x = the amount invested at 7%. Then $8000 - x$ = the amount invested at 5% and we have

	Amount in dollars	X	Percent	=	Income in dollars
At 7%	x		0.07		$0.07x$
At 5%	$8000 - x$		0.05		$0.05(8000 - x)$
Combined investments					510

Therefore

$$0.07x + 0.05(8000 - x) = 510$$

Multiplying both sides of this equation by 100 we get

$$7x + 5(8000 - x) = 51,000$$
$$7x + 40,000 - 5x = 51,000$$
$$2x = 11,000$$
$$x = 5500$$

The amount invested at 7% is $5500 and the amount invested at 5% is 8000 - 5500, or $2500.

Check: Income from 7% investment = 0.07(5500) = $385.
Income from 5% investment = 0.05(2500) = $125.

$$\$385 + \$125 = \$510$$

Example 5: The tens' digit of a two-digit number is 3 times the units' digit. When the digits are reversed, the number obtained is 36 less than the original number. What is the original number?

Solution: Clearly the basic equation is

Number with digits reversed = original number - 36

But here letting x = original number will not do us any good, as it doesn't say anything about the digits! We can, however, let u = the units' digit of the original number so that $3u$ is the tens' digit. Then recalling that, for example, 73 is a shorthand for $(7 \times 10) + 3$ we can represent the original number by $(3u \times 10) + u = 30u + u = 31u$. Then our number with digits reversed is $(u \times 10) + 3u = 10u + 3u = 13u$ and our basic equation becomes

$$13u = 31u - 36$$

from which we get $u = 2$ and $3u = 6$,

Check: Original number is 62; reversing the digits we get 26, which is 36 less than the original number.

EXERCISES 3.8

1. One boy can shovel snow from a sidewalk in 12 minutes and a second boy can do the same job in 18 minutes. How long should it take them to shovel the snow from the sidewalk working together?

2. A farmer can plow a field in 12 hours and his son can plow it in 36 hours. If they work together, using two plows, how long will it take them to plow the field?

3. A water tank can be filled by one pipe in 3 hours and by a second pipe in 2 hours. How many hours will it take the two pipes together to fill the tank?

4. If Bill can paint a house in 4 days and Roger can do it in 6 days, how long will it take them when working together?

5. Printing press A can do a certain job in 3 hours and press B can do the same job in 2 hours. If both presses work together at the same time, in how many hours can they complete the job?

6. A dairy has 400 pounds of milk containing 5% butterfat. How many pounds of milk containing 2% butterfat must be added to produce milk containing 4% butterfat?

7. Some United States silver coins are $\frac{9}{10}$ pure silver or " $\frac{9}{10}$ fine." How much pure silver must be melted with 200 grams of silver $\frac{3}{5}$ fine to make it a standard fineness for coinage?

8. How many cubic centimeters of water must be added to 20 cubic centimeters of a 30% solution of sulphuric acid to make a 15% solution?

9. How many pounds of salt must be added to 20 pounds of a 10% salt solution to make a $33\frac{1}{3}\%$ solution?

10. A chemist wishes to dilute 240 cubic centimeters of pure hydrochloric acid so that he will have a 24% solution. Into how much water should the acid be poured?

11. Jimmez goes 8 miles per hour and José goes 32 miles per hour in the opposite direction. How many hours has Jimmez traveled when they are 112 miles apart?

12. George and Patricia, who are 400 kilometers apart, get in their automobiles and start driving toward each other. George drives at 60 kilometers per hour and Patricia drives at 55 kilometers per hour. How soon will they meet if George has an accident and is delayed an hour?

13. A boat left a marina at 10 miles per hour. Two hours later another boat left the same marina at 14 miles per hour in the same direction as the first boat. How long did it take the second boat to overtake the first boat? How far were they from the marina when they met?

14. A passenger train traveling at 50 miles per hour leaves a station 6 hours after a freight train and overtakes it in 4 hours. Find the speed of the freight train.

15. Two trains start at the same time from towns 605 kilometers apart and meet in 5 hours. If the speed of one train is 11 kilometers per hour less than the speed of the other, what is the speed of each?

16. A widow invests a certain amount of money at 4% annual interest and twice as much at 5%. If her annual income from both investments is $84, how much does she invest at each rate?

17. A farmer invested part of his money at 8% and $500 more than this at 5%. If his income from both these investments is $545, how much is invested at each rate?

18. Mr. Clark invested some money at 10% and an amount $1300 less at 8%. If the first investment produces $160 more interest per year than the second, how much is invested at each rate?

19. Ms. Smith receives $661.50 per year from three investments. She has invested half her money at 7%, $\frac{1}{3}$ at 9%, and the rest at 6%. How much money does, she have invested at each rate?

20. How can $7200 be invested, one part at 8% and the other part at 10%, so that both investments produce the same income?

21. The sum of the digits of a two-digit number is 9 and the number is 12 times the tens' digit. Find the number.

22. The ones' digit of a two-digit number exceeds twice the tens' digit by 1. If the sum of the digits is 7, what is the number?

23. The tens' digit of a two-digit number is 5 more than the ones' digit. Find the number if it is equal to 8 times the sum of its digits.

24. The sum of the digits of a two-digit number is 8. If the number with its digits reversed is 7 times the tens' digit of the original number, what is the original number?

25. A two-digit number is 4 times the sum of the digits. The ones' digit is 3 more than the ten's digit. Find the number.

26. A machine can do a certain job in 12 hours. To step up production, a more modern machine is put into operation and both machines now take 4 hours to do the job. If the old machine should break down, how long would it take the new machine to do the job alone?

27. At 12:00 noon, two river steamers are 120 miles apart on the Mississippi River, one traveling north and one traveling south. If they meet at St. Louis at 6:00 P.M. and the northbound boat steams at 9 miles per hour, how fast is the southbound boat traveling?

28. How much water must be added to 2 liters of a disinfectant solution which contains 30% of the active ingredient to form a solution which contains 20% of the active ingredient?

29. A woman lends some money at 8% and an equal amount at 5%. If the income from the 5% loan is $150 less than the 8% loan, how much money was loaned at each rate?

30. A woman can paint her house in 6 days. When her neighbor helps her, they can do the job in 4 days. How long would it have taken the neighbor to do it alone?

31. A freight train left Paris at 5:00 A.M. traveling at 48 kilometers per hour. At 7:00 A.M. an express train traveling at 80 kilometers per hour left Paris in the same direction. When did the express train overtake the freight train?

32. A man has twenty-one $1000 bonds. Some are $4\frac{1}{2}$% bonds and the others are 6% bonds. His income from the $4\frac{1}{2}$% bonds is the same as the income from the 6% bonds. How many of each kind did he have?

33. How much milk testing 3% butterfat must be mixed with cream testing 20% butterfat to make 80 gallons of a mixture testing 14% butterfat?

34. A swimming pool can be filled in $2\frac{1}{2}$ hours if water enters through a pipe. It can be filled in 4 hours using a fire hose. How long will it take to fill the pool using both the pipe and the hose?

35. An automobile radiator contains 16 quarts of a 20% antifreeze solution. How much must be drained off and replaced by pure antifreeze so that the radiator mixture will be 30% antifreeze?

36. A plane left Chicago for New York flying at a rate of 420 miles per hour. Forty-five minutes later a second plane followed it at 510 miles per hour. How soon will the second plane overtake the first?

37. One computer works twice as fast as another. When operating at the same time, they can complete a complicated job in 4 hours. In how many hours could the faster machine do the job by itself?

38. The sum of the digits of a two-digit number is 6. The number with its digits reversed is 3 times the tens' digit of the original number. Find the original number.

39. A wholesaler in Brazil blends his own coffee. He mixes one brand selling at $1.38 a kilogram with a premium brand selling at $2.38 a kilogram. If he wants to make 70 kilograms of a blend that sells for $2.00 per kilogram, how much of each kind should he use?

40. A certain jet travels 4 times as fast as a small private plane. If both planes start from the same place at the same time traveling in the same direction, and are 6405 kilometers apart in 5 hours, what is the speed of each plane?

41. If a dairyman has 80 pounds of milk testing 3% butterfat, how many pounds of skim milk (which contains no butterfat) must he remove to have milk testing 3.6% butterfat?

42. One bottle cap machine can cap 1200 bottles an hour and a second older machine can cap 900 bottles an hour. If the faster machine starts an hour before the slower machine, how long will it take the machines working together to cap 8200 bottles?

43. You have $25,000 to invest. You decide to invest $10,000 in bonds yielding 8%. You want to invest part of the remainder at 5% and the rest at 7% so that you will have a total income of $6\frac{1}{2}$% on your total investment. How much should you invest at 7%?

44. A woman made a trip of 496 kilometers by automobile in Europe averaging 32 kilometers per hour inside towns and 80 kilometers per hour outside towns. If it took her 8 hours for the trip, how much of her trip was inside towns and how much outside?

45. A swimming pool has two inlet pipes. One pipe fills the pool in 3 hours and the other pipe in 6 hours. The drain pipe can empty the pool in 4 hours. By accident the drain pipe is left open when the pool is being filled. How many hours will it take to fill the pool if both inlet pipes are being used?

46.   A man's age is 2 years less than 3 times his son's age. Each age has the same digits, whose sum is 10. How old is the man and how old is his son?

47.   A man made two investments totaling $9500. On the first he lost 10%, while on the second he gained 20%. If he lost $500 on the two investments, how much was each investment?

48.   A small plane leaves an airport and travels east at 325 miles per hour. At the same time a westbound jet leaves the same airport and travels at 625 miles per hour. In how many hours will the planes be 3325 miles apart?

49.   How much alcohol must be added to 16 liters of a 30% mixture of alcohol and water in order to make a 50% mixture?

50.   One pipe can empty a tank in 10 hours. After it had been open 2 hours, a second pipe was opened and the two pipes emptied the tank in 3 more hours. How long would it take the second pipe alone to empty the entire tank?

3.9   Formulas

Let us return now to the equation of Chapter 1 relating temperature in Fahrenheit to temperature in Celsius:

(1) $$F = \frac{9}{5}C + 32$$

We said in Chapter 1 that the "algebra machine" would enable us to go easily from (1) to

(2) $$C = \frac{5}{9}(F - 32)$$

Now we can see what happens inside the "algebra machine." We have

$$F = \frac{9}{5}C + 32$$

$$5F = 9C + 160 \qquad \text{Why?}$$

$$9C = 5F - 160 \qquad \text{Why?}$$

$$9C = 5(F - 32) \qquad \text{Why?}$$

$$C = \frac{5}{9}(F - 32) \qquad \text{Why?}$$

An equation in two or more variables relating various measurements is often called a *formula*. In a formula it is frequently necessary to solve for one variable in terms of the other. Thus in the temperature scale formula (1) we were given F in terms of C and then solved for C in terms of F to get (2).

Here is another example: The formula

(3) $$I = Prt$$

relates the interest I obtained by investing a principal P at interest rate r for a time t. From (3), by using the division property, we can get

$$P = \frac{I}{rt}, \qquad r = \frac{I}{Pt}, \qquad t = \frac{I}{Pr}$$

That is, we can solve (3) for P, for r, or for t.

As a final example, consider the following formula from thermodynamics:

$$E = \frac{RT}{J(k - 1)}$$

To solve for R we first use the multiplication property of equality to get

(4) $$EJ(k - 1) = RT$$

and then the division property of equality to get

$$R = \frac{EJ(k - 1)}{T}$$

To solve for J we again use the division property of equality on (4) to get

$$J = \frac{RT}{E(k - 1)}$$

To solve for k we apply the distributive property to (4) to get

$$EJk - Ej = RT$$

The addition property of equality then gives us

$$EJk = RT + EJ$$

and, finally, the division property of equality gives us

$$k = \frac{RT + EJ}{EJ}$$

EXERCISES 3.9

odd ones

In each exercise solve the given formula for the indicated letter.

1. $C = 3.1416d$ (Geometric formula.) Solve for d.
2. $R = 0.379G$ (Railroad construction formula.) Solve for G.
3. $C = np$ (Cost formula.) (a) Solve for n. (b) Solve for p.

4. $d = rt$ (Distance formula.)   (a) Solve for r.   (b) Solve for t.

5. $V = lwh$ (Geometric formula.)   (a) Solve for l.   (b) Solve for w.
    (c) Solve for h.

6. $F = DgV$ (Formula for Archimedes' principle.)   (a) Solve for D.
    (b) Solve for V.

7. $M = \frac{W}{E}$ (Mechanical advantage formula.)   (a) Solve for W.   (b) Solve for E.

8. $R = \frac{E}{I}$ (Electrical circuit formula.)   (a) Solve for E.   (b) Solve for I.

9. $S = \frac{1}{2}at^2$ (Motion formula.)   Solve for a.

10. $W = \frac{V^2}{R}$ (Electrical power formula.)   Solve for R.

11. $C = 5p + 5$ (Postage formula.)   Solve for p.

12. $R.P.M. = \frac{12S}{d}$ (Formula for speed of a milling cutter.)   Solve for S.

13. $T = \frac{2}{3}fw$ (Formula for torque on a flat pivot.)   (a) Solve for f.
    (b) Solve for w.

14. $T = \frac{SJ}{C}$ (Torsion formula for round shafts.)   (a) Solve for S.   (b) Solve for J.   (c) Solve for C.

15. $K = \frac{wV^2}{2g}$ (Energy formula.)   (a) Solve for w.   (b) Solve for g.

16. $H = \frac{0.4\pi NI}{L}$ (Magnetic intensity formula.)   (a) Solve for N.   (b) Solve for I.   (c) Solve for L.

17. $I = \frac{E - e}{R}$ (Electrical circuit formula.)   (a) Solve for E.   (b) Solve for R.

18. $W = \frac{2PR}{R - r}$ (Differential pulley formula.)   (a) Solve for P.   (b) Solve for r.

19. $C = \frac{Kab}{b - a}$ (Temperature conversion formula.)   (a) Solve for K.
    (b) Solve for a.

20. $S = \frac{rl - a}{b - a}$ (Geometric progression formula.)   (a) Solve for r.   (b) Solve for l.   (c) Solve for a.

21. $V = \frac{V_t + V_o}{2}$ (Motion formula in physics.)   (a) Solve for V_t.
    (b) Solve for V_o.

22. $S = T - \frac{1.299}{N}$ (Formula for tap drill size for U.S. standard thread.)
    (a) Solve for T.   (b) Solve for N.

23. $V^2 = V_o^2 + 2gh$ (Motion formula in physics.)   (a) Solve for g.   (b) Solve for h.

24. $P = 2l + 2w$ (Geometric formula for perimeter.)   (a) Solve for l.
    (b) Solve for w.

25. $C = \frac{5}{9}(F - 32)$ (Temperature conversion formula.)   Solve for F.

26. $A = \frac{m}{t}(p + t)$ (Formula relating to the thickness of a pipe.)   (a) Solve for m.   (b) Solve for p.   (c) Solve for t.

27. $P = \frac{A}{1 + ni}$ (Business formula for present value.)   (a) Solve for A.
    (b) Solve for n.

28. $A = \frac{h(b_1 + b_2)}{2}$ (Formula for the area of a trapezoid.)   (a) Solve for h.
    (b) Solve for b_1

29. $R = \frac{l^2}{6d} + \frac{d}{2}$ (Formula for the radius of the curvature of a lens.)
    Solve for l^2.

30. $S = \pi r^2 + \pi r l$ (Geometric formula for surface area.)   Solve for l.

31. $P = \frac{n + 2}{od}$ (Formula for the pitch of gears.)   (a) Solve for n.   (b) Solve for d.

32. $i = \frac{2E}{R + 2r}$ (Formula for electrical current.)   (a) Solve for E.   (b) Solve for R.   (c) Solve for r.

33. $I = \frac{E}{r + \frac{R}{n}}$ (Formula for electrical current.)   (a) Solve for E.
    (b) Solve for r.   (c) Solve for R.

34. $V = \frac{v}{1 - \frac{v}{c}}$ (Formula relating to the frequency of sound waves.)   (a) Solve for c.   (b) Solve for v.

3.10   Solution of Linear Inequalities in One Variable

Consider the inequality

$$5x - 7 < 3x + 4$$

Unlike the corresponding equation, $5x - 7 = 3x + 4$, the solution set of the inequality has more than one member. Thus

   5 is a solution since $(5 \cdot 5) - 7 = 18 < (3 \cdot 5) + 4 = 19$

   2 is a solution since $(5 \cdot 2) - 7 = 3 < (3 \cdot 2) + 4 = 10$

and

   0 is a solution since $(5 \cdot 0) - 7 = -7 < (3 \cdot 0) + 4 = 4$

Using the properties of inequalities given in Section 2.11, we can solve inequalities in much the same fashion as equations. Thus, beginning with

(1) $5x - 7 < 3x + 4$

we have

$(5x - 7) + 7 < (3x + 4) + 7$ (adding 7 to both sides of the inequality)

$5x < 3x + 11$

$5x - 3x < (3x + 11) - 3x$ (subtracting $3x$ from both sides of the inequality)

$2x < 11$

(2) $x < \dfrac{11}{2} = 5\dfrac{1}{2}$ $\left(\text{multiplying both sides by } \dfrac{1}{2}\right)$

By reversing the steps, from $x < 5\frac{1}{2}$ we can obtain $5x - 7 < 3x + 4$. Thus (1) and (2) are equivalent inequalities.

The *set builder* notation is very useful to describe this situation. For the solution set of (1) we can write

$$\{ x : 5x - 7 < 3x + 4 \}$$ which we read as

"the set of all x such that $5x - 7$ is less than $3x + 4$." Similarly, we can write the solution set of (2) as

$$\left\{ x : x < 5\frac{1}{2} \right\}$$

which we read as "the set of all x such that x is less than $5\frac{1}{2}$."

To say, then, that (1) and (2) are equivalent inequalities is to say that

$$\{ x : 5x - 7 < 3x + 4 \} = \left\{ x : x < 5\frac{1}{2} \right\}$$

When an inequality is equivalent to one of the form $x < k$ (or $x \leqslant k$, $x > k$, or $x \geqslant k$) for some number k, we say that the inequality is a *linear* inequality. We will consider only linear inequalities here.

We have said nothing about the domain of the variable in the inequality. As with equations, we assume, unless otherwise stated, that the domain is the set of real numbers. If, however, we have a more restricted domain, we will specifically say so. Thus, in considering the inequality $5x - 7 < 3x + 4$, we might state that the domain of x is the set of natural numbers. Then, if this were the case, we would have

$$\left\{ x : x \text{ is a natural number and } x < 5\tfrac{1}{2} \right\}$$

for the solution set and could further note that this solution set is equal to $\{ 1, 2, 3, 4, 5 \}$.

There is no way, of course, to list all the members of $\{ x : x < 5\tfrac{1}{2} \}$ if the domain of x is the set of real numbers. It is useful, however, to picture the solution set on a number line as shown in Figure 3.1.

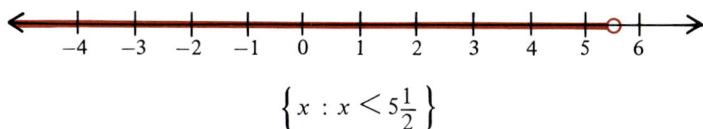

$$\left\{ x : x < 5\tfrac{1}{2} \right\}$$

Figure 3.1

Note that we use the "open circle" at the $5\tfrac{1}{2}$-point to indicate that $5\tfrac{1}{2}$ is *not* in the solution set. On the other hand, if we had $\{ x : x \leqslant 5\tfrac{1}{2} \}$, we would use a "closed circle" to indicate that $5\tfrac{1}{2}$ is in the solution set, as shown in Figure 3.2.

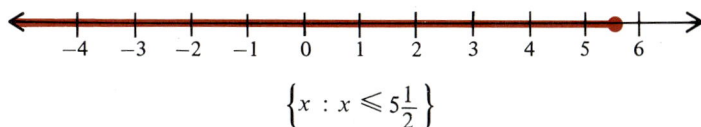

$$\left\{ x : x \leqslant 5\tfrac{1}{2} \right\}$$

Figure 3.2

Finally, the graph of $\{ x : x \text{ is a natural number and } x < 5\tfrac{1}{2} \}$ is shown in Figure 3.3. The solution set is $\{1, 2, 3, 4, 5\}$.

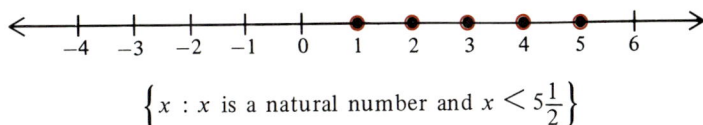

$$\left\{ x : x \text{ is a natural number and } x < 5\tfrac{1}{2} \right\}$$

Figure 3.3

We conclude this section with two more examples of the solution of linear inequalities.

Example 1:

$$5 - 6x > 23$$

$$(5 - 6x) - 5 > 23 - 5 \qquad \text{Why?}$$

$$-6x > 18$$

$$\left(-\frac{1}{6}\right)(-6x) < \left(-\frac{1}{6}\right)(18) \qquad \text{Why?}$$

$$x < -3$$

(Note the reversal in sense — from $>$ to $<$ — because we multiplied both sides of the inequality by a negative number.)

Thus the solution set is $\{\, x : x < -3 \,\}$ and the graph is as shown in Figure 3.4.

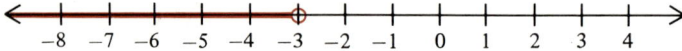

Figure 3.4

Example 2:

$$3x + 5 \leqslant 7 - 2x$$

$$3x \leqslant 2 - 2x \qquad \text{Why?}$$

$$5x \leqslant 2 \qquad \text{Why?}$$

$$x \leqslant \frac{2}{5} \qquad \text{Why?}$$

Thus the solution set is $\{\, x : x \leqslant \frac{2}{5} \,\}$ and the graph is as shown in Figure 3.5.

Figure 3.5

EXERCISES 3.10

Solve the following inequalities. Express your answer in (a) set builder notation and (b) as a graph on a number line.

1. $3x + 7 < 28$

2. $5y + 4 < 19$

3. $6n - 8 > 4$

4. $2x - 7 > 5$

5. $\frac{y}{3} + 2 \leqslant 8$

6. $\frac{x}{8} + 9 \leqslant 24$

7. $\frac{2n}{3} \geqslant 8$

8. $\frac{4k}{5} \geqslant 4$

9. $8n + 6 < 5n + 27$

10. $6y + 5 < 2y + 13$

11. $3x + 18 + x > 28 + 2x$

12. $9y - 12 - y > 5y + 15$

13. $5n - 30 \geqslant 2n$

14. $7y + 40 \geqslant 12y$

15. $7x - 12 < 28 + 5x$

16. $22 - 6z > 7 + 3z$

17. $5n + 18 > 4n + 7$

18. $4x - 11 < 8x + 29$

19. $36 - 7x \leqslant 2x - 18$

20. $9y - 14 \geqslant 5y - 19$

21. $4(n + 3) < 13$

22. $7(x - 5) < -34$

23. $3(2x + 1) > 25$

24. $5(3y - 4) > -5$

25. $6(x - 7) + 30 < 0$

26. $16y - (7y - 5) - 11 > 0$

27. $3(x + 4) - 2(x + 2) > 3(x - 2)$

28. $5(n - 4) < 3(4n - 7) - 2(3n + 4)$

29. $\frac{2}{3}x - 5 \leqslant \frac{3}{4}x + 5$

30. $\frac{1}{3}y + \frac{1}{4} \geqslant \frac{1}{5}y + \frac{1}{6}$

31. $\frac{x}{7} - 6 < \frac{x}{3} + \frac{1}{2}$

32. $\frac{5y}{3} > 7y + 2$

33. $\frac{4x}{7} > \frac{7x}{4} + 5$

34. $\frac{x + 3}{4} < \frac{x - 5}{6}$

35. $\frac{2x + 6}{3} > \frac{3x - 2}{6}$

36. $\frac{1}{2}(x - 2) \leqslant \frac{1}{2}(7 - x)$

37. $\frac{3}{2}(y - 1) > \frac{1}{2}(y - 3) - \frac{7}{10}$

38. $\frac{x + 2}{4} - \frac{x + 2}{2} \leqslant 2$

39. $3 - x < \frac{7 - x}{4} - \frac{2x - 3}{3}$

40. $\frac{x + 3}{5} + \frac{2x - 2}{6} - 6 \leqslant 0$

*41. $0.027x > 1.5538$

*42. $-0.018y < 1234.4$

*43. $0.194w - 3.005 < 2.084$

*44. $0.5583 - 1.64x > 1.5477$

*45. $4.594x + 7.821 \geqslant x - 10.497$

*46. $38.88w - 2.285 > 2.030w + 3.376$

*47. $\frac{2.389x - 1.5276}{1.57} \leqslant 3.062$

*48. $\frac{5.297 - 4.41y}{2.8} \leqslant 23.90$

REVIEW EXERCISES

Section 3.1

In exercises 1–10 state whether the given statement is true or false. For each false statement change the $=$, $<$, or $>$ to make a true statement.

1. $6 + 8 < 13$

2. $9 + 7 > 7 + 9$

3. $3 \times 6 = 15 + 3$

4. $\frac{1}{2} + \frac{3}{4} > \frac{2}{3} + \frac{1}{6}$

5. $\left(\frac{2}{3} \times 6\right) + 5 < \left(\frac{4}{5} \times 20\right) - 5$

6. $3\frac{3}{4} + 4\frac{1}{5} > 6 + \frac{3}{2} + \frac{18}{40}$

7. $4.14 \div 300 = 828 \div 1500$

8. $\frac{23}{45} \cdot 9 > \frac{6}{11} \cdot 8$

9. 6% of $120 = 12\%$ of 600

10. $\frac{1}{4}\%$ of $100 < 25\%$ of 1

In exercises 11–20 find the solution set of the given equation or inequality if the domain of the variable is $\{\,1, 2, 3, 4, 5, 6, 7, 8, 9\,\}$.

11. $x + 2 = 5$

12. $x + 2 < 5$

13. $x + 2 > 5$

14. $6n + 3 < 15$

15. $\frac{y}{5} + 3 > 6$

16. $p - 6 \neq 3$

17. $2x - 4 > 10$

18. $x^2 < 20$

19. $15 > \frac{x}{3} + 12$

20. $2x < 3x - 1$

Section 3.2

In exercises 21–50 find the solution set of the given equation. The domain of each variable is the set of real numbers.

21. $3x + 4 = 16$

22. $5x - 6 = 14$

23. $x + 48 = 9x$

24. $7t + 40 = 12t$

25. $6x + 11 = 2x + 9$

26. $4y + 17 = 5$

27. $6x - 10 = -34$

28. $8a - 7 = a + 42$

29. $2n + 7 = 11 - 6n$

30. $4k - 12 = 65 - 3k$

31. $4x - 11 = 8x + 29$

32. $2h + 27 = 17 - 3h$

33. $9a - 14 = 5a - 19$

34. $3x + 4 = x + 10$

35. $5x - 2 = 2x + 1$

36. $3y + 7 = 25 + y$

37. $4x + 20 = x - 4$

38. $7t - 10 = 11t - 2$

39. $4(n + 3) = 44$

40. $7(a - 5) = 28$

41. $3(2x + 1) = 27$

42. $10(2n + 5) - 8n = 56$

43. $18 - 4(2s - 5) = 14$

44. $7x - 8 - 3(x - 10) = 0$

45. $3(2t - 11) - 4(3t + 12) = -141$

46. $5(6x - 3) - 2(18 - 3x) + 99 = 0$

47. $\frac{1}{3}x = \frac{2}{7}x - 1$

48. $\frac{7}{2}y - \frac{4}{3}y = \frac{11}{6} - \frac{2}{5}y$

49. $\frac{1}{3}(x - 2) = \frac{1}{7}(4 - x)$

50. $\frac{1}{3}(x + 5) - 4 = \frac{1}{4}(x - 10)$

Section 3.3

In exercises 51–59 find a common fraction in lowest terms which is equal to the given repeating decimal.

51. $0.\overline{7}$

52. $0.\overline{1}$

53. $0.\overline{12}$

54. $0.\overline{35}$

55. $6.\overline{2}$

56. $3.\overline{76}$

57. $5.\overline{123}$

58. $7.\overline{9}$

59. $4.1\overline{658}$

Section 3.5

In exercises 60–75 give a mathematical phrase equivalent to the given verbal phrase. Use any letter or letters you wish for the variables.

60. Three times a number.

61. Eleven more than a number.

62. Five less than a number.

63. The product of a number and 3.

64. The product of two numbers.

65. The sum of two numbers.

66. Seven less than 3 times a number.

67. Eight more than twice a number.

68. Three less than $\frac{1}{2}$ a number.

69. Five more than $\frac{2}{3}$ a number.

70. Thirteen times the sum of two numbers.

71. The value in pennies of a certain number of dimes.

72. The value in dollars of a certain number of nickels.

73. One-third of a number which is 6 less than N.

74. A three-digit numeral whose tens' digit is 2 less than the hundreds' digit and whose hundreds' digit is twice a given number.

75. The annual salary of a man who earns a certain number of dollars per week.

Sections 3.6 — 3.8

In exercises 76–95 solve each problem and check your answer.

76. One number is 7 times another and their difference is 78. Find each number.

77. Find three consecutive integers whose sum is 528.

78. The perimeter of a rectangle is 360 feet. If the width is $\frac{2}{3}$ the length find the length and width.

79. The length of a playground exceeds twice its width by 8 feet. If 260 feet of fencing are needed to enclose the playground, what is its length and width?

80. A collection of nickels and dimes consisting of 26 coins amounts to $1.75. Find the number of each kind.

81. Fifty-two coins, all dimes or quarters, amount to $6.40. Find the number of each.

82. George and Ed together have $100. George gives $20 to Ed, after which Ed gives $6 to George. They now have the same amount of money. How much money did each have at first?

83. In the year 1897 a man paid a certain sum for a horse and $\frac{7}{10}$ as much for a carriage. If the horse had cost $70 less and the carriage $50 more, the price of the horse would have been $\frac{4}{5}$ that of the carriage. What was the cost of each?

84. Two persons are 92 kilometers apart and start at the same time to walk toward each other. If A walks at 5 kilometers per hour and B walks at 6.5 kilometers per hour, how far will each have traveled when they meet? If they start at 7:00 A.M., at what time will they meet?

85. A tank can be filled by one pipe in 9 hours and emptied by another in 21 hours. In what time will the tank be filled if the valves on both pipes are left open and the tank is empty to begin with?

86. A man buys two pieces of cloth. They are the same width but·one is 6 yards longer than the other. For the longer piece he paid $7 for 10 yards and for the shorter piece he paid $5 for 3 yards. He sold both pieces together at 9 yards for $11 and made $5 on the transaction. How many yards were there in each piece?

87. A woman walked a certain distance at 5 kilometers per hour and returned immediately in an automobile at 70 kilometers per hour. If the entire trip took 5 hours, how long did she walk?

88. You have $3000 invested at 8% and $4000 at 7%. How much must you invest at 5% to make your total income 6% of your total investment?

89. The manager of a candy store has on hand 100 kilograms of candy which has been selling at $1.80 a kilogram. She decides to mix with it some candy which sells at $3.00 a kilogram. How many kilograms of the latter must she use so that the mixture will be worth $2.10 a kilogram?

90. How much alcohol must be added to a 500 cubic centimeters of a 10% solution of iodine to make a 6% solution?

91. A milk dealer has a 10 gallons of cream containing 30% butterfat. How much milk containing 4% butterfat must be mixed with it to make a mixture containing 10% butterfat?

92. You can do a piece of work in 10 days, while it takes your friend 15 days to do the same job. After your friend works alone for 3 days he asks you to help him. How many days will it take to finish the job if you help him?

93. A motorist traveling 60 miles per hour is being pursued by a state patrol car traveling at 69 miles per hour. If the patrol car is 6 miles behind the motorist at the start of the pursuit, how long will it take the patrol car to overtake the motorist?

94. How many tons of 8% copper ore and how many tons of 3% copper ore must be mixed to make 300 tons of 6% ore?

95. In a two-digit number the ones' digit is 6 and the number is 4 times the sum of the digits. Find the number.

Section 3.9

In exercises 96–111 solve each of the formulas for the letter indicated.

96. $Q = \frac{WL}{T}$ Solve for L.

97. $X = \frac{1}{2\pi fc}$ Solve for c.

98. $T = \frac{1}{a} + t$.   Solve for t.

99. $F = \frac{wv}{gR}$ Solve for R.

100. $F = \frac{KmM}{d^2}$ Solve for M.

101. $V = \frac{1}{3}\pi r^2 h$ Solve for h.

102. $V = \frac{1}{6}h(b + 4M + B)$ Solve for h.

103. $S = \frac{a}{1 - r}$ Solve for r.

104. $D = \frac{NP}{\pi}$ Solve for N.

105. $E = Mc^2$ Solve for M.

106. $v = \frac{V}{bjd}$ Solve for j.

107. $H = \frac{Eit}{4.18}$ Solve for t.

108. $l = d + \frac{2r^3}{5d}$ Solve for r^3.

109. $S = \frac{ab}{a + b}$ Solve for b.

110. $S = \frac{a - b}{ab}$ Solve for a.

111. $f = \left(\frac{w}{k} - 1\right)\frac{1}{k}$ Solve for w.

Section 3.10

In exercises 112–131 solve the given inequality. Express your answer in (a) set builder notation and (b) as a graph on a number line.

112. $2x + 5 < 13$

113. $5x - 6 > -14$

114. $3x + 2 \geqslant -7$

115. $\frac{1}{3}y - 9 \leqslant 12$

116. $9x < 7x + 18$

117. $6x + 11 < x + 31$

118. $4x - 3 \leqslant 8x + 33$

119. $5x + 9 > 14 - 2x$

120. $\frac{n}{5} + 2 \geqslant 22$

121. $\frac{3k}{5} > -6$

122. $7x - 29 \leqslant 16x - 17$

123. $13 - 6a \geqslant 13a - 6$

124. $\frac{2}{3}t + \frac{5}{2} > 0$

125. $\frac{5}{6} > \frac{1}{2}x + \frac{1}{3}$

126. $x + \frac{x}{2} - \frac{3x}{5} > 9$

127. $\frac{5x}{3} - \frac{3x}{5} + \frac{11}{6} \leqslant 0$

128. $2(5x + 1) - 4 < 3(x - 7) - 16$

129. $10y + (3y + 2) < 9y + (5y - 4)$

130. $4x + \frac{8x - 12}{7} \leqslant \frac{9x}{2}$

131. $\frac{5z}{3} + \frac{2z - 2}{9} \geqslant z - 2$

CUMULATIVE REVIEW A

1. $-\frac{1}{8} \cdot \frac{1}{3} = ?$

2. $\frac{-3}{7} + \frac{6}{5} = ?$

3. $\frac{7}{9} - \frac{7}{8} + \frac{2}{3} = ?$

4. $\frac{3}{7} - \left| \frac{3}{4} - \frac{5}{7} \right| = ?$

5. $\dfrac{3\frac{2}{7}}{\frac{1}{2} - 4} = ?$

6. $8a + 3a + a = ?$

7. $4x + 7x + 2x = ?$

8. $(6x) \cdot (7xy) = ?$

9. $(5ab) \cdot (3b) = ?$

10. $3(4 + y) = ?$

11. $a(3 + b) = ?$

12.   Rewrite $pq + pr$ using the distributive property.

13. $(4a + 3b) + (7a + 8b) = ?$

14. $4xy(2x + 8y + 9z) = ?$

15. $2x^2(5y^2 + 4yz) = ?$

16. $(3 + 4) + 5 = 3 + (4 + 5)$ is an example of the___?___ property of addition.

17. $[3 \cdot (4 + 8)] \cdot 4 = [(4 + 8) \cdot 3] \cdot 4$ is an example of the___?___ property of ___?___ .

18. If a and b are integers $(ab \neq 0)$, then $\frac{1}{ab}$ is called the___?___of ab.

19. If x, y, and z have the values 3, 4, and 0, respectively, what is the value of $2x \cdot 3y + (4z + x)$?

20. Find the decimal equivalent of $\frac{5}{9}$.

21. Find the common fraction in lowest terms equal to $0.\overline{14}$.

22. Find the common fraction in lowest terms equal to $0.1\overline{23}$.

In exercises 23–34 find the solution set of the given equation.

23. $8 - 3x = 20$

24. $5a + 2 = 7a$

25. $9x - 7 = 4x - 12$

26. $6y + 5 = 7 + 8y$

27. $5x - 8 + 3x = -3x - 4 + 9x$

28. $3a - 5 = 2a - 5$

29. $x + 2 = x + 5$

30. $3x + (4x - 2) = 6x + 13$

31. $2 - (x - 1) = 8$

32. $5y - 2(y - 6) = 9$

33. $12 + 3x = 5x - 3(x - 4)$

34. $2(x - 4) - 3(4 - x) = 1$

In exercises 35–42 find the solution set of the given inequality and graph it on a number line.

35. $4x + 6 > -18$

36. $2y - 7 > 7$

37. $16 - 3x \leqslant 4 + 3x$

38. $x - 3(x + 2) < -8$

39. $8x + 12 \geqslant 2(6x + 4)$

40. $6x + (2x + 12) < 12 - 4x$

41. $-3x - 7 < 8 + 2x$

42. $(6x + 8) - (2x - 3) \leqslant 18 - (x - 3)$

In exercises 43–50 (a) write an equation for the given situation and (b) find the solution for the problem.

43. The sum of 3 times a number and 8 is 71. What is the number?

44. If 6 times a number is subtracted from 10 times the same number, the difference is 36. What is the number?

45. If the length of a rectangle is 6 meters more than 3 times the width, and if its perimeter is 10 times the width, what are the dimensions of the rectangle?

46. John left his home at 4 P.M. and traveled south on the freeway at 50 miles per hour. If Bill left 30 minutes later in the same direction and traveled at 55 miles per hour, at what time did Bill catch up with John?

47. A small cake and a large cake together make 15 servings. Ten of the small cakes and 6 of the large cakes serve exactly 110 people. How many servings are there in each type of cake?

48. A chemistry student wishes to reduce 60 cubic centimeters of a 40% carbonate solution to a 15% carbonate solution by adding water. How much water must be added?

49. How many pounds each of $1.80 per pound and $2.25 per pound candy must be mixed to have 160 pounds of candy selling at $1.95 per pound?

50. A total of $8000 is invested in two companies, *A* and *B*. Company *A* pays 7.5% on the amount invested, and company *B* pays 9%. How much is invested in each company if the total return is $705?

4

SYSTEMS OF LINEAR EQUATIONS AND INEQUALITIES

4.1 Linear Equations in Two Variables

Consider the equation

$$x - y = 5$$

It is an example of a linear equation in two variables, x and y. Since $6 - 1 = 5$, we say that the *ordered pair* (6, 1) is in the solution set of the equation. Similarly, as you should check, (7, 2), (−1, −6), and $(5\frac{1}{2}, \frac{1}{2})$ are also in the solution set. On the other hand, since $1 - 6 \neq 5$, (1, 6) is not in the solution set. Note that (6, 1) and (1, 6) both involve the same numbers, 6 and 1, but in a different order, and that the order makes a difference. This is why we speak of *ordered* pairs.

As in the study of linear equations in one variable, the study of linear equations in two variables involves the concepts of domain and equivalence of equations. Thus, if the domain of x and y is the set of natural numbers, then the solution set of the equation $x + y = 5$ is { (1, 4), (2, 3), (4, 1), (3, 2) }. On the other hand, if the domain of x and y is the set of integers or the set of real numbers, then the solution set is infinite. Unless otherwise specified, we will consider the domain of all variables to be the set of real numbers.

For the concept of equivalence we again use the definition that two equations are equivalent if and only if they have the same solution set. Thus, the equations

$$x - y = 5, \qquad 2x - 2y = 10, \qquad \text{and } x = 5 + y$$

are all equivalent equations. The easiest way to see that this is so is to solve for y (or x) in each equation to obtain in each case

$$y = x - 5 \text{ (or } x = 5 + y)$$

Thus from $x - y = 5$ we have, by adding $-x$ to both sides,

$$-y = 5 + (-x) = 5 - x$$

and then

$$(-1)(-y) = (-1)(5 - x) \qquad \text{(multiplying both sides by } -1)$$

so that

$$y = -5 + x = x - 5$$

From $2x - 2y = 10$ we get, by dividing both sides by 2, $x - y = 5$ so, again, $y = x - 5$. Finally, from $x = 5 + y$ we get, by adding -5 to both sides,

$$x - 5 = y$$

On the other hand, $2x + y = 3$ and $6x + 2y = 9$ are not equivalent equations since the equation $2x + y = 3$ has $(1, 1)$ in its solution set $(2 \cdot 1 + 1 = 3)$ whereas $(1,1)$ is not in the solution set of the equation $6x + 2y = 9$ because $(6 \cdot 1) + (2 \cdot 1) = 8 \neq 9$.

Any equation in two variables which is equivalent to an equation $ax + by = c$ where a, b, and c are any real numbers with $a \neq 0$ and $b \neq 0$ is called a *linear equation in two variables.*

EXERCISES 4.1

In exercises 1-15 find the solution set (set of ordered pairs, (x, y)) of the given equation if the domain of x is $\{-2, -1, 0, 1, 2\}$.

Example (a): $y = 2x$

Solution:

x	$2x$	y
-2	$2 \cdot -2$	-4
-1	$2 \cdot -1$	-2
0	$2 \cdot 0$	0
1	$2 \cdot 1$	2
2	$2 \cdot 2$	4

Solution set $= \{(-2, -4), (-1, -2), (0, 0), (1, 2), (2, 4)\}$.

Example (b): $x - y = 2$

Solution:

Solving for y, we have

$-y = 2 - x$

$y = x - 2$

x	$x - 2$	y
-2	-2 - 2	-4
-1	-1 - 2	-3
0	0 - 2	-2
1	1 - 2	-1
2	2 - 2	0

Solution set = $\{(-2, -4), (-1, -3), (0, -2), (1, -1), (2, 0)\}$.

1.	$x + y = 2$	9.	$y = x - 1$
2.	$y = x + 2$	10.	$x - y = 5$
3.	$y = x - 2$	11.	$x + y = 5$
4.	$x - y = 0$	12.	$x - y = 0$
5.	$y = x$	13.	$y = 2x$
6.	$x + y = 1$	14.	$x = 3y$
7.	$x - y = 1$	15.	$y = x + 2$
8.	$y = x + 1$		

In exercises 16–30 find the solution set of the given equation if the domain of x is $\{-5, -2, 0, \frac{3}{2}, 7\}$.

16.	$y = x - 2$	24.	$x - y = 2$
17.	$x + y = 1$	25.	$3y = x$
18.	$x - y = 1$	26.	$2x + y = 3$
19.	$x + y = 0$	27.	$y = \frac{1}{2}x$
20.	$2x + y = 0$	28.	$\frac{y}{5} = 2x$
21.	$2x - y = 6$	29.	$\frac{y}{6} = x$
22.	$y = -2x$	30.	$\frac{3x}{2} + y = 1$
23.	$y = -3x$		

In exercises 31–40 indicate whether the two given equations are equivalent or not equivalent. If the equations are not equivalent, find an ordered pair of numbers that is in the solution set of one equation but not in the solution set of the other.

Example (a):
$$x + y = 5$$
$$2x + 2y = 15$$

Solution: These are not equivalent equations; $(2, 3)$ is a solution of $x + y = 5$ but not of $2x + 2y = 15$ $(2 \cdot 2 + 2 \cdot 3 = 4 + 6 = 10 \neq 15)$

Example (b):
$$2x + y = 12$$
$$4x + 2y = 24$$

Solution: These are equivalent equations since from each equation we get $y = 12 - 2x$.

31. $x - y = 1$
$2x - 2y = 2$

36. $x - y = 0$
$2x - 3y = 0$

32. $y = 2x$
$2y = 4x$

37. $x = y + 5$
$y = x - 5$

33. $y = x + 1$
$2y = x + 2$

38. $\frac{1}{2}x + \frac{1}{3}y = 1$
$3x + 2y = 6$

34. $x + y = 2$
$y = 2 - x$

39. $x + y = 3$
$x + y = 5$

35. $x = y + 4$
$x + 2 = y + 6$

40. $2x = 3y$
$x = \frac{3}{2}y$

In exercises 41–50 each pair of equations is an equivalent pair. What property or properties of equality can be applied to the first equation to arrive at the second?

Example (a):
$$x + y = 3$$
$$2x + 2y = 6$$

Solution: Multiplication property of equality. (Each side of the first equation has been multiplied by 2 to obtain the second equation.)

Example (b):
$$2y - x = 6$$
$$y = \frac{x + 6}{2}$$

Solution: Addition and multiplication properties of equality. (x was added to both sides of the equality and then both sides of the equation were multiplied by $\frac{1}{2}$.)

41. $y = x + 2$
    $y + 3 = x + 5$

42. $y - 2 = 3x$
    $y = 3x + 2$

43. $x - y = 6$
    $2x - 2y = 12$

44. $x - y = 10$
    $x = 10 + y$

45. $x + y = 0$
    $x = -y$

46. $3y + 5 = x$
    $y = \dfrac{x - 5}{3}$

47. $2x + 6y = 8$
    $x + 3y = 4$

48. $2x + y = 6$
    $x = \dfrac{1}{2}(6 - y)$

49. $\dfrac{3}{4}y - \dfrac{1}{2}x = \dfrac{2}{3}$
    $9y - 6x = 8$

50. $\dfrac{5x}{3} + y = 3$
    $x = \dfrac{3(3 - y)}{5}$

4.2 Coordinate Planes

Figure 4.1 shows a *coordinate plane* — also called a *rectangular coordinate system.* It consists of two number lines at right angles to each other and intersecting at the point, called the *origin,* that corresponds to 0 on each. The lines are usually drawn horizontally and

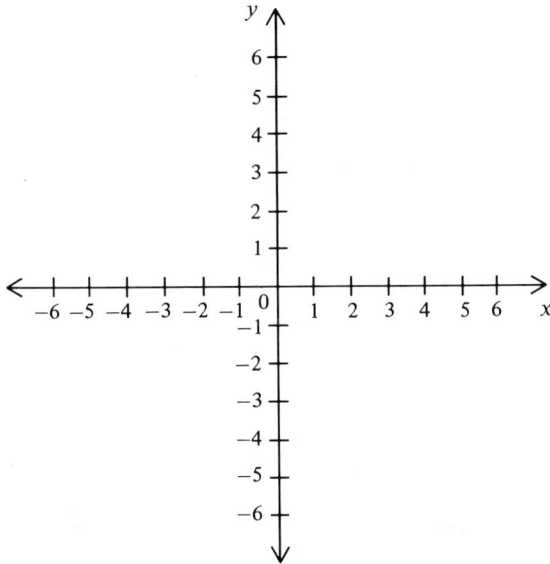

Figure 4.1

vertically; the horizontal line is called the *x-axis*, unless otherwise specified, and the vertical line is called the *y-axis*, unless otherwise specified. Together the two lines are referred to as the *coordinate axes*. A coordinate plane is also sometimes referred to as a *cartesian plane*, in honor of René Descartes, a seventeenth-century French mathematician who was the first to use it.

Just as a number line enables us to picture any number geometrically, so a coordinate plane enables us to picture any ordered *pair* of numbers. Thus, in Figure 4.2, the point corresponding to (3,2) is 3 units to the right of the *y*-axis and 2 units above the *x*-axis; the point corresponding to (-4,3) is 4 units to the left of the *y*-axis and three units above the *x*-axis; the point corresponding to (-3,-2) is 3 units to the left of the *y*-axis and 2 units below the *x*-axis; the point corresponding to (2,0) is on the *x*-axis; and the point corresponding to (0,1) is on the *y*-axis.

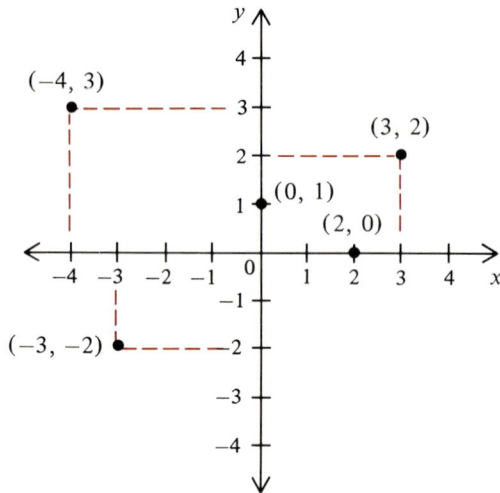

Figure 4.2

The first number of an ordered pair of numbers is called the *x-coordinate* or *abscissa* of the point corresponding to the ordered pair, and the second member is called the *y-coordinate* or *ordinate*. Together

the two numbers are referred to as the *coordinates* of the point. Thus
the point corresponding to (3,2) has abscissa 3, ordinate 2, and coordinates 3 and 2. (For brevity we will write "(3,2)-point" rather than "the
point corresponding to (3,2).")

The coordinate axes divide the plane into four *quadrants*, as shown
in Figure 4.3. The (x, y)-point is in the

first quadrant if $x > 0$ and $y > 0$;

second quadrant if $x < 0$ and $y > 0$;

third quadrant if $x < 0$ and $y < 0$;

fourth quadrant if $x > 0$ and $y < 0$.

Thus the (2, 3)-point is in the first quadrant; the (-2, 3) point is in the
second quadrant; the (-2, -3)-point is in the third quadrant; and the
(2, -3)-point is in the fourth quadrant. When $x = 0$ the (x, y)-point is
on the y-axis, and when $y = 0$ the (x, y)-point is on the x-axis. Points
on either the x-axis or the y-axis do not belong to any quadrant.

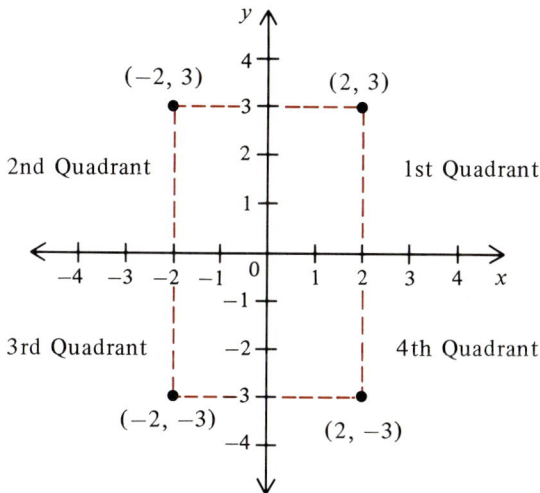

Figure 4.3

EXERCISES 4.2

In exercises 1–20 list the ordered pair of numbers that is associated with each numbered point in the following coordinate plane.

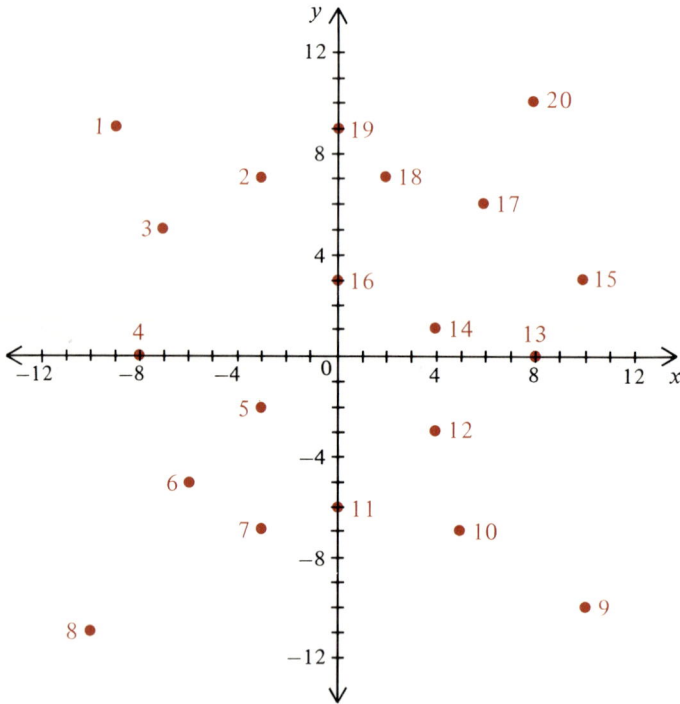

In exercises 21–35 graph on a rectangular coordinate system the points with the given coordinates. You may use the same coordinate system for all points, but label each point with the letter associated with its coordinates.

21. *A* (3, 7)	29. *I* (0, 0)
22. *B* (-3, 7)	30. *J* (4, 0)
23. *C* (3, -7)	31. *K* (0, -5)
24. *D* (-3, -7)	32. *L* (-3, 0)
25. *E* (6, 6)	33. *M* (0, 7)
26. *F* (-4, 4)	34. *N* $\left(1\frac{1}{2}, -3\frac{1}{2}\right)$
27. *G* (-8, 6)	35. $\left(-\frac{8}{3}, -\frac{9}{4}\right)$
28. *H* (-7, -7)	

In exercises 36-44 refer to the points you graphed in exercises 21-35. Select a replacement for each ___?___ to make a true sentence.

36. The points *A* and *E* lie in quadrant ___?___ .
37. The points *B, F,* and *G* lie in quadrant ___?___ .
38. The points *C* and *N* lie in quadrant ___?___ .
39. The points *D* and *H* lie in quadrant ___?___ .
40. The point *I* is called the ___?___ of the rectangular coordinate system.
41. The point *J* lies on the ___?___ (positive, negative) half of the ___?___ axis.
42. The point *K* lies on the ___?___ (positive, negative) half of the ___?___ axis.
43. The point *L* lies on the ___?___ (positive, negative) half of the ___?___ axis.
44. The point *M* lies on the ___?___ (positive, negative) half of the ___?___ axis.
45. If the (*x, y*)-point is in the second quadrant, *x* is a ___?___ number and *y* is a ___?___ number.

In exercise 46 plot each point on the same cartesian plane and connect each point with the next by a line segment in the order given.

46. *A* (5, 0)   *M* (-2, -6)
  B (-2, 0)   *N* (0, -6)
  C (-1, 1)   *0* (0, -3)
  D (-3, 3)   *P* (4, -3)
  E (-4, 6)   *Q* (4, -6)
  F (-5, 3)   *R* (6, -6)
  G (-4, 2)   *S* (6, -1)
  H (-7, -1)   *T* (8, 1)
  I (-6, -2)   *U* (5, 4)
  J (-5, -3)   *V* (7, 1)
  K (-3, -1)   *A* (5, 0)
  L (-2, -2)

47. Make up a design of your own on a coordinate plane and list, as in exercise 46, coordinates of enough points to describe the design.

4.3 Graphs of Linear Equations

Suppose that we locate (*plot*) on a coordinate plane some of the points corresponding to ordered pairs of numbers in the solution set of the equation $x - y = 5$. Some of the points are (6, 1), (7, 2), (-1, -6), (2, -3), (3, -2) and $(5\frac{1}{2}, \frac{1}{2})$. These points are shown in Figure 4.4. It appears that all of these points lie on the straight line shown in the

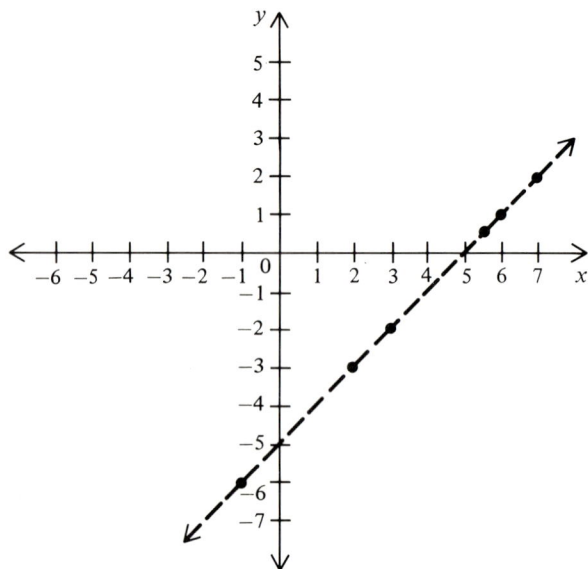

Figure 4.4

figure. This is indeed the case and, in fact, all of the points corresponding to the ordered pairs of real numbers in the solution set of the equation lie on this line, called the *graph* of the equation.

Furthermore, the graph of any linear equation, $ax + by = c$, is a straight line. That is, the set of all ordered pairs of real numbers in the solution set of the equation all lie on a straight line. Only two points are needed, of course, to determine a straight line, but as a check, it is best to plot at least three points.

In the equation $ax + by = c$, when $b = 0$, we have an equation of the form $ax = c$ as, for example, $2x = 5$. It may seem at first glance that the graph of the equation $2x = 5$ is just the point $(\frac{5}{2}, 0)$ on the x-axis. However, we can write the equation $2x = 5$ as $2x + 0 \cdot y = 5$ and then see that any value of y together with $x = \frac{5}{2}$ gives a solution of this equation. For example:

$$2 \cdot \frac{5}{2} + 0 \cdot 1 = 5, \quad 2 \cdot \frac{5}{2} + 0 \cdot (-1) = 5,$$

$$\text{and} \quad 2 \cdot \frac{5}{2} + 0 \cdot \frac{9}{7} = 5$$

Thus the graph of $2x = 5$ is the vertical line shown in Figure 4.5.

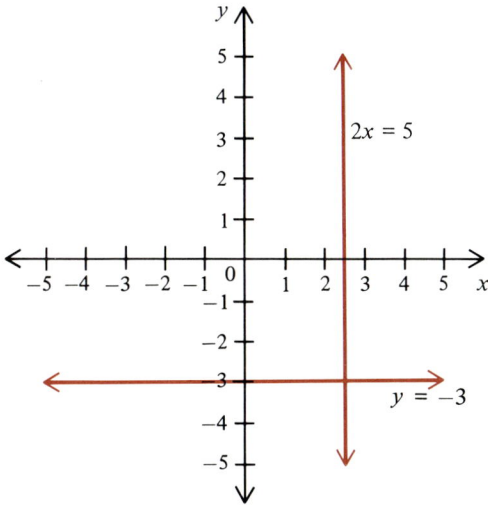

Figure 4.5

Similarly, if $a = 0$, $b = 1$, and $c = -3$ in $ax + by = c$, we get $y = -3$ whose graph, as shown in Figure 4.5, is a horizontal line.

EXERCISES 4.3

In exercises 1–25 draw a graph of the given equation.

Example: Graph $x + 2y = 6$

Solution: $y = \dfrac{-x}{2} + 3$ is equivalent to $x + 2y = 6$.

x	$\dfrac{-x}{2} + 3$	y
-2	$\dfrac{-(-2)}{2} + 3$	4
0	$\dfrac{-0}{2} + 3$	3
2	$\dfrac{-2}{2} + 3$	2

Plot this partial solution set on a coordinate plane and draw the line passing through the three points.

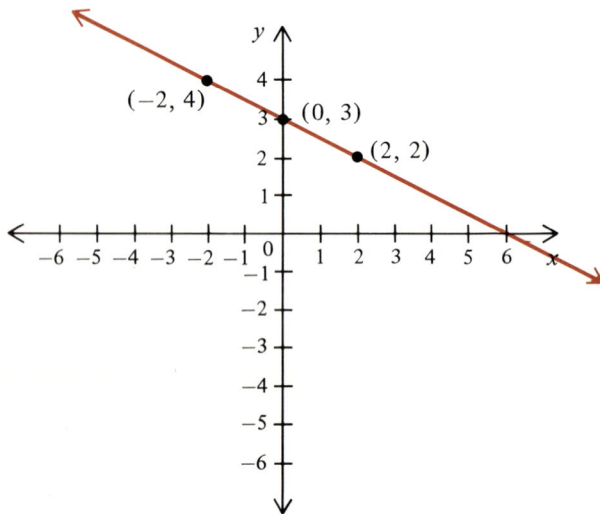

1.	$y = 3x$	14.	$3x - 4y = 7$
2.	$y = \frac{1}{2}x$	15.	$2x - 3y = 12$
3.	$y = -2x$	16.	$2x + 3y = 0$
4.	$y = -\frac{1}{2}x$	17.	$y = 2$
5.	$x + y = 3$	18.	$y = -2$
6.	$x + y = 5$	19.	$x = -6$
7.	$x - y = 2$	20.	$x = 7$
8.	$x - y = 3$	21.	$\frac{3}{4}x + y = 0$
9.	$2x - y = 1$	22.	$\frac{3}{4}x - y = 0$
10.	$3x - y = 2$	23.	$\frac{1}{3}x + \frac{1}{2}y = 2$
11.	$x + 2y = 6$	24.	$\frac{1}{4}x - \frac{1}{2}y = 0$
12.	$x - 2y = 6$	25.	$1.8x - 0.6y = 2.5$
13.	$3x + 2y = 6$		

Each of exercises 26–34 involves two equations. Graph each pair of equations on one set of coordinate axes.

26. $x - y = 4$
    $y = 3x$

27. $x + y = 5$
    $x + y = 3$

28. $x + y = 3$
    $2x + 2y = 6$

29. $x + y = 1$
    $x - y = 1$

30. $x + 2y = 3$
    $2x + y = 3$

31. $4x + y = 6$
    $2x + y = 6$

32. $2x + 5y = 3$
    $7x + 6y = -24$

33. $y = x$
    $y = -x$

34. $4x + 3y = 12$
    $4x + 3y = 9$

The questions in exercises 35–39 refer to the graphs you drew in exercises 26–34.

35. In exercises 26–34 some pairs of equations have ordered pairs of numbers, (x, y), which are common to the solution sets of both equations. For each of the exercises that have common solutions, list a number pair that is common to the solution sets of both equations. List more than one pair, if possible.

36. Which exercises have equations that have at least one ordered pair of numbers that is a common solution to both equations?

37. Which exercises have equations that have only one ordered pair of numbers that is a common solution to both equations? What can you say about the lines which are the graphs of these two equations?

38. Which exercises have equations that have more than one ordered pair of numbers that are common solutions to both equations? What can you say about the lines which are the graphs of these two equations?

39. Which exercises have equations which have no ordered pairs as a common solution to the two equations? What can you say about the lines that are the graphs of the two equations?

40. An electronics manufacturer has the following data on the total cost y of filling an order, including packaging, for x transistor radios.

Number of radios x	1	2	5	8	10	15	20	25
Total cost (dollars) y	13	26	55	88	90	135	180	200

   (a) Display this information graphically by using a coordinate plane and graphing the eight points — $(1, 13)$, $(2, 26)$, etc.
   (b) Join the points on the graph by means of a smooth curve. Is this curve a straight line?

(c) On the same coordinate plane graph the equation $y = 9x$.

(d) Could the equation $y = 9x$ be used to approximate the data given by the electronics manufacturer?

41. In Chapter 1 the relationship between the temperature and the number of chirps a cricket makes was expressed by the equation $t = \frac{n}{4} + 40$. (a) Draw a graph of this equation. (b) Are there values for t and n which appear on the graph but which are not practical or reasonable in the physical world?

42. In Chapter 1 the relationship between the Fahrenheit and Celsius temperature scales was given by the equation $F = \frac{9}{5} C + 32$.

(a) Draw a graph of this equation. (b) Are there values for F and C which appear on the graph but which are not practical or reasonable in the physical world?

4.4 Graphical Solution of Systems of Linear Equations

Consider the following problem: The perimeter of a rectangle is 14 meters and its length is 3 meters less than 4 times its width in meters. Find the rectangle's dimensions.

Now this problem can certainly be solved by the techniques described in Chapter 3. It is also possible, and sometimes easier, to solve such problems by the use of equations in two variables. Thus, if we let x = the length of the rectangle in meters and y = the width in meters, we can translate the given conditions as

$$2x + 2y = 14$$

(perimeter is 14 meters) and

$$x = 4y - 3$$

(length is 3 meters less than 4 times its width in meters).

It follows that finding a solution to this problem is equivalent to finding a solution of the equation $2x + 2y = 14$, which is *also* a solution of the equation $x = 4y - 3$. Finding such a common solution is known as solving a *system of linear equations*, which we write as

$$\begin{cases} 2x + 2y = 14 \\ \quad\; x = 4y - 3 \end{cases}$$

Now we know that the points corresponding to solutions of a given linear equation all lie on the same line. Thus if we graph the two equations of a system, we will find that one of the three following things will happen:

1. The graphs will be two distinct but parallel lines, which means that the solution set of the system is the empty set.

2.   The graphs will be the same line, which means that the two equations of our system are equivalent and any solution of one is a solution of the other.

3.   The graphs will be two lines intersecting at a single point.

In Figure 4.6 we show the graphs of the two equations of our system. Since the two lines appear to intersect at the point (5, 2), we conjecture that the solution set of our system is { (5, 2) }. To check our conjecture we note that

$$(2 \cdot 5) + (2 \cdot 2) = 14 \qquad \text{and} \qquad 5 = (4 \cdot 2) - 3$$

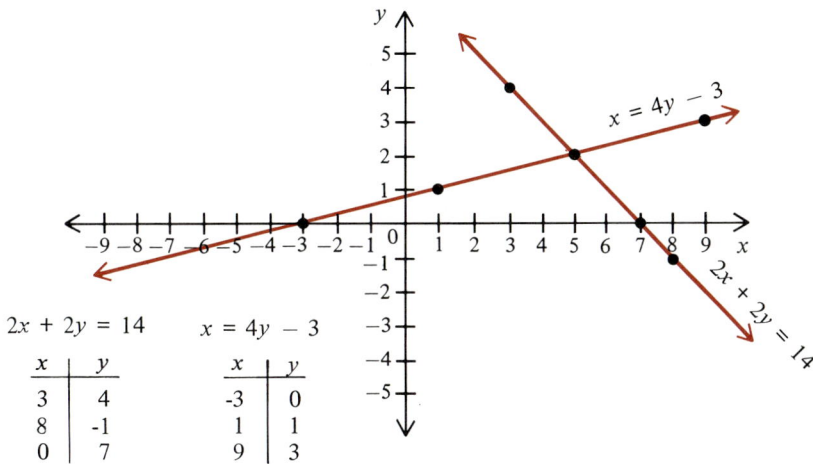

$2x + 2y = 14$

x	y
3	4
8	-1
0	7

$x = 4y - 3$

x	y
-3	0
1	1
9	3

Figure 4.6

Thus the rectangle has a width of 2 meters and a length of 5 meters.

Let us compare the graphical solution of this system with the graphical solution of the following two systems:

$$(1) \quad \begin{cases} 4x - y = 9 \\ 8x - 2y = 6 \end{cases} \quad \text{and (2)} \quad \begin{cases} 3x + 2y = 6 \\ 6x + 4y = 12 \end{cases}$$

Figure 4.7 shows the graph of the equations of (1) as two parallel lines so that the solution set of the system is the empty set. On the other hand, Figure 4.8 shows the graph of each of the two equations of (2) as one and the same line. Hence the solution set of the system is the same as the (infinite) solution set of either of the two equivalent equations of the system.

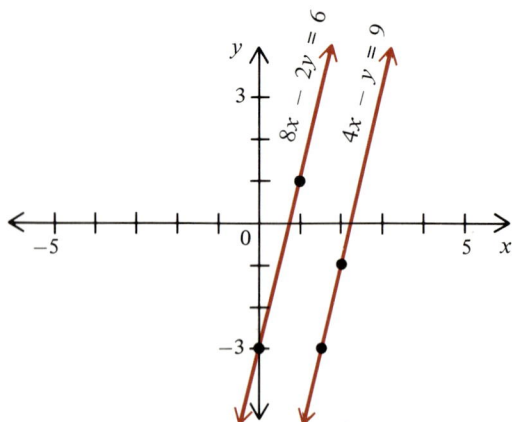

Figure 4.7

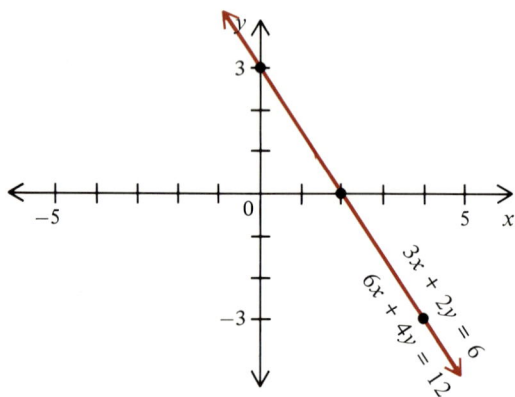

Figure 4.8

EXERCISES 4.4

Find graphically the solution set of each of the following systems of equations. Be sure to check your solutions.

Example (a): $\begin{cases} x + y = 5 \\ x - y = 1 \end{cases}$

Solution: $x + y = 5$ $\qquad\qquad\qquad\qquad$ $x - y = 1$

$\qquad\qquad\qquad y = 5 - x$ $\qquad\qquad\qquad\qquad$ $-y = -x + 1$

$\qquad\qquad\qquad\qquad\qquad\qquad\qquad\qquad\qquad$ $y = x - 1$

x	$5 - x$	y
-2	$5 - (-2)$	7
0	$5 - 0$	5
2	$5 - 2$	3

x	$x - 1$	y
-3	$-3 - 1$	-4
0	$0 - 1$	-1
1	$1 - 1$	0

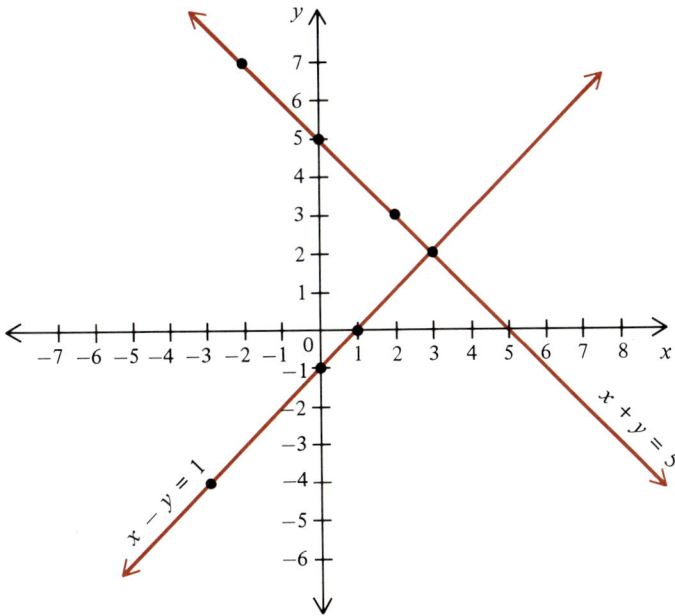

Solution set = $\{(3, 2)\}$.

Check: $3 + 2 = 5$ $\quad$ and $\quad$ $3 - 2 = 1$

Example (b): $\begin{cases} y = x + 2 \\ y = x \end{cases}$

Solution: $\qquad\qquad\qquad$ $y = x + 2$ $\qquad\qquad\qquad$ $y = x$

x	$x + 2$	y
-4	$-4 + 2$	-2
0	$0 + 2$	2
2	$2 + 2$	4

x	y
-3	-3
0	0
3	3

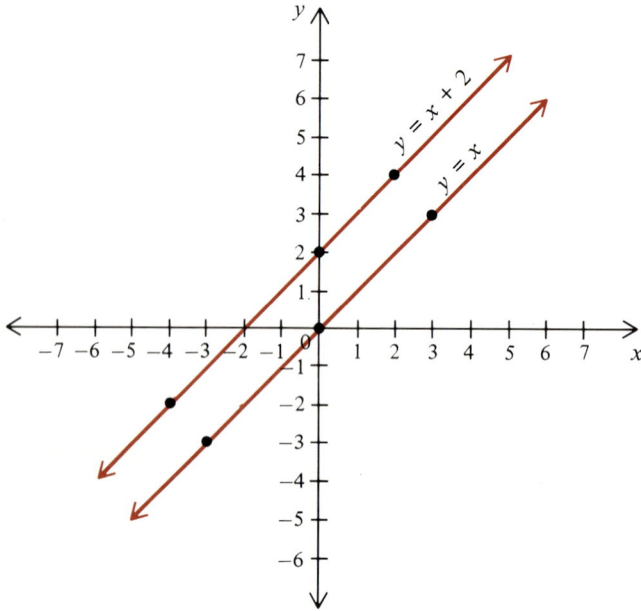

Solution set = { } (empty set).

1. $\begin{cases} x - y = 4 \\ x + y = 6 \end{cases}$

2. $\begin{cases} y = 3 - x \\ y = 1 + x \end{cases}$

3. $\begin{cases} x + y = 4 \\ x - y = 1 \end{cases}$

4. $\begin{cases} y = x + 5 \\ 2x - 2y = -10 \end{cases}$

5. $\begin{cases} 4x + y = 8 \\ 4x - y = 0 \end{cases}$

6. $\begin{cases} x + y = 1 \\ x - y = -3 \end{cases}$

7. $\begin{cases} y = 5 - x \\ y = 3 + x \end{cases}$

8. $\begin{cases} y = 2x + 4 \\ x + y = 1 \end{cases}$

9. $\begin{cases} x - y = 1 \\ y = 2x + 1 \end{cases}$

10. $\begin{cases} 3x - y = 4 \\ x - y = 2 \end{cases}$

11. $\begin{cases} y = 2x + 3 \\ x + y = 3 \end{cases}$

12. $\begin{cases} 3x + y = 1 \\ x - y = 3 \end{cases}$

13. $\begin{cases} x - y = 0 \\ x + y = -4 \end{cases}$

14. $\begin{cases} 5x - y = 6 \\ x - y = 2 \end{cases}$

15. $\begin{cases} 5x + y = 3 \\ x - y = 3 \end{cases}$

16. $\begin{cases} x - y = 0 \\ x + y = -6 \end{cases}$

17. $\begin{cases} x + 2y = 5 \\ x + y = 0 \end{cases}$

18. $\begin{cases} x - y = 0 \\ x + y = 4 \end{cases}$

19. $\begin{cases} 2x - y = 4 \\ x - 3y = -3 \end{cases}$

20. $\begin{cases} x - y = 3 \\ -3x - 3y = -9 \end{cases}$

21. $\begin{cases} 2x + y = -3 \\ x - y = -9 \end{cases}$

31. $\begin{cases} 5x + 3y = 7 \\ 2\frac{1}{2}x + 1\frac{1}{2}y = 3\frac{1}{2} \end{cases}$

22. $\begin{cases} y = 4x \\ y = -2x - 6 \end{cases}$

32. $\begin{cases} 2x + \frac{1}{4}y = 3 \\ x - \frac{1}{8}y = \frac{1}{2} \end{cases}$

23. $\begin{cases} x + 2y = 3 \\ 2x + 4y = 6 \end{cases}$

33. $\begin{cases} \frac{1}{2}x - \frac{1}{3}y = 1 \\ 2x - 1\frac{1}{3}y = 4 \end{cases}$

24. $\begin{cases} x + 2y = 5 \\ x = 3y \end{cases}$

34. $\begin{cases} 0.2x - 0.3y = 0 \\ x + 2y = 7 \end{cases}$

25. $\begin{cases} 2y - 4x = 6 \\ y = 3x \end{cases}$

35. $\begin{cases} 2x + y = -2 \\ y = 2x - 4 \end{cases}$

26. $\begin{cases} 2x + 3y = 7 \\ x = 2y \end{cases}$

36. $\begin{cases} 3x + 2y = 8 \\ x - 4y = 5 \end{cases}$

27. $\begin{cases} 3x + 2y = 8 \\ x - 4y = 5 \end{cases}$

37. $\begin{cases} 3x + 4y = 3.5 \\ x - y = 0 \end{cases}$

28. $\begin{cases} 2x + y = 13 \\ 3x - y = 2 \end{cases}$

38. $\begin{cases} x + y = 5 \\ x + y = 3 \end{cases}$

29. $\begin{cases} 4x + 6y = 19 \\ x + y = 6 \end{cases}$

39. $\begin{cases} 6x + 6y = -3 \\ 9x + 9y = 1 \end{cases}$

30. $\begin{cases} 2x - 5y = 6 \\ 4x - 10y = 8 \end{cases}$

40. $\begin{cases} 16x = 5 \\ 2y = 9 \end{cases}$

4.5 Algebraic Solution of Systems of Linear Equations

You probably have noticed that only integers, or fractions with a denominator of 2, appeared in the solution sets of the systems of equations of the previous section. This was no accident! The problems were so constructed because it is difficult to determine noninteger answers by graphical methods; $\frac{2}{3}$, for example, would be hard to read from a graph.

Fortunately, algebraic procedures for finding solutions are available. To illustrate one such procedure let us consider again the system

(1) $$\begin{cases} x + y = 5 \\ x - y = 1 \end{cases}$$

for which the graphical procedure yielded the solution set { (3, 2) }.

Now there are many systems of linear equations *equivalent* to (1), i.e., having the same solution set as (1). One of these systems is, as shown by Figure 4.9, the system

(2)
$$\begin{cases} 2x + 3y = 12 \\ x - 2y = -1 \end{cases}$$

Both (1) and (2) have the solution set $\{ (3, 2) \}$.

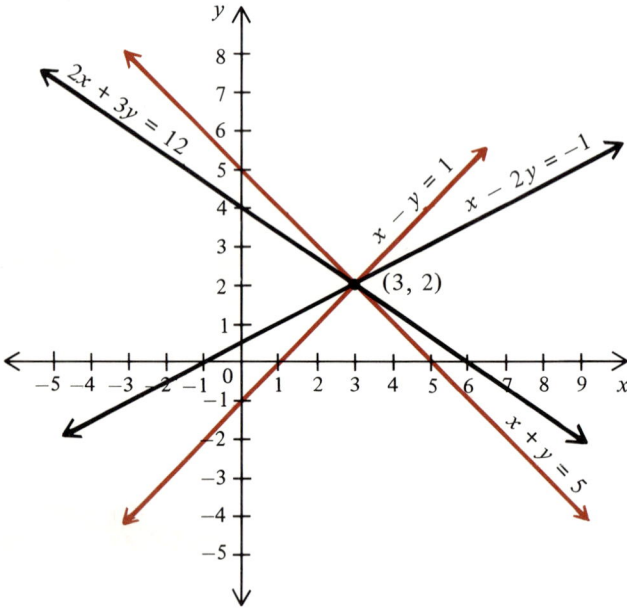

Figure 4.9

Also, of course, as illustrated by Figure 4.10, the system

(3)
$$\begin{cases} x = 3 \\ y = 2 \end{cases}$$

is another system equivalent to (1) and is the simplest system equivalent to (1).

What we need to do, then, is to learn how to get a simple system like (3) equivalent to a given system such as (1). In the case of (1), it follows by the properties of equality that

since $x + y = 5$ and $x - y = 1$, $(x + y) + (x - y) = 5 + 1$

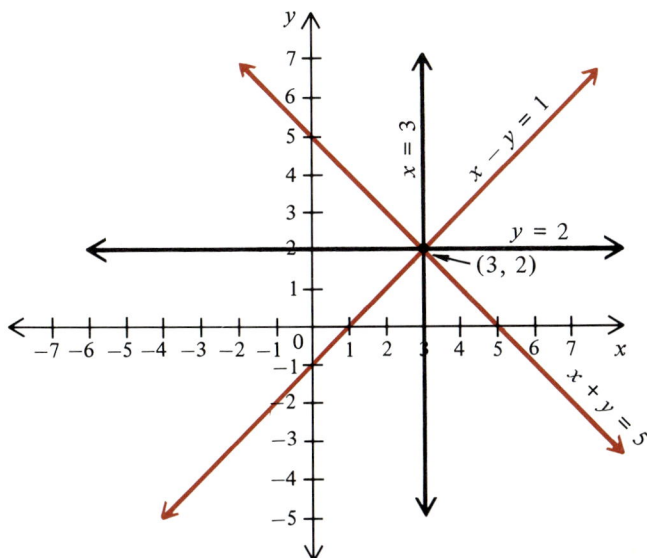

Figure 4.10

Thus

$$2x + (-y + y) = 6$$

$$2x + 0 = 6$$

$$2x = 6$$

$$x = 3$$

Substituting for x in the equation $x + y = 5$, we have

$$3 + y = 5$$

$$y = 2$$

We have thus obtained the simpler equivalent system $\begin{cases} x = 3 \\ y = 2 \end{cases}$ without resorting to graphing.

Now consider the system

(4)
$$\begin{cases} 3x + 2y = 12 \\ 7x - 4y = 2 \end{cases}$$

Here it is no longer possible to "eliminate" one of the variables by simple addition or subtraction. We can however, multiply the first equation of the system by 2 to obtain the equivalent system

(5)
$$\begin{cases} 6x + 4y = 24 \\ 7x - 4y = 2 \end{cases}$$

Then we can proceed with (5) to get

$$(6x + 4y) + (7x - 4y) = 24 + 2$$
$$13x = 26$$

so that we have

(6)
$$\begin{cases} 3x + 2y = 12 \\ 13x = 26 \end{cases}$$
or
(7)
$$\begin{cases} 3x + 2y = 12 \\ x = 2 \end{cases}$$

From (7) we get $(3 \cdot 2) + 2y = 12$ so that $y = 3$ and our solution is (2, 3).

Check: $(3 \cdot 2) + (2 \cdot 3) = 6 + 6 = 12$ and $(7 \cdot 2) - (4 \cdot 3) = 14 - 12 = 2$.

Now let us consider the system

(8)
$$\begin{cases} 7x + 4y = 43 \\ 5x + 6y = 26 \end{cases}$$

If we multiply the first equation by 3 and the second equation by 2, we obtain the equivalent system

(9)
$$\begin{cases} 21x + 12y = 129 \\ 10x + 12y = 52 \end{cases}$$

Then, since

$$(21x + 12y) - (10x + 12y) = 11x = 129 - 52 = 77$$

we get $x = 7$ and, from the first equation of (8),

$$(7 \cdot 7) + 4y = 43$$

so that

$$y = \frac{-6}{4} = -\frac{3}{2}$$

Thus the solution set of the system is $\{ (7, -\frac{3}{2}) \}$.

Check: $(7 \cdot 7) + 4(-\frac{3}{2}) = 49 - 6 = 43$ and $(5 \cdot 7) + 6(-\frac{3}{2}) = 35 - 9 = 26$.

In this problem, if you wished to obtain a system equivalent to (8) in which one of the equations involved only the variable y, what numbers would you use as multipliers for the two equations of (8)? Carry out the solution using these multipliers.

The final two examples show two special cases. Study them carefully. Consider the system

(10)
$$\begin{cases} x + y = 5 \\ x + y = 3 \end{cases}$$

If we multiply each term of the second equation by -1, and then apply the addition property of equality, we obtain

(11)
$$\begin{cases} x + y = 5 \\ -x - y = -3 \end{cases}$$
and
$$\begin{aligned} (x + y) + (-x - y) &= 5 + (-3) \\ (x - x) + (y - y) &= 2 \\ 0 + 0 &= 2 \\ 0 &= 2 \quad \text{(False)} \end{aligned}$$

Because we arrive at a false statement, let us examine the graphs of these two equations as shown in Figure 4.11. The graphs show that the lines are parallel and thus that there is no solution to this system of equations. Such a system is called *inconsistent* because the two given equations have no common solution.

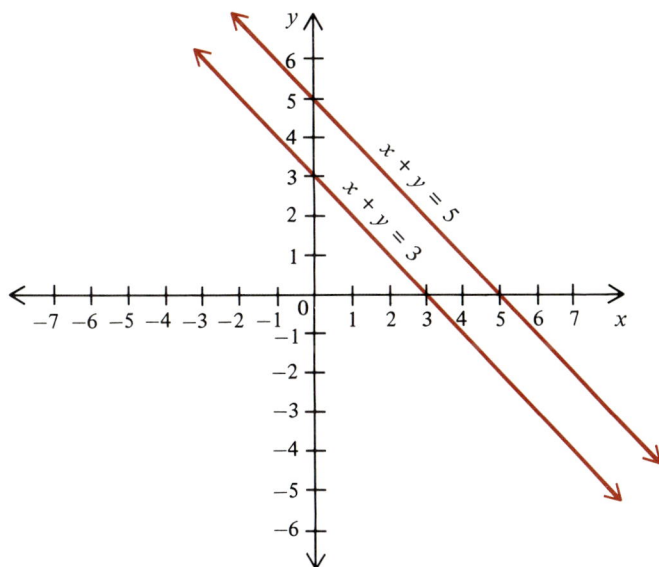

Figure 4.11

Now let us consider the system

(12)
$$\begin{cases} x - y = 3 \\ 2x - 2y = 6 \end{cases}$$

If we attempt to solve this system by multiplying the first equation by -2, and then applying the addition property of equality, we have

(13)
$$\begin{cases} -2x + 2y = -6 \\ 2x - 2y = 6 \end{cases}$$
and
$$(-2x + 2y) + (2x - 2y) = -6 + 6$$
$$(-2x + 2x) + (2y - 2y) = 0$$
$$0 + 0 = 0$$

Although we obtain the true statement $0 + 0 = 0$, this does not indicate any solution. The graph of the two equations is shown in Figure 4.12. The equations of (12) are equations of the same line and therefore have an infinite number of common solutions. Such equations are called *dependent* equations. From these two examples we see that for a system of linear equations to have one and only one solution, the graphs of the equations must not be parallel nor may they be the same line. For two linear equations to have one common solution they must

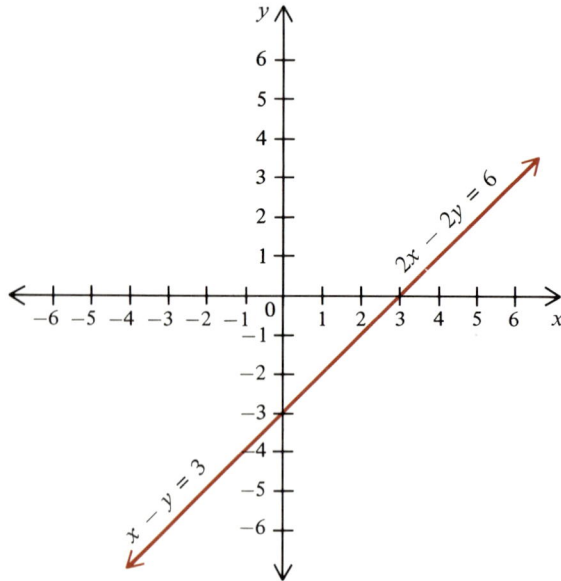

Figure 4.12

be both *consistent* (i.e., *not* inconsistent) and *independent* (i.e., *not* dependent).

When a calculator is used to solve certain systems of equations, different procedures for solving the system can result in (small) differences in the answer obtained. For this reason a specific procedure is given for such problems.

Example: $\begin{cases} 3.12x + 0.43y = 18 \\ 1.26x - 0.34y = 16 \end{cases}$

Solve for x by eliminating y and then find y by substituting for x in the first equation.

Solution: Multiply both sides of the first equation by 0.34 and both sides of the second equation by 0.43; then add to get

$$1.0608x + 0.1462y = 6.12$$
$$\underline{0.5418x - 0.1462y = 6.88}$$
$$1.6026x \qquad\qquad = 13$$
$$x = 8.1118182$$

Substituting 8.1118182 for x in the first equation, we have

$$(3.12)(8.1118182) + 0.43y = 18$$
$$25.308872 + 0.43y = 18$$
$$0.43y = -7.308872$$
$$y = -16.997376$$

Solution set = { (8.1118182, -16.997376) }.

EXERCISES 4.5

Find the solution sets of the following systems of equations and check your solutions in both equations. (Note that exercises 49–58 are to be solved using a calculator.)

Example (a): Solve for a and b: $\begin{cases} 2a - 3b = 8 \\ 3a - 2b = 17 \end{cases}$

Solution: To eliminate the variable a, multiply the first equation by 3, the second equation by -2, and add.

$$6a - 9b = 24$$
$$-6a + 4b = -34$$
$$\overline{\qquad -5b = -10}$$
$$b = 2$$

Substitute 2 for b in either of the original equations and solve for a.

$2a - 3b = 8$	$3a - 2b = 17$
$2a - 3(2) = 8$ or	$3a - 2(2) = 17$
$2a = 14$	$3a = 21$
$a = 7$	$a = 7$

Solution set = $\{(7, 2)\}$.

Check in *both* equations:

$2a - 3b = 8$	$3a - 2b = 17$
$2(7) - 3(2) \overset{?}{=} 8$	$3(7) - 2(2) \overset{?}{=} 17$
$14 - 6 \overset{?}{=} 8$	$21 - 4 \overset{?}{=} 17$
$8 = 8$	$17 = 17$

(The symbol "$\overset{?}{=}$" is used here to ask the question "Is the left side equal to the right side?" We use it in each step until we can see by inspection whether or not equality holds.)

Example (b): Solve for x and y:
$$\begin{cases} x = 2y \\ x + 3y = 10 \end{cases}$$

Solution: Sometimes a system of equations can be solved easily by eliminating one variable by substitution. In this example we substitute $2y$ for x in the second equation and solve for y.

$$x = 2y$$
$$x + 3y = 10$$
$$2y + 3y = 10$$
$$5y = 10$$
$$y = 2$$

Substitute 2 for y in the first equation and hence find x.

$$x = 2y$$
$$x = 2(2)$$
$$x = 4$$

Solution set = $\{(4, 2)\}$.

Check:

$$x = 2y \qquad\qquad x + 3y = 10$$
$$4 \stackrel{?}{=} 2(2) \qquad\qquad 4 + 3(2) \stackrel{?}{=} 10$$
$$4 = 4 \qquad\qquad 4 + 6 \stackrel{?}{=} 10$$
$$10 = 10$$

1. $\begin{cases} x + y = 8 \\ x - y = 2 \end{cases}$

2. $\begin{cases} x - y = 5 \\ x + y = 13 \end{cases}$

3. $\begin{cases} 3x + y = 9 \\ 3x - y = 3 \end{cases}$

4. $\begin{cases} x - 4y = -2 \\ x + 4y = 14 \end{cases}$

5. $\begin{cases} 2a - b = 20 \\ 2a + b = 48 \end{cases}$

6. $\begin{cases} 4r - 5t = 31 \\ 2r + 5t = -7 \end{cases}$

7. $\begin{cases} 2x + 4y = 10 \\ x + 2y = 5 \end{cases}$

8. $\begin{cases} 3x - 2y = 3 \\ x - y = -1 \end{cases}$

9. $\begin{cases} 5a - b = -15 \\ a + 3b = 13 \end{cases}$

10. $\begin{cases} 4n + 3d = 7 \\ n + 2d = -2 \end{cases}$

11. $\begin{cases} 6x - y = 15 \\ 2x + 5y = 21 \end{cases}$

12. $\begin{cases} 6 = x - 3y \\ 20 = 2x - 6y \end{cases}$

13. $\begin{cases} x - y = 73 \\ 2x - 7y = 21 \end{cases}$

14. $\begin{cases} 6a - 7b = 44 \\ 4a - b = 22 \end{cases}$

15. $\begin{cases} x + 7y = 9 \\ 4x + y = -18 \end{cases}$

16. $\begin{cases} y = 4 - x \\ y - x = 1 \end{cases}$

17. $\begin{cases} 3r - 6s = 12 \\ 4r - 3s = 1 \end{cases}$

18. $\begin{cases} y = 2x + 6 \\ x + y = 1 \end{cases}$

19. $\begin{cases} 2x - 4y - 1 = 0 \\ x = 1 + 2y \end{cases}$

20. $\begin{cases} 2h - 8j - 12 = 0 \\ 7h + 4 = 5j \end{cases}$

21. $\begin{cases} 2x - 9y = 0 \\ 3x - 6y = -5 \end{cases}$

22. $\begin{cases} 4a - 7b = 8 \\ 5a + 9b = 81 \end{cases}$

23. $\begin{cases} 5x + y = 1 \\ x - y = 3 \end{cases}$

24. $\begin{cases} 2c - 5d = -39 \\ 3c + 4d = 22 \end{cases}$

25. $\begin{cases} 4x + 6y = 19 \\ x + y = 6 \end{cases}$

26. $\begin{cases} y = 2x \\ x + y = 3 \end{cases}$

27. $\begin{cases} x = 2y + 3 \\ 3x - 6y = 12 \end{cases}$

28. $\begin{cases} x + y = 12 \\ x = 3y \end{cases}$

29. $\begin{cases} 5x + 3y = 7 \\ 2x - 4y = -15.4 \end{cases}$

30. $\begin{cases} 12x + 16y = 0 \\ 5x - 5y = 15 \end{cases}$

31. $\begin{cases} 5a + 3b = 14 \\ 2a - 7b = 22 \end{cases}$

32. $\begin{cases} 5x - 3y = 7 \\ 3x - 5y = 2 \end{cases}$

33. $\begin{cases} 3x + 8y = -43 \\ 7x - 2y = 3 \end{cases}$

34. $\begin{cases} 6x + 5y = 11 \\ 5x - 3y = -41 \end{cases}$

35. $\begin{cases} 8r - 15s = 26 \\ 4r - 9s = 10 \end{cases}$

36. $\begin{cases} x = y \\ x + y = 10 \end{cases}$

37. $\begin{cases} 2x + y = 7 \\ y = x - 5 \end{cases}$

38. $\begin{cases} x - 8 = y \\ y = 5x \end{cases}$

39. $\begin{cases} \dfrac{2}{3}x + \dfrac{1}{5}y = 6 \\ \dfrac{1}{6}x - \dfrac{1}{2}y = -4 \end{cases}$

40. $\begin{cases} \dfrac{1}{2}x + y = \dfrac{2}{3} \\ \dfrac{1}{3}x + 4y = -\dfrac{1}{2} \end{cases}$

41. $\begin{cases} \dfrac{1}{2}x + \dfrac{1}{3}y = 5 \\ 3x + 2y = 30 \end{cases}$

42. $\begin{cases} 2\dfrac{1}{2}x - \dfrac{3}{4}y = -14 \\ \dfrac{1}{3}x + \dfrac{2}{3}y = -8 \end{cases}$

43. $\begin{cases} \dfrac{7}{8}x + \dfrac{1}{4}y = 3\dfrac{1}{8} \\ \dfrac{2}{3}x - \dfrac{1}{3}y = 1\dfrac{1}{3} \end{cases}$

44. $\begin{cases} 0.5x - 0.3y = 7 \\ 0.3x - 0.4y = 2 \end{cases}$

45. $\begin{cases} 0.4x + 0.6y = 8 \\ 0.1x - 0.2y = -0.1 \end{cases}$

46. $\begin{cases} 0.35x + 0.7y = 0.105 \\ 6x + 5y = 1.1 \end{cases}$

47. $\begin{cases} 0.2x + y = 0.1 \\ 2x - 0.01y = 1 \end{cases}$

48. $\begin{cases} 0.4x + 0.6y = 8 \\ 0.1x + 0.2y = -0.1 \end{cases}$

*49. $\begin{cases} 1.35x + 1.7y = 1.105 \\ 1.6x + 2.5y = 1.1 \end{cases}$
(Solve for x by eliminating y, and then find y by substituting for x in the first equation.)

*50. $\begin{cases} 0.06x = 0.02y + 0.75 \\ 0.03y = 0.84 - 0.04x \end{cases}$
(Proceed as in exercise 49.)

*51. $\begin{cases} 4.15x + 37y = 429 \\ 0.005x + 0.17y = 3.67 \end{cases}$
(Proceed as in exercise 49.)

*52. $\begin{cases} 2.10x + 5.46y = 4.46 \\ 21.4x + 2.38y = 5.56 \end{cases}$
(Proceed as in exercise 49.)

*53. $\begin{cases} 0.012x + 0.37y = 5.36 \\ 37.67x + 4.308y = 0.567 \end{cases}$
(Solve for y by eliminating x, and then find x by substituting for y in the first equation.)

*54. $\begin{cases} 7200x + 8100y = 79,200 \\ 10,000x - 4000y = 61,200 \end{cases}$
(Proceed as in exercise 53.)

*55. $\begin{cases} 999x + 298y = 1 \\ 578x + 9.37y = 251.6 \end{cases}$
(Proceed as in exercise 49.)

*56. $\begin{cases} 2.37x + 4.06y = 7.62 \\ 0.003x + 1.03y = -0.0045 \end{cases}$
(Solve for y by eliminating x, and then find x by substituting for y in the second equation.)

*57. $\begin{cases} 275x + 385y = 7652 \\ 27.5x + 5.07y = 15.06 \end{cases}$
(Proceed as in exercise 56.)

*58. $\begin{cases} -0.2121x + 0.0707y = 0.4324 \\ 0.9899x - 0.114y = 0.9192 \end{cases}$
(Proceed as in exercise 56.)

4.6 Another Approach to Word Problems

Sometimes the use of two variables makes the solving of word problems easier. If you do use two variables, then you must have two equations; thus two basic equations are required.

Example 1: Find two numbers whose sum is 108 and whose difference is 36.

Solution: Basic equations:

First number + second number = 108
First number – second number = 36

Let x = first number.

Let y = second number.

Then

$$\begin{cases} x + y = 108 \\ x - y = 36 \end{cases}$$

Solving the system of equations for x by the addition property of equality:

$$2x = 144$$

$$x = 72$$

Substituting 72 for x in one of the original equations and solving for y:

$$x + y = 108$$
$$72 + y = 108$$
$$y = 36$$

Solution: The two numbers are 72 and 36.

Check: $72 + 36 \overset{?}{=} 108$ $72 - 36 \overset{?}{=} 36$

$108 = 108$ $36 = 36$

Example 2: Two planes took off from an airport and flew in opposite directions. One plane traveled 220 kilometers per hour faster than the other. At the end of $2\frac{1}{2}$ hours the planes were 2800 kilometers apart. Find the average speed of each plane.

Solution: Basic equations:

> Speed of first plane = speed of second plane + 220
>
> Distance first plane traveled + distance second plane traveled = total distance apart

Let x = speed of first plane in kilometers per hour.

Let y = speed of second plane in kilometers per hour.

Then $\quad x = y + 220 \quad$ (first equation)

We know that $d = rt$ (distance equals rate times time) and that both planes traveled for $2\frac{1}{2}$ hours. Thus the distance the first plane traveled was $2\frac{1}{2}x$ kilometers and the distance the second plane traveled was $2\frac{1}{2}y$ kilometers.

Therefore: $\quad 2\frac{1}{2}x + 2\frac{1}{2}y = 2800 \quad$ (second equation)

Our two equations are:

$$\begin{cases} x = y + 220 \\ 2\frac{1}{2}x + 2\frac{1}{2}y = 2800 \end{cases}$$

Solving for x and y by first substituting $y + 220$ for x in the second equation:

$$2\frac{1}{2}(y + 220) + 2\frac{1}{2}y = 2800$$
$$2\frac{1}{2}y + 550 + 2\frac{1}{2}y = 2800$$
$$5y = 2250$$
$$y = 450$$

Solving for x:

$$x = y + 220$$
$$x = 450 + 220$$
$$x = 670$$

The speed of the first plane was 670 kilometers per hour and the speed of the second plane was 450 kilometers per hour.

Check: $\quad 670 \overset{?}{=} 450 + 220 \qquad 2\frac{1}{2}(670) + 2\frac{1}{2}(450) \overset{?}{=} 2800$

$\qquad\qquad\qquad 670 = 670 \qquad\qquad\qquad 1675 + 1125 \overset{?}{=} 2800$

$\qquad\qquad\qquad\qquad\qquad\qquad\qquad\qquad\qquad 2800 = 2800$

Example 3: Mr. Knutson invested a total of $10,500. Part of his money was invested at 6% and the rest at 9%. The total interest received was $810. How much money did Mr. Knutson invest at each rate of interest?

Solution: Basic equations:

Money invested at 6% + money invested at 9% = $10,500

Interest from 6% investment + interest from 9% investment = $810

Let x = amount of money invested at 6%.

Let y = amount of money invested at 9%.

Then $x + y = 10,500$ (first equation)

Interest = Principal × Rate

Interest from 6% investment is $0.06x$.

Interest from 9% investment is $0.09y$.

Therefore: $0.06x + 0.09y = 810$ (second equation)

Solve the system:
$$\begin{cases} x + y = 10,500 \\ 0.06x + 0.09y = 810 \end{cases}$$

Multiply first equation by 9: $\qquad 9x + 9y = 94,500$

Multiply second equation by -100: $\qquad -6x - 9y = -81,000$

Eliminate y by the addition $\qquad\qquad 3x = 13,500$

property of equality: $\qquad\qquad x = 4500$

Substituting in first equation: $\qquad x + y = 10,500$

$$4500 + y = 10,500$$

$$y = 6000$$

Therefore $4500 was invested at 6% and $6000 was invested at 9%.

Check: $4500 + 6000 \overset{?}{=} 10,500 \qquad 0.06(4500) + 0.09(6000) \overset{?}{=} 810$

$$10,500 = 10,500 \qquad\qquad\qquad 270 + 540 \overset{?}{=} 810$$

$$810 = 810$$

EXERCISES 4.6

Solve the following word problems using two variables and two equations.

1. The sum of two numbers is 50 and their difference is 26. Find the two numbers.

2. The difference between two numbers is 6. If the first number is added to twice the second, the sum is 33. What are the numbers?

3. The sum of two numbers is 36. Five less than 5 times the smaller is 4 more than 4 times the larger. What are the numbers?

4. The sum of two numbers is 52 and the first number is $\frac{5}{8}$ as large as the second. Find the numbers.

5. Twice the base of a rectangle is 8 meters longer than 4 times its height. If the perimeter is 50 meters, what are the rectangle's dimensions?

6. The perimeter of a triangle is 41 inches. The second side is 9 inches longer than $\frac{1}{3}$ of the first and the third side is 1 inch shorter than $\frac{1}{2}$ the first. Find the length of each side.

7. Ms. Rogers paid $2400 cash for a new car. On her old car she was allowed $250 more than $\frac{1}{3}$ of the value of the new car. What was the cost of the new car?

8. Last year the Fresno Fanatics won 2 more than 3 times as many games as the Salinas Sabers. Between them they won 150 games. How many games did each win?

9. A meat market charged $1\frac{1}{3}$ times as much for sirloin steak as round steak. A restaurant manager ordered 14 pounds of round steak and 18 pounds of sirloin steak and the bill was $42.18. What was the cost per pound of each kind of steak?

10. A clothing merchant finds that his operating expenses average 20% of his sales. He plans to sell fur coats which he buys for $264 so as to make a profit of 25% of the selling price. What must be the selling price?

11. A woman has 20 coins that are quarters or dimes. The value of the coins is $3.05. How many of each kind of coin does she have?

12. Tickets for a school basketball game were sold for 75¢ when purchased at school and $1.00 when purchased at the gate. If 340 tickets were sold and the receipts from their sale amounted to $287.50, how many tickets were sold at the school and how many at the gate?

13. A man bought 13 stamps at one price and 32 stamps at a higher price. For all the stamps he paid $1.41. If he had bought twice as many of the lower-priced stamps and 6 fewer of the higher-priced stamps, he would have spent 11¢ less. What was the demonination of each type of stamp he purchased?

14. Mrs. Garcia invested a total of $4000 in securities. Part of the money was invested at 10% and part at 8%. The annual income from both investments totaled $360. What amount of money was invested at each rate?

15. The sum of Nancy's automobile license number and her telephone number is 8,689,430. Her telephone number decreased by 100 times her license number is 7,716,800. What is Nancy's phone number?

16. A candy dealer wishes to mix chocolates selling for $2.00 per kilogram and bonbons selling at $3.00 per kilogram. How many kilograms of each should he use to obtain a mixture of 35 kilograms which he can sell for $2.20 per kilogram?

17. A wholesale grocer decided to mix two grades of coffee which sold at 75¢ and $1.20 a pound, respectively, so as to obtain a "blend" which she can sell for 96¢ per pound. How many pounds of each should she use if she desires 45 pounds of the mixture?

18. The owners of a nut shop mixed pecan nuts selling at $3.60 a kilogram with cashew nuts selling at $2.40 a kilogram. How many kilograms of each did they use to obtain a mixture of 100 kilograms which they could sell at $2.70 a kilogram?

19. Mr. Clock received $4000 when his endowment policy matured. He invested a part of this sum in a 10% mortgage and the remainder in 8% bonds. His annual income from the two investments is $370. How much money did he invest at each rate?

20. Ms. Thompson put $2200 into two investments which yielded 6% and 8% respectively. If she received twice as much interest from the 6% investment as she did from the 8%, how much money did she have invested at each rate?

21. Mr. Clark invested part of $21,000 at 4% and the remainder at 7%. His annual income from these two was equivalent to a return of 6% on the entire sum invested. How much money was invested at each rate?

22. Mrs. Downs purchased two gasoline stations. For one she paid $5000 and for the other $7000. At the end of the year her total net income from these investments amounted to $2680. She found that the net income from the second station was $680 more than the first. What was her rate of return from each investment?

23. Two airplanes start from St. Louis and fly in opposite directions. One travels 200 miles an hour faster than the other. At the end of 5 hours they are 4640 miles apart. How fast is each plane traveling?

24. Two couples on a European trip traveled in separate cars. One couple averaged 72 kilometers per hour and the other 56 kilometers per hour. The faster couple traveled 48 kilometers farther than the slower one and made the trip in 2 hours less time. How long did each couple travel?

25. A freighter 132 nautical miles from New York City was steaming toward New York at a speed of 12 knots (12 nautical miles per hour). At 9 A.M. it radioed that an injured seaman should be flown to a hospital. A half hour later, a helicopter left New York for the ship, traveling at 72 knots. At what time will the helicopter reach the ship?

26. Two airplanes pass each other at 12:00 noon flying in opposite directions. Plane *A* has a speed of 480 kilometers per hour and plane *B* has a speed of 360 kilometers per hour. They continue to travel at the same speed and

in the same direction, with plane B landing a half hour earlier than plane A and 1920 kilometers from where plane A landed. At what time did plane A land?

27. Two ships sailing in opposite directions pass a buoy. One ship travels at 20 knots, the other at 15 knots. After both sail the same distance, they return to the buoy, with the faster ship arriving a half hour before the slower ship. How far from the buoy in nautical miles did each travel?

28. On an automobile trip in the mountains, Mr. Jenkins averaged 45 kilometers per hour. On the return trip he averaged 50 kilometers per hour. If the trip both ways took 7 hours and 36 minutes, how far into the mountains did Mr. Jenkins travel?

29. For every hour that Ms. Barker works, her daughter takes $1\frac{1}{2}$ hours to do the same amount of work. If when they work together it takes them 6 hours to hay a certain field, how long would it take each of them to do it alone?

30. It takes an older machine just twice the time to do a certain job as it does a new model. If both machines working together can do a job in 4 hours, how long would it take each machine to do it alone?

31. Ray and Jane can do a piece of work in 9 hours. After working together 7 hours, Jane finishes the job in 5 more hours. In how many hours could each do the work alone?

32. Jill and Grace can do a piece of work in $8\frac{3}{4}$ days; but if Jill had worked $\frac{5}{6}$ as fast and Grace $\frac{3}{2}$ as fast, they would have done it in $7\frac{7}{8}$ days. In how many days could each do the work alone?

33. On her birthday Mrs. Jenkins received a bouquet from her husband which contained a flower for each year of her age. It was made up of roses at $6.00 a dozen and carnations at $4.50 a dozen. There were half as many roses as carnations. If the cost of the bouquet was $15.00, how old was Mrs. Jenkins?

34. The sum of the digits of a two-digit number is 7. If the digits are reversed, the new number is 9 more than the original number. What was the original number?

35. A man bought 26 stamps at one price and 64 stamps at a higher price. For all of the stamps he paid $2.82. If he had bought twice as many of the lower-priced stamps and 12 fewer of the higher-priced stamps, he would have spent 22¢ less. What was the denomination of each type of stamp that he bought?

36. The perimeter of an isosceles (2 sides of the same length) triangle is 52 centimeters. The base is 4 centimeters longer than one of the two congruent sides of the triangle. Find the dimensions of the triangle.

37. In a set of coins consisting of pennies, nickels, and dimes, half the coins are pennies. The value of the collection is $1.26. There are 4 more nickels than dimes. How many coins of each type are in the collection?

38. At the first home basketball game, 940 student and 560 adult tickets were sold. At the next game 640 adult and 1000 student tickets were sold. If the receipts were $2060 and $2280, respectively, what was the cost of each type of ticket?

39. Pat sets out to walk to a town 20 kilometers away at a rate of 4 kilometers per hour. One hour afterwards Dick starts to follow him. When Dick overtakes Pat, he turns back and reaches the starting point at the same time Pat reaches his destination. If Dick walks at half again the speed Pat walks, how far from the starting point did Dick overtake Pat?

40. If Jerry gives Betty $12, Betty will have $\frac{9}{5}$ as much money as Jerry, but if Betty gives Jerry $12, Jerry will have $\frac{4}{3}$ as much money as Betty. How much money has each?

41. A resolution was passed by 10 votes; but if $\frac{1}{4}$ of those who voted for it had voted against it, the resolution would have been defeated by 6 votes. How many voted for the resolution and how many voted against it?

42. Divide 59 into two parts such that $\frac{2}{3}$ of the smaller shall be 4 less than $\frac{4}{7}$ of the larger.

43. A girl left a service station at 8:00 A.M., traveling at 60 kilometers per hour. Fifteen minutes later her boyfriend left the station, following the same route as the girl and traveling at 72 kilometers per hour. How long did it take him to overtake her?

44. A woman pays two bills with a $5 bill. One bill was $\frac{6}{7}$ of the other, and the change she received after paying both bills was 7 times the difference of the bills. How much did she owe in each case?

45. If 5 is added to the numerator of a certain fraction it is equal to $\frac{5}{3}$, and if 5 is subtracted from its denominator it is equal to $\frac{5}{2}$. Find the fraction.

4.7 Linear Inequalities in Two Variables

Let us begin by considering the graph of the inequality

(1) $$3x - 4y < 7$$

If $x = 2$ and $y = 1$, then $3x - 4y < 7$ because

$$3 \cdot 2 - 4 \cdot 1 = 2 < 7$$

Similarly, if $x = 0$ and $y = 2$, then $3x - 4y < 7$ because

$$3 \cdot 0 - 4 \cdot 2 = -8 < 7$$

Thus we can conclude that both (2, 1) and (0, 2) are in the solution set of (1).

On the other hand, if $x = 1$ and $y = -1$, then $3x - 4y = 3 \cdot 1 - 4 \cdot (-1) = 7$ so that we conclude that $(1, -1)$ is not in the solution set of (1). Similarly, since $3 \cdot 1 - 4 \cdot (-3) = 15 > 7$, we know that $(1, -3)$ is not in the solution set of (1).

How can the solution set of (1) be pictured geometrically? To answer this question we first form an inequality equivalent to (1) and an equation equivalent to the equation $3x - 4y = 7$ as follows:

$$3x - 4y < 7 \qquad\qquad\qquad 3x - 4y = 7$$

$$-4y < 7 - 3x \qquad\qquad\qquad -4y = 7 - 3x$$

$$\left(-\frac{1}{4}\right)(-4y) > \left(-\frac{1}{4}\right)7 + \left(-\frac{1}{4}\right)(-3x) \qquad \left(-\frac{1}{4}\right)(-4y) = \left(-\frac{1}{4}\right)7 + \left(-\frac{1}{4}\right)(-3x)$$

$$y > \frac{3}{4}x - \frac{7}{4} \qquad\qquad\qquad y = \frac{3}{4}x - \frac{7}{4}$$

Now we draw the graph of $y = \frac{3}{4}x - \frac{7}{4}$ as shown in Figure 4.13. For every (x, y)-point on the line we have $y = \frac{3}{4}x - \frac{7}{4}$. Therefore every point above the line will have a y-coordinate such that $y > \frac{3}{4}x - \frac{7}{4}$. As a check, consider the points $(0, 0)$, $(1, 2)$, and $(-2, -1)$.

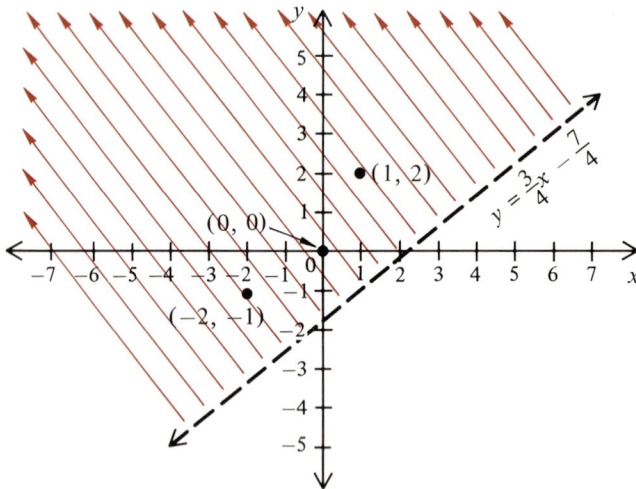

Figure 4.13

For $x = 0$ and $y = 0$, $\quad \dfrac{3}{4}x - \dfrac{7}{4} = 0 - \dfrac{7}{4} = -\dfrac{7}{4} \quad$ and $\quad 0 > -\dfrac{7}{4}$

For $x = 1$ and $y = 2$, $\quad \dfrac{3}{4}x - \dfrac{7}{4} = \dfrac{3}{4} - \dfrac{7}{4} = -1 \quad$ and $\quad 2 > -1$

For $x = -2$ and $y = -1$, $\dfrac{3}{4}x - \dfrac{7}{4} = \dfrac{-6}{4} - \dfrac{7}{4} = -\dfrac{13}{4}$ and $-1 > -\dfrac{13}{4}$

    The graph of the inequality $3x - 4y < 7$, then, is the set of all points that lie in the *half plane above* the graph of $3x - 4y = 7$, as indicated in Figure 4.13. (Note that we might have expected from the "$<$" in our given inequality that the graph would be *below* the corresponding line.)

    In general, the same line of argument will show that the graph of the inequality $ax + by < c$ will consist either of all the points in the half plane lying above the line whose equation is $ax + by = c$ or of all the points in the half plane lying below the line. To decide whether it is the half plane above or the half plane below, we need only check on the coordinates of any points not on the line. For example, for the inequality

$$2x + 3y < 8$$

we can check that $(1, 1)$ is in the solution set since $2 \cdot 1 + 3 \cdot 1 = 5 < 8$. Thus, as shown in Figure 4.14, every point in the graph of $2x + 3y < 8$ lies on the same side of the line as the $(1, 1)$-point.

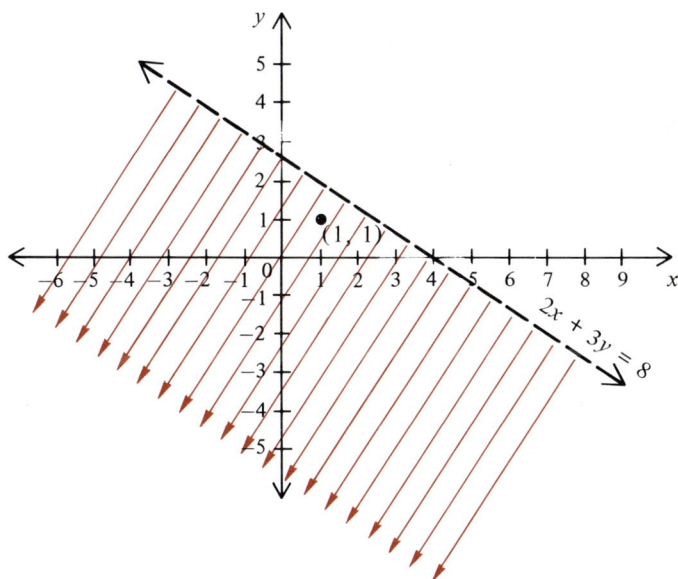

Figure 4.14

If we were considering the graph of $2x + 3y \leqslant 8$, for example, the line corresponding to the equation $2x + 3y = 8$ would be included in the graph and drawn as a solid line. This is because $2x + 3y \leqslant 8$ means that either $2x + 3y < 8$ or $2x + 3y = 8$.

We illustrate these techniques of graphing inequalities in the following three examples.

Example 1: Graph $x + y > 5$.

1. For $x + y = 5$ we have the following table:

x	y
2	3
0	5
5	0

2. Graph $x + y = 5$ using a dashed line.

3. Choose a point not on the line whose equation is $x + y = 5$, say the $(0, 0)$-point. Then $0 + 0 < 5$ and so $(0, 0)$ is *not* in the solution set of $x + y > 5$.

4. Because $(0, 0)$ is not in the solution set, and the $(0, 0)$-point is below the line, the graph of $x + y > 5$ is the half plane above the line, as shown in Figure 4.15.

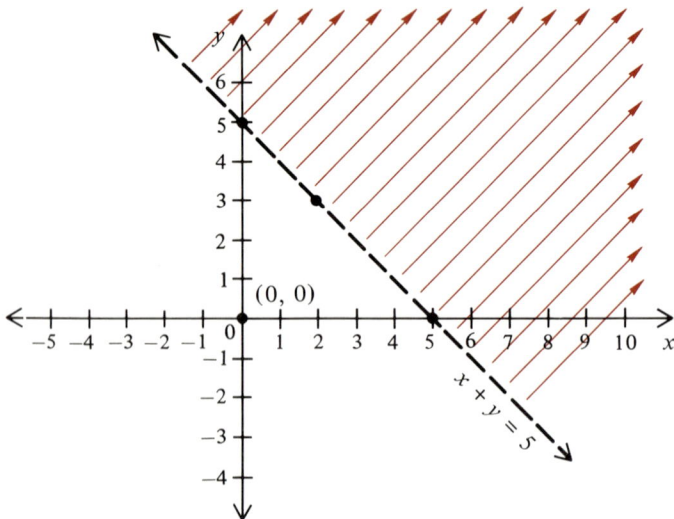

Figure 4.15

Example 2: Graph $y - 2x < 0$.

1. For $y - 2x = 0$ we have the following table:

x	y
-3	-6
0	0
3	6

2. Graph $y - 2x = 0$ using a dashed line.

3. Choose the $(1, 1)$-point, which is not on the line. We have $y - 2x = 1 - 2 \cdot 1 = -1$, and since $-1 < 0$ we conclude that $(1, 1)$ is in the solution set of $y - 2x < 0$.

4. Since the $(1, 1)$-point is below the line, we conclude that the graph of $y - 2x < 0$ is the half plane below the line, as shown in Figure 4.16.

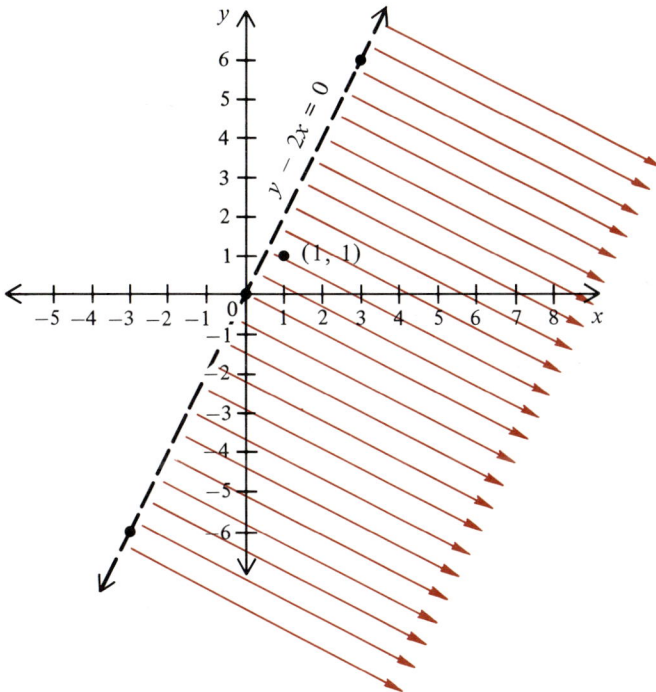

Figure 4.16

Example 3: Graph $x + y \leqslant 2$.

1. For $x + y = 2$ we have the following table

x	y
2	0
1	1
0	2

2. Graph $x + y = 2$ using a solid line.

3. Choose the (2, 2)-point, which is not on the line. Since $2 + 2 = 4 > 2$, (2, 2) is not in the solution set of $x + y \leqslant 2$.

4. The graph is the line $x + y = 2$ and the half plane that does not contain the (2, 2)-point. Since the (2, 2)-point is above the line, this half plane is the one below the line, as shown in Figure 4.17.

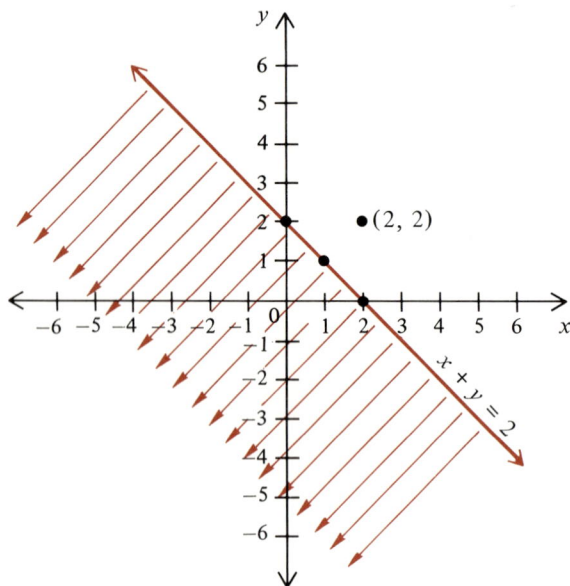

Figure 4.17

EXERCISES 4.7

Draw a graph of each of the following inequalities.

1. $x + y < 3$
2. $x + y > 3$
3. $x + y \leqslant 3$
4. $x + y \geqslant 3$
5. $x - y > 3$
6. $x - y < 3$
7. $y > 2x$
8. $y \leqslant -3x$
9. $2x - y < 1$
10. $2x + y > 4$
11. $x \leqslant 2y + 4$
12. $3x + 2y < 6$
13. $3x - 4y > 7$
14. $3y \leqslant x$
15. $2x > 3y$
16. $2x - 5y < 6$
17. $y + 3x > 0$
18. $x + y \leqslant 1$
19. $y - 2x < -1$
20. $3x + 6y \leqslant 1$

21. $3x + y > -2$
22. $\frac{1}{2}x - \frac{2}{3}y < \frac{1}{4}$
23. $0.3x + 0.4y \geqslant 1$
24. $2y < 3$
25. $x < 0$
26. $150x + 200y \geqslant 1000$
27. $5y < 0$
28. $2\frac{1}{2}x - 4\frac{1}{3}y > 7$
29. $x + y < 1$
30. $3\frac{1}{2}x - y < 10$
31. $5x + 15y > 30$
32. $x > y$
33. $3x \leqslant 2y$
34. $7 + 5y \leqslant x$
35. $\frac{1}{2}y < 10$
36. $5x \leqslant y - 7$
37. $y \leqslant 3x + 2$
38. $y \geqslant \frac{1}{2}x + 4$
39. $y < -x$
40. $-y \geqslant -x$

4.8 Systems of Linear Inequalities

If we have a system of inequalities such as

(1) $\qquad \begin{cases} x + 2y > 4 \\ 2x - y \geqslant 3 \end{cases}$ or (2) $\begin{cases} x + 2y > 4 \\ 2x - y \geqslant 3 \\ x + y \leqslant 9 \end{cases}$

we can solve the system graphically by graphing each member of the system and considering the *intersection* of the solution sets; i.e. the solutions common to all of the members of the system.

In system (1), then, we have the solution set pictured by the cross-hatched area shown in Figure 4.18.

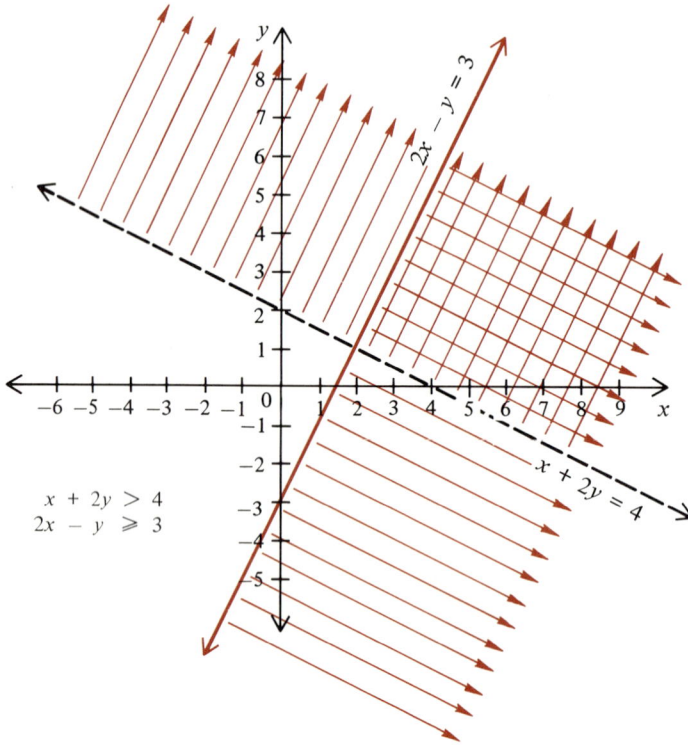

$$x + 2y > 4$$
$$2x - y \geqslant 3$$

Figure 4.18

For the system (2), the solution set is pictured by the shaded area in Figure 4.19.

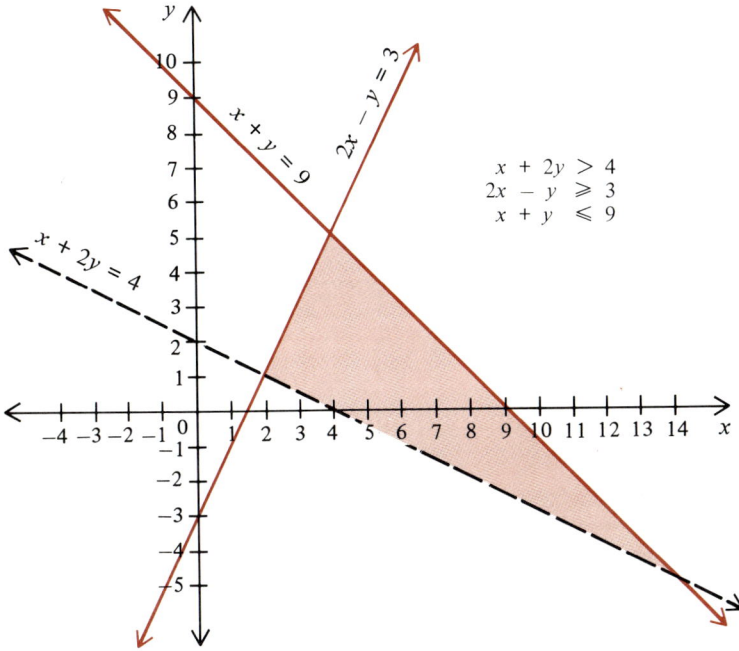

Figure 4.19

EXERCISES 4.8

The lines with equations $x + y = 2$, $x - 2y = 10$, and $10x + 3y = -30$ separate the plane into regions A-G, as shown on page 148. In exercises 1-10 state what regions show solutions of the specified system.

Example (a): $x + y > 2$ Regions C, D, and E

Example (b): $\begin{cases} x + y > 2 \\ 10x + 3y > -30 \end{cases}$ Regions D and E

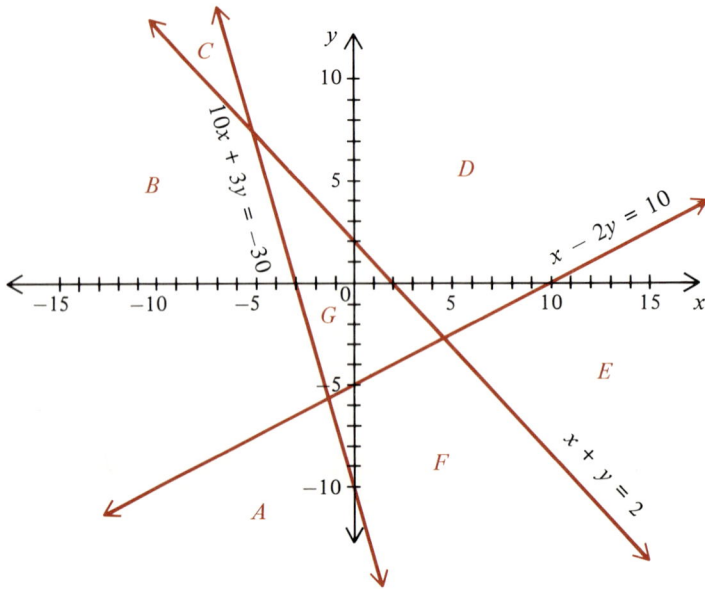

1. $x - 2y < 10$

2. $x + y < 2$

3. $10x + 3y < -30$

4. $\begin{cases} x + y < 2 \\ x - 2y < 10 \end{cases}$

5. $\begin{cases} x - 2y > 10 \\ 10x + 3y < -30 \end{cases}$

6. $\begin{cases} x - 2y > 10 \\ 10x + 3y > -30 \end{cases}$

7. $\begin{cases} 10x + 3y < -30 \\ x + y < 2 \end{cases}$

8. $\begin{cases} x + y > 2 \\ x - 2y > 10 \end{cases}$

9. $\begin{cases} 10x + 3y < -30 \\ x - 2y < 10 \\ x + y > 2 \end{cases}$

10. $\begin{cases} 10x + 3y < -30 \\ x - 2y < 10 \\ x + y < 2 \end{cases}$

In exercises 11–16 write a system of inequalities to describe the region indicated.

Example: Region made up of D and E.

Solution: $\begin{cases} x + y > 2 \\ 10x + 3y > -30 \end{cases}$

11. A, F, and E

12. A, B, and C

13. C and D

14. D

15. F

16. G

In exercises 17–51 draw a graph of the given system. Be careful to draw as solid lines only those lines that belong on the graphs. Shade the portion of the coordinate plane that is the graph of the given system.

17. $\begin{cases} x + y \geqslant 2 \\ x - y < 3 \end{cases}$

18. $\begin{cases} y - x \leqslant 0 \\ y + x \leqslant 0 \end{cases}$

19. $\begin{cases} y \geqslant 2x \\ x + y \leqslant 6 \end{cases}$

20. $\begin{cases} y - x < 8 \\ y + x > 4 \end{cases}$

21. $\begin{cases} x - y \geqslant -8 \\ y > 3x \end{cases}$

22. $\begin{cases} x - y < -6 \\ x + y > 4 \end{cases}$

23. $\begin{cases} y \leqslant 2x \\ x + y < 6 \end{cases}$

24. $\begin{cases} x + y \geqslant 5 \\ 2x - y \geqslant 7 \end{cases}$

25. $\begin{cases} 2x + 5y < 20 \\ x < y \end{cases}$

26. $\begin{cases} 3x + y \leqslant 4 \\ 2x - y < 6 \end{cases}$

27. $\begin{cases} 4x - y < 8 \\ 4x - 3y > 8 \end{cases}$

28. $\begin{cases} 4x - 3y < -1 \\ 6x + y > -7 \end{cases}$

29. $\begin{cases} x + y < 5 \\ 2x + y > -3 \end{cases}$

30. $\begin{cases} 3x + y - 6 > 0 \\ x - 3y - 1 < 0 \end{cases}$

31. $\begin{cases} 2x \leqslant y \\ x + y < 7 \end{cases}$

32. $\begin{cases} 2x \geqslant y \\ x + y > 7 \end{cases}$

33. $\begin{cases} 2x - y < 5 \\ 2x + y > 5 \end{cases}$

34. $\begin{cases} x > y \\ y > x \end{cases}$

35. $\begin{cases} y \leqslant -3x + 4 \\ y - 2x \geqslant 0 \end{cases}$

36. $\begin{cases} 3x - 4y \leqslant 1 \\ x \leqslant 5 \end{cases}$

37. $\begin{cases} 3x - 4y \geqslant 1 \\ x \geqslant 5 \end{cases}$

38. $\begin{cases} 3x - 4y \geqslant 1 \\ x \leqslant 5 \end{cases}$

39. $\begin{cases} 3x - 4y \leqslant 1 \\ y \geqslant 5 \end{cases}$

40. $\begin{cases} 3x + 2y \leqslant 1 \\ 2x - y > 4 \end{cases}$

41. $\begin{cases} 5x + 2y < 0 \\ 2x - y > 4 \end{cases}$

42. $\begin{cases} -3x < y \\ -y > x - 7 \\ y > 2 \end{cases}$

43. $\begin{cases} x < 3 \\ y > -3 \\ x > -3 \end{cases}$

44. $\begin{cases} x \leqslant 2 \\ x - 2y > 0 \\ 3x + y \geqslant 0 \end{cases}$

45. $\begin{cases} x \geqslant 0 \\ y > x \\ y < x + 2 \end{cases}$

46. $\begin{cases} 5x + 7y \leqslant 35 \\ 5x - 7y \geqslant -35 \\ y \geqslant 0 \end{cases}$

47. $\begin{cases} x + y > 4 \\ x + y < 6 \\ 2x - y < 7 \end{cases}$

48. $\begin{cases} x \leqslant 2 \\ x - 2y > 0 \\ 3x + y \geqslant 0 \end{cases}$

49. $\begin{cases} x + y < 0 \\ x - y < 4 \\ y \leqslant 0 \end{cases}$

50. $\begin{cases} x + y \leqslant 8 \\ x + y \geqslant -8 \\ x - y \leqslant 8 \\ x - y \geqslant -8 \end{cases}$

51. $\begin{cases} x - y \leqslant -4 \\ 2x - y \geqslant -3 \\ 2x - y \leqslant 3 \\ y \geqslant 1 \end{cases}$

REVIEW EXERCISES

Section 4.1

In exercises 1–16 indicate whether or not the two equations are equivalent.

1. $x + y = 0$
   $2x + 2y = 0$

2. $x + 5 = y$
   $3x + 15 = 3y$

3. $x - y = 6$
   $y = x - 6$

4. $x - 2y = 10$
   $5x - 10y = 15$

5. $2x - y = 13$
   $y = 2x - 13$

6. $x + y = 7$
   $7x = 49 - 7y$

7. $x = y + 4$
   $x + 3 = y + 7$

8. $2x - y = 8$
   $4x - 2y = 16$

9. $x = 3y - 4$
   $3x + 4 = 9y - 8$

10. $x - 2y = 10$
    $5x - 10y = 50$

11. $x = 2y$
    $4y - 2x = 0$

12. $x = y + 1$
    $2x - 2y = 2$

13. $4x + 8y = 12$
    $x + 2y = 3$

14. $3x - y = 9$
    $9x - 3y = 12$

15. $y = x - 5$
    $y + 5 = x$

16. $x + y = 2$
    $y = x - 2$

Section 4.2

In exercises 17–31 list the ordered pair of numbers that is associated with each numbered point on the graph below.

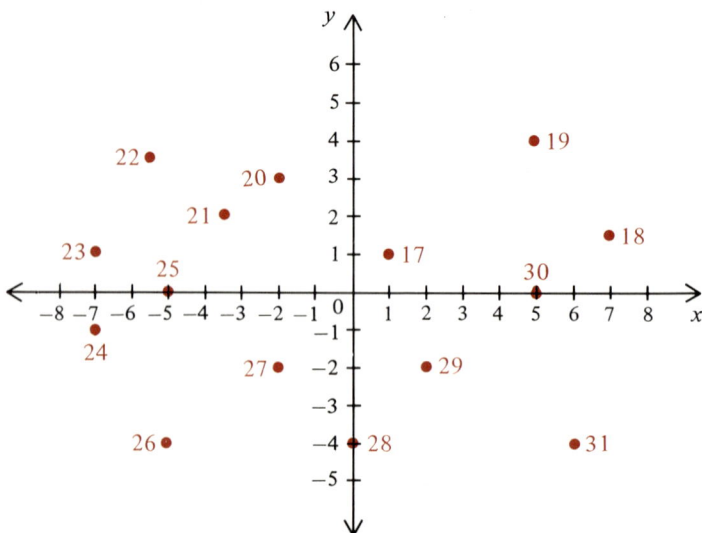

In exercises 32–34 plot, on the same coordinate plane, each point whose coordinates are given, and connect them in order by line segments.

32. A (3, -5)   E (0, 7)   I (-3, -5)
    B (1, -5)   F (-4, -3)   J (-3, -6)
    C (1, -3)   G (-1, -3)   K (3, -6)
    D (4, -3)   H (-1, -5)   A (3, -5)

33. A (-1, 1) $G \left(3\frac{1}{2}, 5\frac{1}{2}\right)$ M (4, 2)

    B (-4, 0) $H \left(3\frac{1}{2}, 4\frac{1}{2}\right)$ N (0, -3)

    C (0, 4) $I \left(2\frac{1}{2}, 5\frac{1}{2}\right)$ O (0, -7)

    D (1, 3) $J \left(3\frac{1}{2}, 5\frac{1}{2}\right)$ P (1, -8)

    E (1, 6)   K (5, 4)   Q (-1, -8)
    F (2, 7)   L (3, 3)   A (-1, 1)

34. $A \left(-1, -3\frac{1}{2}\right)$ $H \left(1, 3\frac{1}{2}\right)$ O (2, -2)

    B (-1, -5)   I (1, 2)   P (1, -2)

    C (-2, -5)   J (2, 1) $Q \left(\frac{3}{4}, -2\frac{1}{4}\right)$

    D (-2, 1)   K (1, 1) $R \left(-\frac{1}{2}, -2\right)$

    E (-3, 1)   L (1, 0) $S \left(\frac{1}{2}, -3\right)$

    F (-2, 2)   M (0, 0) $T \left(0, -3\frac{1}{2}\right)$

    $G \left(-2, 3\frac{1}{2}\right)$ $N \left(\frac{1}{2}, 1\right)$ $U \left(-1, -3\frac{1}{2}\right)$

Section 4.3

In exercises 35–52 use the properties of equality to solve the given equation for y in terms of x.

35. $2x - y = 8$ 38. $2x + y = 12$ 41. $x - 3y = 8$
36. $x - 3y = 9$ 39. $3x + 4y = 24$ 42. $6x + 4y = -12$
37. $2x - 3y = 12$ 40. $3x + y = -6$ 43. $3y = 2x$

44. $2x - y - 13 = 0$

45. $x + 3y + 4 = 0$

46. $4x = 6y$

47. $\frac{1}{6}x + \frac{1}{2}y = 2$

48. $6x = 5y + \frac{1}{3}$

49. $\frac{3}{4}x + \frac{2}{5}y = 4$

50. $x + \frac{1}{9}y = -\frac{1}{3}$

51. $\frac{1}{4}x = 4y$

52. $\frac{2}{3}x = 5y$

In exercises 53–68 draw a graph of the given equation.

53. $x - y = 6$
54. $x - 2y = 8$
55. $3x - y = 9$
56. $3x - 2y = 12$
57. $y = 2x - 5$
58. $x + y = 8$

59. $x + 2y = 12$
60. $5x = y$
61. $4x + 3y = 12$
62. $y = 7 - 3x$
63. $y - x = 6$
64. $2x - y = 10$

65. $x = 2y$
66. $y = 5$
67. $x = 3$
68. $x = -2$

Section 4.4

In exercises 69–88 solve graphically the given system of equations and check your solution.

69. $\begin{cases} x + y = 5 \\ y - x = 3 \end{cases}$

70. $\begin{cases} x = y + 1 \\ x + y = -7 \end{cases}$

71. $\begin{cases} y = 2x + 7 \\ x + 2y = -11 \end{cases}$

72. $\begin{cases} x + 2y = 0 \\ y = -5x + 9 \end{cases}$

73. $\begin{cases} x + 3y = 6 \\ 3x - y = 8 \end{cases}$

74. $\begin{cases} x - y = 7 \\ 2x + y = 5 \end{cases}$

75. $\begin{cases} 2x = y \\ x + y = 9 \end{cases}$

76. $\begin{cases} x + 3y = 9 \\ 2x - y = 11 \end{cases}$

77. $\begin{cases} 5x + y = -5 \\ 2x - y = 12 \end{cases}$

78. $\begin{cases} x + 4y = 14 \\ 3x - y = 3 \end{cases}$

79. $\begin{cases} 3x + 2y = 12 \\ 6x + 4y = -12 \end{cases}$

80. $\begin{cases} 2x - y = 10 \\ x + 2y = 5 \end{cases}$

81. $\begin{cases} y + 2x = 3 \\ x - 2y = 9 \end{cases}$

82. $\begin{cases} 2x - y = 5 \\ 6x - 3y = -6 \end{cases}$

83. $\begin{cases} x - 4y = 10 \\ x + 2y = 7 \end{cases}$

84. $\begin{cases} 2x + y = -5 \\ 3y - 4x = 15 \end{cases}$

85. $\begin{cases} 5x - 2y = 8 \\ 2x + y = -4 \end{cases}$

86. $\begin{cases} 3x - y = 9 \\ 9x - 3y = 12 \end{cases}$

87. $\begin{cases} 2x - y = 8 \\ 4x - 2y = 16 \end{cases}$

88. $\begin{cases} 2x - y = 13 \\ x + 3y = -4 \end{cases}$

Section 4.5

In exercises 89–110 solve the given system of equations algebraically and check your solution.

89. $\begin{cases} x + 2y = 6 \\ x + 3y = 4 \end{cases}$

90. $\begin{cases} 4x - 3y = -6 \\ 5x - y = 20 \end{cases}$

91. $\begin{cases} a + 5b = 13 \\ 5a + 4b = 2 \end{cases}$

92. $\begin{cases} 2x + 3y = 8 \\ 3x + 2y = -3 \end{cases}$

93. $\begin{cases} 4r + 5s = 0 \\ 3r + 2s = 23 \end{cases}$

94. $\begin{cases} 8x + y = 8 \\ 3x - 2y = -16 \end{cases}$

95. $\begin{cases} 3a + 7b = -15 \\ 2a + 8b = -10 \end{cases}$

96. $\begin{cases} 9w - 4z = -5 \\ 6w + 5z = 12 \end{cases}$

97. $\begin{cases} 8x - y = 0 \\ 4x + 2y = 15 \end{cases}$

98. $\begin{cases} 8m - 9n = -2 \\ 6m + 3n = 5 \end{cases}$

99. $\begin{cases} 10x + 8y = 1 \\ 4x - 12y = -11 \end{cases}$

100. $\begin{cases} 5a + 4b = 1 \\ 4a + 5b = -1 \end{cases}$

101. $\begin{cases} \dfrac{1}{2}x + \dfrac{1}{4}y = 8 \\ \dfrac{2}{3}x - \dfrac{3}{4}y = 2 \end{cases}$

102. $\begin{cases} 8x = 3y - 4 \\ \dfrac{4}{3}x + \dfrac{3}{2}y = 2 \end{cases}$

103. $\begin{cases} \dfrac{1}{2}a + \dfrac{5}{8}b = 0 \\ \dfrac{2}{5}a + \dfrac{3}{4}b = 4 \end{cases}$

104. $\begin{cases} 3x = 6y + 1 \\ 3x = 4y \end{cases}$

105. $\begin{cases} 7x + y + 1 = 0 \\ y + 10x = 2 \end{cases}$

106. $\begin{cases} a - 9b = 4 \\ a - 15b = 0 \end{cases}$

107. $\begin{cases} x = y + 1 \\ 3x - 2y = 0 \end{cases}$

108. $\begin{cases} 5m + n = 1 \\ m + n = 5 \end{cases}$

109. $\begin{cases} x - 2y = -4 \\ \dfrac{1}{2}x + y = -4 \end{cases}$

110. $\begin{cases} 12x + y - 1 = 0 \\ 3y + 10x + 10 = 0 \end{cases}$

Section 4.6

In exercises 111–135 solve the given problem using two equations and two variables.

111. The sum of two numbers is 128 and their difference is 48. Find the numbers.

112. Half the sum of two numbers is 40 and 3 times their difference is 18. Find the numbers.

113. You want to invest $14,000 in two enterprises which together will give you an income of $1484. If one enterprise yields 11% and the other 10%, how much should you invest in each?

114. A woman invested $18,000 in two enterprises, one yielding 8% and the other yielding 10%. Had the amount invested at 10% been twice as much, her income would have been $110 more. How much did she invest at each rate?

115. You have a solution of 60% acid. You add some distilled water, reducing the concentration to 40%. You then add 1 liter more of distilled water, reducing the concentration to 30%. How much of the 30% solution do you now have?

116. A chemist has two bottles of alcohol solutions of different strengths. The first is 25% alcohol; the second is 45% alcohol. How many ounces of solution must he draw from each bottle to make a mixture of 24 ounces that will be $33\frac{1}{3}$ % alcohol?

117. Two cars start at the same time from the same place and travel in opposite directions. One car travels at a rate that is 25 kilometers per hour more than twice the rate of the other car. What is the rate of each car if they are 825 kilometers apart in 5 hours?

118. You start to walk to a town 10 miles away at the rate of 3 miles per hour. After walking part of the way you obtain a ride from a passing motorist who is traveling at an average rate of 24 miles per hour. If you reach the town 1 hour after you started, how long did you walk?

119. Three kilograms of tea and 6 kilograms of coffee cost $25.32, and 5 kilograms of tea and 4 kilograms of coffee cost $24.92. What is the price per kilogram of the tea and the price per kilogram of the coffee?

120. As a candy dealer you want to make a mixture of 100 pounds of candy to sell at $2.40 a pound using some candy worth $2.00 a pound and some worth $3.20 per pound. How many pounds of each should you use?

121. If it takes Jack 3 hours to row 20 kilometers downstream and 6 hours to row back to where he started, find the average rate Jack rows in still water and the rate of the current of the river.

122. An outboard-powered boat travels upstream a distance of 12 miles in 2 hours and 15 minutes. If the current had been twice as strong, the trip would have taken 3 hours. How long should it take to make the trip downstream in the slower current?

123. Five times the smaller of two numbers is 4 times the larger. The difference between 3 times the smaller and twice the larger is 4. What are the numbers?

124. The perimeter of a rectangular farm is 4 kilometers. The length of the farm is 200 meters more than twice the width. What is the length and width of the farm?

125. A girl has a pile of coins, consisting of dimes and quarters, worth $14.40. She observes that if the dimes were quarters and the quarters were dimes she would have $2.70 more. How many of each coin does she have?

126. A certain two-digit number is 4 times the sum of the digits. If 27 is added to the number, the digits will be reversed. Find the number.

127. Six years ago Bob was 12 times as old as Jim was then. In 3 years he will be 3 times as old as Jim will be then. How old is each now?

128. George has $91 in $5 bills and $2 bills. If he has 23 bills in all, how many $5 bills does he have?

129. Eight guilders and 7 francs are worth $4.88, and 9 guilders and 70 francs are worth $20.40. What is the value of a guilder and the value of a franc?

130. If the greater of two numbers is divided by the smaller, the quotient is 3 and the remainder is 4. If 5 times the smaller is divided by the greater, the quotient is 1 and the remainder is 10. What are the numbers?

131. A man owns two stores. One year one of the stores returned 10% on his investment while the other store lost 5%, resulting in a gain of $12,500. The next year the first store earned 8% and the second 6%, resulting in a gain of $19,000. What is the value of each store?

132. June works for 8 days and Wally for 15 days to complete a certain job. If June works for 12 days and Wally for 10 days, they can complete the same job. How long would it take working alone for each one to do the same job?

133. A tank has a pipe that drains it continuously. If the pipe that fills the tank is turned on full force, the tank is full in 6 hours. If the filling valve is turned on only halfway, it takes 24 hours to fill the tank. How long does it take the drain to empty a full tank?

134. An instructor gives 3 points for each correct answer and subtracts 2 points for each incorrect answer. Don received a score of 3, but said, "If the instructor had given five points for each correct answer, I could have missed one more and had a score of fourteen." How many problems did he miss originally?

135. A merchant has two casks of wine, which together contain 224 liters. He pours from the first into the second as much as the second contained to begin with; he then pours from the second into the first as much as was left in the first; and again from the first into the second as much as was left in the second. There is now $\frac{3}{4}$ as much in the first as in the second. How many liters did each contain to begin with?

Section 4.7

In exercises 136–155 draw a graph of the given inequality.

136. $5x + 2y < 6$

137. $x - 3y \geqslant 3$

138. $x + y < 1$

139. $\frac{2}{3}x - \frac{1}{2}y > \frac{1}{2}$

140. $x \leqslant 0$

141. $0.4y - 0.3x \geqslant 1$

142. $200x + 150y \leqslant 1000$

143. $5y < 0$

144. $2\frac{1}{4}x + 1\frac{1}{2}y < 3$

145. $x - y < 1$

146. $x + y \geqslant 2$

147. $x - 3\frac{1}{2}y < 10$

148. $0.5x + 1.5y < 0.3$

149. $2x > 3y$

150. $y > x$

151. $5y + 7 < x$

152. $\frac{1}{2}y > \frac{1}{2}$

153. $x \leqslant 5y - 7$

154. $y \geqslant 3x + 4$

155. $y \leqslant \frac{1}{2}x + 8$

Section 4.8

In exercises 156–167 draw a graph of the given system.

156. $\begin{cases} x + 2y < -5 \\ 2x - y > 0 \end{cases}$

157. $\begin{cases} x + y < 4 \\ x - y > 2 \end{cases}$

158. $\begin{cases} 3x - y > -5 \\ x + 2y > 3 \end{cases}$

159. $\begin{cases} 3x - 2y \leqslant 3 \\ 2x + 3y \leqslant 2 \end{cases}$

160. $\begin{cases} x + 4y > 1 \\ x + 6y < 3 \end{cases}$

161. $\begin{cases} y < 2x - 4 \\ x > 1 - y \end{cases}$

162. $\begin{cases} x + 3y + 6 \geqslant 0 \\ y \geqslant 4x - 2 \end{cases}$

163. $\begin{cases} 2x + 4y < 8 \\ 3x + 6y > 30 \end{cases}$

164. $\begin{cases} 2x + 4y > 8 \\ 3x + 6y < 30 \end{cases}$

165. $\begin{cases} 3y \leqslant 4x + 2 \\ 2x + 5y \geqslant 12 \end{cases}$

166. $\begin{cases} x + y \leqslant 5 \\ x + y \geqslant -5 \\ x - y \leqslant 5 \\ x - y \geqslant -5 \end{cases}$

167. $\begin{cases} x < 5 \\ y < 5 \\ x > -5 \\ y > -5 \end{cases}$

CUMULATIVE REVIEW B

1. $\dfrac{6}{7} - \dfrac{6}{9} = ?$

2. $\left(\dfrac{-6}{-7}\right)\left(\dfrac{-3}{10}\right) = ?$

3. $\dfrac{7}{2} \cdot \left(\dfrac{3}{4} + \dfrac{1}{8}\right) = ?$

4. The fact that $(4 + 5)(y + 2) = (4 + 5)y + (4 + 5) \cdot 2$ illustrates the __?__ property.

5. If P and Q have the values 7 and 8, respectively, what is the value of $(2P + 5Q) \cdot 3 + 2$?

6. $3x(y) + 2(x + y) + 5(xy) = ?$

7. $6a(3b) + 4ab + b(8a) + (2 + 5)(ab) = ?$

8. Find the fraction in lowest terms equal to $0.\overline{75}$.

9. What is the decimal equivalent of $\frac{5}{12}$?

In exercises 10–22 find the solution set of the given equation or inequality. Graph the solution set of each inequality on a number line.

10. $8x + 2 - 3x = 4x - 12 + 2x$

11. $3x - 8 = x + 7 + 2x$

12. $2x + 3 = 3x + 4 - x$

13. $x + 2(x - 3) = x + 2$

14. $5 - (2x + 1) = 8$
15. $2x - 4 = 6x + 2(x - 2)$
16. $x - 9 = 9 - x$
17. $2x - 10 < -6 + 3x$
18. $6x - 8 < -x - 85$
19. $12 - 6x > -6 + 3x$
20. $x + 3(x + 1) < 2x + 17$
21. $7x - 6 > 3(5x + 6)$
22. $x(8 + x) > x(x + 2) + 24$
23. List five members of the solution set of the equation $2x + 3y = 10$.
24. Are the equations $3x + 5y + 6 = 0$ and $\frac{5}{2}y + \frac{3}{2}x = -3$ equivalent equations?

In exercises 25 and 26 solve the given equation for y in terms of x.

25. $2(x + y) + 3(x + 2y) = 5 - 2y$
26. $5yx - 7 = 9 - 3yx$
27. In the formula $V = \dfrac{Pr^4}{8lk}$ solve for k.

In exercises 28–35 graph the given equation.

28. $x + y = 5$ 31. $x = y - 5$ 34. $y = 2x$
29. $x - y = 5$ 32. $x = y$ 35. $y = 8$
30. $y = x + 5$ 33. $x = 8$

In exercises 36 and 37 solve the given system of equations by graphing.

36. $\begin{cases} 2x + 3y = 9 \\ 4x - 5y = 7 \end{cases}$ 37. $\begin{cases} 4x + 5y = 8 \\ 3x - 4y = 6 \end{cases}$

In exercises 38–43 solve the given system of equations algebraically.

38. $\begin{cases} 3x - y = 1 \\ 5x - 3y = 7 \end{cases}$ 40. $\begin{cases} 5x + y + 2 = 0 \\ x + 2y = 5 \end{cases}$ 42. $\begin{cases} 6x - 5y = -10 \\ 5x - 2y = -17 \end{cases}$
39. $\begin{cases} 3a = 4b \\ 4a - 5b = 2 \end{cases}$ 41. $\begin{cases} 3x + 2y = 17 \\ 4x + y = 16 \end{cases}$ 43. $\begin{cases} 3x + 7y = 2 \\ 7x + 8y = -2 \end{cases}$

In exercises 44–48 solve the given problem.

44. Jack has 50 coins, all nickels or dimes, amounting to $4. How many of each kind does he have?

45. Three pipes fill a tank. One can fill it in 4 minutes, another in 6 minutes, and a third in 12 minutes. How long would it take the three pipes together to fill the tank?

46. Two automobiles start at the same time from the same place and travel in opposite directions. The first travels at 54 miles per hour and the second travels at 45 miles per hour. In how many hours will they be 231 miles apart?

47. Mr. Johansson has 3 times as much money invested in 4% bonds as he does in 8% bonds. How much does he have invested in each if his total annual return is $1400?

48. A druggist must make 300 cubic centimeters of a 25% barbiturate solution from her 15% and 30% solutions. How much of each should she use?

In exercises 49 and 50 solve the given system by graphing.

49. $\begin{cases} y \geqslant 2x + 1 \\ 2x + y \leqslant 9 \end{cases}$

50. $\begin{cases} y > x \\ x < 5 \end{cases}$

5

RADICALS AND EXPONENTS

5.1 Positive Integral Exponents

A very convenient abbreviation for $2 \cdot 2$ is 2^2; for $2 \cdot 2 \cdot 2, 2^3$; for $a \cdot a \cdot a \cdot a, a^4$. In general, for any real number a and any positive integer n we say that

$$a^1 = a \text{ and } a^n = a \cdot a \cdot \ldots \cdot a \text{ (n factors of a) if } n > 1$$

In a^n, a is called the *base*, n is called the *exponent*, and a^n is called a *power* of a. Thus

$$2^5 \quad \nearrow \text{the } exponent$$
$$\searrow \text{the } base \qquad \text{is a } power \text{ of } 2$$

We read a^n as "a to the nth" but we also read a^2 as "a squared" and a^3 as "a cubed" because the area of a square of side a is $a \cdot a$ and the volume of a cube of edge a is $a \cdot a \cdot a$.

Consider now the following examples of work with exponents:

1. $a^2 \cdot a^3 = (a \cdot a) \cdot (a \cdot a \cdot a) = a^5 = a^{2+3}$

2. $\dfrac{a^5}{a^2} = \dfrac{a \cdot a \cdot a \cdot a \cdot a}{a \cdot a} = \dfrac{a \cdot a}{a \cdot a} \cdot \dfrac{a \cdot a \cdot a}{1} = 1 \cdot \dfrac{a \cdot a \cdot a}{1}$

    $= a \cdot a \cdot a = a^3 = a^{5-2} \quad (a \neq 0)$

3. $\dfrac{a^2}{a^5} = \dfrac{a \cdot a}{a \cdot a \cdot a \cdot a \cdot a} = \dfrac{a \cdot a}{a \cdot a} \cdot \dfrac{1}{a \cdot a \cdot a} = 1 \cdot \dfrac{1}{a \cdot a \cdot a}$

    $= \dfrac{1}{a \cdot a \cdot a} = \dfrac{1}{a^3} = \dfrac{1}{a^{5-2}} \quad (a \neq 0)$

4. $(ab)^3 = (ab) \cdot (ab) \cdot (ab) = (a \cdot a \cdot a) \cdot (b \cdot b \cdot b) = a^3 b^3$

5. $\left(\dfrac{a}{b}\right)^3 = \dfrac{a}{b} \cdot \dfrac{a}{b} \cdot \dfrac{a}{b} = \dfrac{a \cdot a \cdot a}{b \cdot b \cdot b} = \dfrac{a^3}{b^3} \ (b \neq 0)$

6. $(a^2)^3 = (a \cdot a) \cdot (a \cdot a) \cdot (a \cdot a) = a^6 = a^{2 \cdot 3}$

These examples can easily be generalized to give us the following basic properties of positive integral exponents.

For any real numbers a and b and positive integers n and m

P-1 $a^m a^n = a^{m+n}$

P-2 $\dfrac{a^m}{a^n} = a^{m-n}$ if $m > n$ and $a \neq 0$

P-3 $\dfrac{a^m}{a^n} = \dfrac{1}{a^{n-m}}$ if $n > m$ and $a \neq 0$

P-4 $(ab)^n = a^n b^n$

P-5 $\left(\dfrac{a}{b}\right)^n = \dfrac{a^n}{b^n}$ if $b \neq 0$

P-6 $(a^n)^m = a^{nm}$

These properties can be used to simplify many expressions involving exponents, as the following examples show.

7. $(x^3 y^2)(x^4 y^3) = (x^3 \cdot x^4)(y^2 \cdot y^3) = x^{3+4} y^{2+3} = x^7 y^5$

(Note the use of the commutative and associative properties of multiplication.)

8. $5(2x - 3y)^3 \cdot 4(2x - 3y)^4 = 5 \cdot 4(2x - 3y)^3 (2x - 3y)^4$

$\qquad\qquad\qquad\qquad = 20(2x - 3y)^{3+4} = 20(2x - 3y)^7$

9. $\dfrac{4x^3 y^2}{2xy^3} = \dfrac{2x^{3-1}}{y^{3-2}} = \dfrac{2x^2}{y}$

10. $\dfrac{24xy(2x - y)^3}{(3xy)^4 (2x - y)} = \dfrac{8(3xy)(2x - y)^3}{(3xy)^4 (2x - y)} = \dfrac{8(2x - y)^{3-1}}{(3xy)^{4-1}}$

$\qquad\qquad\qquad\qquad = \dfrac{8(2x - y)^2}{3^3 x^3 y^3} = \dfrac{8(2x - y)^2}{27x^3 y^3}$

11. $\left(\dfrac{3a}{2b}\right)^3 = \dfrac{(3a)^3}{(2b)^3} = \dfrac{3^3 a^3}{2^3 b^3} = \dfrac{27a^3}{8b^3}$

12. $2(x^3y^2)^5 = 2(x^3)^5(y^2)^5 = 2x^{15}y^{10}$

13. $(-3)^2 = (-3)(-3) = 9$
    (Contrast this with $-3^2 = -(3 \cdot 3) = -9$.)

EXERCISES 5.1

In exercises 1–15 write the given expression without exponents.

Example (a): $5x^3y^2 = 5 \cdot x \cdot x \cdot x \cdot y \cdot y$

Example (b): $(2x^2y)(3x^3y^2) = 2 \cdot 3 \cdot x \cdot x \cdot x \cdot x \cdot x \cdot y \cdot y \cdot y$

1. $2x^2y^3$
2. $5ab^4$
3. $7a^4b^3$
4. $(2x^2y)(2xy^2)$
5. $(2a^3b^2)(5ab^3)$
6. $(3a^2b^3)(7a^2b)$
7. $(4x^2y^2)(3xy^3)$
8. $(2xy)(3xy)(5x^2y^3)$

9. $2(x-y)^3(x+y)^2$
10. $5x^3(x-2y^2)^2$
11. $(3x^2y^3)(2x+y)^2$
12. $\dfrac{3x^2y^3}{(x-y)^3}$
13. $\dfrac{(3x^3)^2}{(2y-3)^3}$
14. $\dfrac{(5x^2)^3}{(x+y)^3}$
15. $\dfrac{2(x-y)^2}{[3(x+2y)]^2}$

In exercises 16–49 simplify the given expression by using properties P-1 to P-6.

Example (a): $(2x^2y)(3xy^3) = 6x^3y^4$

Example (b): $\dfrac{16a^5}{8a} = 2a^4$

Example (c): $\dfrac{15a^4b^2c^4}{25a^6b^3c} = \dfrac{3c^3}{5a^2b}$

16. $(10x)(-3x)$
17. $(8xy)(2xy)$
18. $(3x)^3(-2y)$
19. $(3xy^2)(-xy)$
20. $(a^2b)(-2b)^2$

21. $(-5xy^2)^2(-2xy)$
22. $(3pq)(2p^2q)$
23. $(6st^2)(2st)$
24. $(-3abc)^3(3ab^2)^2$
25. $(6xy)(-2x)(-3y)$

26. $(7ab)(3ac)(2bc)$

27. $(9xyz)(6yz)(2z)$

28. $(3xy)(-4x)(3y)(5xy^2)$

29. $(12a^2bc)(ab^2c)(a^2b^2c^4)$

30. $(-3x^2y)^2(-2x^3y^2)(-4x^3y)$

31. $(2x^5y^2)^2(3xy^2)(-5x^2y^3)$

32. $(-2x)(9yz)(-3xz)(-y)$

33. $(3a^2b)(2a^3b^2)^2$

34. $(2xy^2)(3x^2y^2)^3$

35. $(2a^2b)^2(3ab^2)^3(-2a)^3$

36. $(-x)^2(-3xy)^2(-2x)^3$

37. $\dfrac{15a^2b}{5ab}$

38. $\dfrac{10x^3y^2}{2xy}$

39. $\dfrac{32x^3y^2z}{4x^2y^2z}$

40. $\dfrac{46x^3y^4z^5}{23x^2y^4z}$

41. $\dfrac{-30a^4b^5c^2}{6a^2c}$

42. $\dfrac{32x^2y^3z^4}{64x^3y^5z^6}$

43. $\dfrac{27a^4b^2c^3}{81a^6b^3c^4}$

44. $\dfrac{15r^2s^3t^4}{25r^3s^5t}$

45. $\dfrac{32x^5y^4z^2}{20x^2yz^4}$

46. $\dfrac{(3x^2y)^2}{(2xy)(3x^2y)^3}$

47. $\dfrac{(5ab^2)^3}{(2a^2b)(-5ab)^2}$

48. $\dfrac{(12x^2y)(-xy^3)^2}{(5xy)(3x^2y)^3}$

49. $\dfrac{(2a^2b)^3(-5ab)}{(-3ab)^3(-ab^2)}$

5.2 Integral Exponents

How shall we define a^0? If we want P-1 to continue to hold when $n = 0$, we must have

$$a^0 a^m = a^{0+m}$$

But $0 + m = m$ and so we must have

$$a^0 a^m = a^m$$

These considerations lead us to the definition

$$a^0 = 1 \text{ if } a \neq 0$$

(The restriction that $a \neq 0$ is necessary for reasons that are too complicated to go into here.)

Similarly, if we want *P*-2 to hold for all integers *m* and *n* we must have, for example,

$$\frac{a^3}{a^5} = a^{3-5} = a^{-2}$$

But, by *P*-3, we know that

$$\frac{a^3}{a^5} = \frac{1}{a^{5-3}} = \frac{1}{a^2}$$

so that we have

$$a^{-2} = \frac{1}{a^2}$$

These considerations lead us to the definition

If $a \neq 0$, $a^{-n} = \dfrac{1}{a^n}$ for any integer *n*

Example 1: $(xy)^{-3} = \dfrac{1}{(xy)^3} = \dfrac{1}{x^3 y^3}$

Example 2: $3(2x - y)^{-2} = \dfrac{3}{(2x - y)^2}$

Example 3: $\left(\dfrac{a}{b}\right)^{-2} = \dfrac{1}{\left(\dfrac{a}{b}\right)^2} = \dfrac{1}{\dfrac{a^2}{b^2}} = 1 \div \dfrac{a^2}{b^2} = 1 \cdot \dfrac{b^2}{a^2} = \dfrac{b^2}{a^2}$

Example 4: $\dfrac{x^{-2}}{x^{-3}} = \dfrac{\dfrac{1}{x^2}}{\dfrac{1}{x^3}} = \dfrac{1}{x^2} \div \dfrac{1}{x^3} = \dfrac{1}{x^2} \cdot \dfrac{x^3}{1} = \dfrac{x^3}{x^2} = x$

Because of our definitions for a^0 and a^{-n} we can replace *P*-2 and *P*-3 by the single statement

$$\frac{a^m}{a^n} = a^{m-n} \quad \text{if } a \neq 0$$

without any restrictions on *m* and *n*. For *P*-2 tells us the statement is true when $m > n$, whereas if $m = n$ we have the true statement that

$$\frac{a^m}{a^m} = a^{m-m} = a^0 = 1$$

Finally if $m < n$ we have, by P-2 and our definition of a^{-n},

$$\frac{a^m}{a^n} = a^{m-n} = a^{-(n-m)} = \frac{1}{a^{n-m}}$$

which is P-3.

Properties P-1–P-6 were originally stated for positive integral exponents only. They also hold, however, for any integral exponents— positive, negative, or zero. Thus we can do Example 3 more easily by first applying P-5 to get

$$\left(\frac{a}{b}\right)^{-2} = \frac{a^{-2}}{b^{-2}}$$

and then using our definition of negative exponents to get $\frac{b^2}{a^2}$. Similarly, we can apply P-2 to Example 4 to get

$$\frac{x^{-2}}{x^{-3}} = x^{-2-(-3)} = x^1 = x$$

Here are some other examples of the uses of the properties of exponents.

Example 5. $x^3 x^{-2} = x^{3+(-2)} = x^1 = x$

Example 6. $\dfrac{a^0 b^3 c^{-2}}{b^{-2} c^3} = b^{3-(-2)} c^{-2-3} = b^5 c^{-5} = \dfrac{b^5}{c^5}$

Example 7. $\left(\dfrac{x^{-2}}{x^{-3}}\right)^{-2} = \dfrac{(x^{-2})^{-2}}{(x^{-3})^{-2}} = \dfrac{x^4}{x^6} = \dfrac{1}{x^2}$

EXERCISES 5.2

In exercises 1–25 write the given expression in a form that involves only positive exponents.

Example: $x^{-2} y^2 = \dfrac{1}{x^2} \cdot y^2 = \dfrac{y^2}{x^2}$

1. $x^{-1} y^2$

2. $5x^{-3}$

3. $x^2 y^{-2}$

4. $a^{-2} b^{-3}$

5. $(2a^2)^{-3}$

6. $(-3x)^{-2}$

7. $\dfrac{(2a)^{-3}}{b^2}$

8. $\dfrac{5x^2}{y^{-2}}$

9. $3a^0 b^{-3}$

10. $(-2)^{-2}$

11. $(2^{-2})^{-2}$

12. $(-2)^0 x^{-3}$

13. $\dfrac{3x^0 y^{-2}}{z^{-3}}$

14. $\dfrac{3a^{-1}}{b^{-2}}$

15. $\dfrac{3}{(2x)^{-2}}$

16. $\dfrac{2x^{-3}}{x^2}$

17. $\dfrac{x^{-3}y}{x^{-5}y^{-2}}$

18. $\dfrac{a^{-3}b^{-2}z^{-1}}{a^4 b^{-3}z^2}$

19. $\dfrac{cd^{-4}}{c^{-2}d^3}$

20. $\left(\dfrac{ax^{-2}}{y^3}\right)^{-2}$

21. $\dfrac{x^{-2}y^3}{a^{-1}b^{-4}}$

22. $\dfrac{1}{a^{-1}b^{-1}}$

23. $x^5(x^{-3} + x^3)$

24. $x^{-4}(x^7 + x^9)$

25. $x^3(x^{-3} + x^3)$

In exercises 26–40 perform the indicated computations by using the properties of exponents.

Example: $3^{-1} \cdot 3^1 = 3^{-1+1} = 3^0 = 1$

26. $5^1 \cdot 5^{-2}$

27. $2^0 \cdot 2^{-2}$

28. $2^0 \cdot 3^{-2}$

29. $\dfrac{3^{-1}}{3^{-3}}$

30. $5^{-2} \cdot 25$

31. $3^{-2}(9^0)$

32. $(-7)^0 \cdot 7^{-1}$

33. $-7^0 \cdot 3$

34. $(2^2 \cdot 3^3 \cdot 5^4)^0$

35. -1^4

36. -1^5

37. $(-1)^4$

38. $(-1)^5$

39. $\dfrac{1^{15}}{(3.87^2 + 4.765^3)^0}$

40. $\dfrac{7^{-2} \cdot 7^{-5}}{7^{-7}}$

In exercises 41–60 simplify the given expression. Write all answers without negative or zero exponents.

Example (a): $(2a^{-2})(3a^5) = 6a^3$

Example (b): $\dfrac{(x^2 y^{-3})^{-2}}{(x^{-1}y^{-1})^{-3}} = \dfrac{x^{-4}y^6}{x^3 y^3} = \dfrac{y^3}{x^7}$

41. $(2x^2)(3x^{-5})$

42. $(2a)^{-2}(3a^{-2}b)$

43. $(3r^{-2}s^3)^2$

44. $(x^5)(x^{-5})$

45. $(-5a)^0(3a^{-2}b)^{-2}$

46. $(-10x^3)(2x^{-3}y)(10x^3 y)^{-2}$

47. $(x^{-2}y^3)(x^0 y^{-3})(x^2 y^2)^0$

48. $(a^{-3}b^2)(a^2 b^0)(a^2 b^2)$

49. $(3xy)^{-2}(xy^2)$

50. $(2xy)^{-3}(x^2 y)^0$

51. $(rs^2)^{-2}(rs)^{-1}$

52. $(x^2 y)^{-3}(x^2 y)^{-1}$

53. $\dfrac{3x^{-2}}{9x^{-3}y}$

54. $\dfrac{4a^{-3}}{8a^{-2}y}$

55. $\dfrac{r+s}{r^{-2}s^2}$

56. $\dfrac{x^0+y^0}{x^2y^{-1}}$

57. $\dfrac{5a\,(a+b)^2}{3a^{-2}(a+b)^{-2}}$

58. $\dfrac{(x^{-1}y^{-1})^{-2}}{(x^{-2}y^{-2})^{-1}}$

59. $\dfrac{a^{-2}x^{-3}}{(a^{-2}x^{-2})^{-1}}$

60. $\dfrac{(x^{-3}y^{-5})^{-2}}{(x^{-1}y^{-2})^{-2}}$

5.3 Square Roots

The process of squaring a number can, so to speak, be reversed. We have $3^2 = 9$ and say that 9 is the *square* of 3; on the other hand, we call 3 a *square root* of 9. In general, if $x^2 = a$, we call a the square of x and x a square root of a.

Each positive number has two square roots. Thus 9 has the square roots 3 and -3, since $3^2 = 9$ and also $(-3)^2 = 9$. For the *positive* square root we write $\sqrt{9}$ and for the negative square root we write $-\sqrt{9}$. Thus $\sqrt{9} = 3$ and $-\sqrt{9} = -3$. Similarly, $\sqrt{4} = 2$ and $-\sqrt{4} = -2$, $\sqrt{25} = 5$, and $-\sqrt{25} = -5$. We call the symbol $\sqrt{9}$ a *radical* and say that 9 is the *radicand* of the radical $\sqrt{9}$. Similarly, 5 is the radicand of the radical $\sqrt{5}$ and $2x^3y$ is the radicand of the radical $\sqrt{2x^3y}$. The equation $x^2 = a$ where $a > 0$ has, then, the two solutions $\sqrt{a}$ and $-\sqrt{a}$. If $a = 0$, we define $\sqrt{0}$ as 0. In summary: If $a \geqslant 0$, then

$$\overset{\text{radical}}{\sqrt{a}} = x \text{ if } x \geqslant 0 \text{ and } x^2 = a$$

radicand

($\sqrt{9} = 3$ since $3 > 0$ and $3^2 = 9$; $\sqrt{9} \neq -3$ since although $(-3)^2 = 9$ we have $-3 < 0$; but $-\sqrt{9} = -3$.)

Some rational numbers have rational square roots. Thus, for example, $\sqrt{9} = 3$, $\sqrt{\frac{25}{4}} = \frac{5}{2}$, and $\sqrt{0.01} = 0.1$. However, not all rational numbers have rational square roots. For example, $\sqrt{2}$ is not a rational number; that is, there is no rational number x such that $x^2 = 2$, even though we can *approximate* x as a decimal fraction (rational number) to as many decimal places as we like. Thus $(1.4)^2 = 1.96$ and $(1.5)^2 = 2.25$, so that

$$1.4 < \sqrt{2} < 1.5$$

Similarly, $(1.41)^2 = 1.9881$ and $(1.42)^2 = 2.0164$, so that

$$1.41 < \sqrt{2} < 1.42$$

and $(1.414)^2 = 1.999396$ and $(1.415)^2 = 2.002225$, so that

$$1.414 < \sqrt{2} < 1.415$$

How can be be certain that $\sqrt{2}$ is not a rational number? We have
seen that $\sqrt{2} \neq 1.4, 1.5, 1.41, 1.42, 1.414,$ or 1.415, since the square
of none of these numbers is equal to 2. There are, however, infinitely
many more rational numbers to try. How can we be sure that no one
of these is equal to $\sqrt{2}$? We can proceed as follows: Suppose
$\sqrt{2} = \frac{a}{b}$ where a and b are natural numbers. Then

$$(\sqrt{2})^2 = \frac{a^2}{b^2}, \quad 2 = \frac{a^2}{b^2}, \quad 2b^2 = a^2$$

so that

(1) $$a \times a = 2 \times b \times b$$

We will show that (1) is impossible and hence show that we cannot have
$\sqrt{2} = \frac{a}{b}$ where a and b are natural numbers. To do this, we let m be
the number of times that 2 occurs as a factor on the left-hand side,
$a \times a$, of (1) and n the number of times that 2 occurs as a factor on the
right-hand side, $2 \times b \times b$, of (1). Then if (1) is to hold, we must
certainly have $m = n$.

Now what are the possible values of m? We might have $m = 0$, as
would be the case if $a = 17$; we might have $m = 2$, as would be the case
if $a = 18$ ($18 \times 18 = 2 \times 2 \times 9 \times 9$); we might have $m = 4$, as would be
the case if $a = 36$ ($36 \times 36 = 2 \times 2 \times 2 \times 2 \times 9 \times 9$). In any case, m
must be an *even* number. (Remember that 0 is an even number.) But
now, because the right-hand side of (1) is $2 \times b \times b$, a similar analysis
leads to the conclusion that n is an *odd* number. Since no even number
is equal to an odd number, we conclude that $m \neq n$ and hence that
(1) is impossible and that $\sqrt{2} \neq \frac{a}{b}$ for any natural numbers a and b.
That is, $\sqrt{2}$ is not a rational number.

Recall from Chapter 2 that numbers that are not rational are called
irrational. Thus $\sqrt{2}$ is an irrational number. Other irrational numbers
are $\sqrt{3}, \sqrt{5},$ and π (the ratio of the length of the circumference of any
circle to the length of its diameter) and such nonrepeating and
nonterminating decimals as $4.101001000100001 \ldots$ Together, the set
of rational numbers and the set of irrational numbers form the set of
real numbers. To every point on a number line there corresponds a real
number and, conversely, to every real number there corresponds a point
on a given number line.

Notice that in defining $\sqrt{a}$ we have required that a be greater than
or equal to 0. This is because x^2 is nonnegative for any real number x
and so $x^2 = a$ has no real number solution if $a < 0$. For example, if
$x = \sqrt{-4}$ we would need to have $x^2 = -4$. But $2 \times 2 = (-2)(-2) =$

$4 \neq -4$; there is no real number x such that $x^2 = -4$ and hence no real number x such that $x = \sqrt{-4}$. To summarize:

$$\sqrt{a} = x \quad \text{means} \quad \left\{ \begin{array}{l} a \geq 0 \\ x \geq 0 \\ x^2 = a \end{array} \right.$$

EXERCISES 5.3

In exercises 1–20 decide whether the given statement is true or false.

1. $\sqrt{16} = 4$

2. $\sqrt{25} = -5$

3. $\sqrt{9} = 3$

4. $\sqrt{1} = 1$

5. $\sqrt{-1} = 1$

6. $\sqrt{1} = -1$

7. $\sqrt{64} = 8$

8. $\sqrt{0} = 0$

9. $\sqrt{\dfrac{4}{9}} = \dfrac{2}{3}$

10. $\sqrt{\dfrac{9}{4}} = -\dfrac{3}{2}$

11. $\sqrt{\dfrac{-4}{-9}} = \dfrac{2}{3}$

12. $\sqrt{-\dfrac{4}{9}} = \dfrac{2}{3}$

13. $\sqrt{0.1} = 0.1$

14. $\sqrt{0.01} = 0.1$

15. $\sqrt{2} = 1.4$

16. $\sqrt{3} = 1.73$

17. $\sqrt{1.21} = 1.1$

18. $\sqrt{196} = 14$

19. $\sqrt{1.96} = 1.4$

20. $\sqrt{2500} = 500$

In exercises 21–35 solve the given equation.

Example (a): $\sqrt{x} = 5, \quad x = 25$

Example (b): $\sqrt{3x} = 12, \quad 3x = 144$; therefore $x = \dfrac{144}{3} = 48$

21. $\sqrt{x} = 3$

22. $\sqrt{x} = 121$

23. $-\sqrt{x} = -7$

24. $\sqrt{3x} = 81$

25. $\sqrt{5x} = 625$

26. $\sqrt{-x} = 3$

27. $\sqrt{-x} = 5$

28. $\sqrt{x} + 2 = 12$

29. $3 + \sqrt{x} = 28$

30. $\sqrt{\dfrac{x}{3}} = 12$

31. $\sqrt{\dfrac{x}{4}} = 6$

32. $\sqrt{x} + 2 = 13$

33. $\sqrt{x} + 5 = 5$

34. $\sqrt{x} + 12 = 112$

35. $\sqrt{x} = 1.76$

In exercises 36–50 find a rational number equal to the given square root.

36. $\sqrt{49}$

37. $\sqrt{81}$

38. $\sqrt{64}$

39. $\sqrt{169}$

43. $\sqrt{\dfrac{225}{9}}$

47. $\sqrt{0.36}$

40. $\sqrt{225}$

44. $\sqrt{\dfrac{49}{144}}$

48. $\sqrt{0.09}$

41. $\sqrt{\dfrac{9}{16}}$

45. $\sqrt{\dfrac{1}{81}}$

49. $\sqrt{0.0049}$

42. $\sqrt{\dfrac{25}{36}}$

46. $\sqrt{0.25}$

50. $\sqrt{0.000064}$

In exercises 51-60 find, to two decimal places, a rational number approximation to the irrational number given that, to three decimal places, $\sqrt{2} = 1.414$ and $\sqrt{3} = 1.732$.

51. $2\sqrt{3}$

55. $\dfrac{\sqrt{3}}{\sqrt{2}}$

59. $\dfrac{1}{\sqrt{2} + \sqrt{3}}$

52. $3\sqrt{2}$

56. $\dfrac{\sqrt{2}}{\sqrt{3}}$

60. $\dfrac{1}{\sqrt{3} - \sqrt{2}}$

53. $\dfrac{2}{\sqrt{3}}$

57. $\sqrt{2} + \sqrt{3}$

54. $\dfrac{3}{\sqrt{2}}$

58. $\sqrt{3} - \sqrt{2}$

5.4 Multiplication and Division of Square Roots

In developing procedures for computations involving irrational numbers such as square roots, we will assume that the associative and commutative properties of addition and multiplication plus the distributive property of multiplication over addition hold for all real numbers.

For example, to show that

$$\sqrt{a}\sqrt{b} = \sqrt{ab}$$

we first note that, by definition,

$$(\sqrt{a}\sqrt{b})^2 = (\sqrt{a}\sqrt{b})(\sqrt{a}\sqrt{b})$$

Then, by the associative and commutative properties of multiplication, we have

$$(\sqrt{a}\sqrt{b})^2 = (\sqrt{a}\sqrt{b})(\sqrt{a}\sqrt{b}) = (\sqrt{a}\sqrt{a})(\sqrt{b}\sqrt{b})$$

But $\sqrt{a}\sqrt{a} = a$ and $\sqrt{b}\sqrt{b} = b$ by definition, and so it follows that

$$(\sqrt{a}\sqrt{b})^2 = ab$$

Since it is also true that $(\sqrt{ab})^2 = ab$ and $\sqrt{a}\sqrt{b}$ and $\sqrt{ab}$ are both positive numbers, we reach our desired conclusion that

$$\sqrt{a}\sqrt{b} = \sqrt{ab}$$

(Here and elsewhere we will assume that the radicands are nonnegative, i.e., in this case, that $a \geqslant 0$ and $b \geqslant 0$, so that $ab \geqslant 0$.)

Example 1:

$$\sqrt{2}\sqrt{3} = \sqrt{2} \cdot \sqrt{3} = \sqrt{6}; \sqrt{484} = \sqrt{4}\sqrt{121} = 2 \cdot 11 = 22$$

Similarily, if $b \neq 0$ we have

$$\frac{\sqrt{a}}{\sqrt{b}} = \sqrt{\frac{a}{b}}$$

Example 2: $\quad \sqrt{\dfrac{4}{9}} = \dfrac{2}{3} \;$ since $\; \left(\dfrac{2}{3}\right)^2 = \dfrac{4}{9} \;$ and also

$$\sqrt{\frac{4}{9}} = \frac{\sqrt{4}}{\sqrt{9}} = \frac{2}{3}$$

Example 3: $\quad \dfrac{\sqrt{5}}{\sqrt{3}} = \sqrt{\dfrac{5}{3}}$

Any square root can be written in many ways. For example,

$$\sqrt{72} = \sqrt{9 \cdot 8} = \sqrt{9}\sqrt{8} = 3\sqrt{8} = 3\sqrt{4 \cdot 2}$$
$$= 3\sqrt{4}\sqrt{2} = 3 \cdot 2\sqrt{2} = 6\sqrt{2}$$

and

$$\sqrt{\frac{9}{2}} = \frac{\sqrt{9}}{\sqrt{2}} = \frac{3}{\sqrt{2}} = \frac{3\sqrt{2}}{(\sqrt{2})(\sqrt{2})} = \frac{3\sqrt{2}}{2}$$

The last two expressions, $6\sqrt{2}$ and $\frac{3\sqrt{2}}{2}$, are said to be radicals in simplest form. A formal definition of simplest form is as follows:

An expression involving square root radicals is said to be in *simplest form* if:

1. No integral radicand has a factor of the form a^2 where a is an integer greater than 1.

For example, we write

$$\sqrt{32} = \sqrt{16 \cdot 2} = \sqrt{16}\sqrt{2} = 4\sqrt{2} \text{ (Here } a^2 = 4^2 = 16.)$$

and

$$\sqrt{24} = \sqrt{4 \cdot 6} = \sqrt{4}\sqrt{6} = 2\sqrt{6} \text{ (Here } a^2 = 2^2 = 4.)$$

2. A product of radicals is written as one radical.

For example, we write

$$2\sqrt{2}\sqrt{3} = 2\sqrt{6} \text{ and } \sqrt{2}\sqrt{8} = \sqrt{16} = 4$$

3. The radicand is an integer.

For example, we write

$$\sqrt{\frac{2}{9}} = \frac{\sqrt{2}}{\sqrt{9}} = \frac{\sqrt{2}}{3} \ , \ \sqrt{\frac{24}{3}} = \sqrt{8} = 2\sqrt{2} \ ,$$

$$\sqrt{\frac{75}{121}} = \frac{\sqrt{75}}{\sqrt{121}} = \frac{5\sqrt{3}}{11} \ , \text{ and } \sqrt{0.35} = \sqrt{\frac{35}{100}} = \frac{\sqrt{35}}{10}$$

4. No denominator of a fraction involves a radical.

For example, we write

$$\frac{\sqrt{5}}{\sqrt{2}} = \frac{\sqrt{5}}{\sqrt{2}} \cdot \frac{\sqrt{2}}{\sqrt{2}} = \frac{\sqrt{10}}{2} \ , \ \sqrt{\frac{3}{5}} = \frac{\sqrt{3}}{\sqrt{5}} \cdot \frac{\sqrt{5}}{\sqrt{5}} = \frac{\sqrt{15}}{5} \ ,$$

$$\sqrt{\frac{7}{12}} = \frac{\sqrt{7}}{\sqrt{12}} \cdot \frac{\sqrt{3}}{\sqrt{3}} = \frac{\sqrt{21}}{\sqrt{36}} = \frac{\sqrt{21}}{6}$$

Notice that in the last example we saved work by multiplying $\sqrt{12}$ by $\sqrt{3}$ rather than by $\sqrt{12}$. The longer way would involve writing

$$\frac{\sqrt{7}}{\sqrt{12}} \cdot \frac{\sqrt{12}}{\sqrt{12}} = \frac{\sqrt{84}}{12} = \frac{\sqrt{4} \cdot \sqrt{21}}{12} = \frac{2\sqrt{21}}{12} = \frac{\sqrt{21}}{6}$$

When a fraction with a radical in the denominator is rewritten in a form with no radical in the denominator, we say that we have *rationalized the denominator.*

Some of these simplifications make it easier to obtain decimal approximations to expressions involving square roots that are irrational numbers.

Example: Calculate a decimal approximation to $\dfrac{\sqrt{2}}{\sqrt{3}}$ correct to three decimal places.

Hard way: From Table II on page 345, we find that $\sqrt{2} \approx 1.4142$ ("$\approx$" means "approximately equal to") and $\sqrt{3} \approx 1.7321$.

Thus

$$\frac{\sqrt{2}}{\sqrt{3}} \approx \frac{1.4142}{1.7321} \approx 0.8165$$

Easy way: $$\frac{\sqrt{2}}{\sqrt{3}} = \frac{\sqrt{2}}{\sqrt{3}} \cdot \frac{\sqrt{3}}{\sqrt{3}} = \frac{\sqrt{6}}{3}$$

From Table II we have $\sqrt{6} \approx 2.4495$ and so

$$\frac{\sqrt{2}}{\sqrt{3}} = \frac{\sqrt{6}}{3} \approx \frac{2.4495}{3} \approx 0.8165$$

(If a hand calculator is available, the hard way is no longer especially hard — but it certainly is hard without such an aid!)

EXERCISES 5.4

In exercises 1–66 write the given expression in the simplest form according to the definition of simplest form given on pages 170 and 171.

Example (a): $\sqrt{3} \cdot \sqrt{5} = \sqrt{15}$

Example (b): $\sqrt{18} = \sqrt{9 \cdot 2} = \sqrt{9} \cdot \sqrt{2} = 3\sqrt{2}$

Example (c): $\sqrt{\dfrac{2}{5}} = \dfrac{\sqrt{2}}{\sqrt{5}} = \dfrac{\sqrt{2}}{\sqrt{5}} \cdot \dfrac{\sqrt{5}}{\sqrt{5}} = \dfrac{\sqrt{10}}{5}$

1. $\sqrt{3} \cdot \sqrt{7}$	15. $\sqrt{45}$	29. $\sqrt{18} \cdot \sqrt{3}$
2. $\sqrt{5} \cdot \sqrt{6}$	16. $\sqrt{50}$	30. $\sqrt{35} \cdot \sqrt{45}$
3. $2\sqrt{5} \cdot 2\sqrt{3}$	17. $\sqrt{56}$	31. $\sqrt{42} \cdot \sqrt{77}$
4. $3\sqrt{2} \cdot 4\sqrt{3}$	18. $\sqrt{60}$	32. $\sqrt{30} \cdot \sqrt{35}$
5. $2\sqrt{5} \cdot 3\sqrt{7}$	19. $\sqrt{63}$	33. $3\sqrt{10} \cdot 2\sqrt{15}$
6. $3\sqrt{6} \cdot 2\sqrt{5}$	20. $\sqrt{72}$	34. $4\sqrt{6} \cdot 7\sqrt{12}$
7. $(2\sqrt{3})^2$	21. $\sqrt{162}$	35. $3\sqrt{8} \cdot 2\sqrt{2}$
8. $(3\sqrt{5})^3$	22. $\sqrt{147}$	36. $\sqrt{\dfrac{4}{9}}$
9. $(5\sqrt{6})^2$	23. $\sqrt{242}$	37. $\sqrt{\dfrac{27}{3}}$
10. $(2\sqrt{3})^2 \, (3\sqrt{3})^2$	24. $\sqrt{800}$	38. $\sqrt{\dfrac{24}{3}}$
11. $\sqrt{8}$	25. $\sqrt{2025}$	39. $\sqrt{\dfrac{32}{25}}$
12. $\sqrt{12}$	26. $\sqrt{0.2} \cdot \sqrt{10}$	40. $\sqrt{\dfrac{125}{121}}$
13. $\sqrt{20}$	27. $\sqrt{7} \cdot \sqrt{21}$	41. $\sqrt{\dfrac{1}{2}}$
14. $\sqrt{27}$	28. $\sqrt{6} \cdot \sqrt{15}$	42. $\sqrt{\dfrac{5}{7}}$

43. $\sqrt{\dfrac{3}{5}}$

44. $\sqrt{\dfrac{5}{12}}$

45. $\sqrt{\dfrac{5}{2}}$

46. $\dfrac{\sqrt{12}}{\sqrt{3}}$

47. $\dfrac{\sqrt{72}}{\sqrt{2}}$

48. $\dfrac{\sqrt{8}}{\sqrt{3}}$

49. $\dfrac{\sqrt{9}}{\sqrt{3}}$

50. $\dfrac{\sqrt{14}}{\sqrt{7}}$

51. $\dfrac{\sqrt{294}}{\sqrt{6}}$

52. $\dfrac{\sqrt{40}}{\sqrt{5}}$

53. $\dfrac{\sqrt{48}}{\sqrt{3}}$

54. $\dfrac{\sqrt{7}}{\sqrt{10}}$

55. $\dfrac{\sqrt{75}}{\sqrt{8}}$

56. $8\sqrt{\dfrac{5}{18}}$

57. $\dfrac{3\sqrt{18}}{2\sqrt{72}}$

58. $\dfrac{2}{3}\sqrt{\dfrac{9}{27}}$

59. $\dfrac{3}{5}\sqrt{\dfrac{45}{6}}$

60. $\dfrac{2\sqrt{3}}{4\sqrt{7}}$

61. $\dfrac{14}{3}\sqrt{\dfrac{15}{28}}$

62. $\sqrt{\dfrac{3}{5}} \cdot \sqrt{\dfrac{45}{27}}$

63. $\dfrac{2}{3}\sqrt{\dfrac{5}{7}} \cdot \dfrac{7}{5}\sqrt{\dfrac{3}{20}}$

64. $\sqrt{0.9} \cdot \sqrt{0.75}$

65. $\sqrt{2.5} \cdot \sqrt{3.3}$

66. $\sqrt{1.25} \cdot \sqrt{0.45}$

5.5 Simplification of Radicals Involving Variables

Recall that the symbol $\sqrt{a}$ for $a \geqslant 0$ represents the *nonnegative* square root of a, sometimes called the *principal square root*

Thus we have $\sqrt{4} = 2$, even though $(-2)^2 = 4$ as well as $2^2 = 4$. What about $\sqrt{x^2}$? A natural first guess is that $\sqrt{x^2} = x$. This is true for $x \geqslant 0$ as, for example, in $\sqrt{2^2} = \sqrt{4} = 2$. But if $x < 0$ as, for example, in $\sqrt{(-2)^2} = \sqrt{4} = 2 \neq -2$, we do not have $\sqrt{x^2} = x$. What we do have is

$$\sqrt{x^2} = |x|$$

where $|x|$ is the absolute value of x as defined in Section 2.12. Thus

$$\sqrt{7^2} = \sqrt{49} = |7| = 7 \quad \text{and also}$$
$$\sqrt{(-7)^2} = \sqrt{49} = |-7| = 7$$

In this introductory treatment, however, we will avoid the complications of absolute value by assuming that the replacement set of each variable is the set of nonnegative real numbers. Under this assumption, then,

$$\sqrt{x^2} = x, \quad \sqrt{x^2 y^2} = \sqrt{x^2}\sqrt{y^2} = xy,$$
$$\sqrt{x^3} = \sqrt{x^2 \cdot x} = \sqrt{x^2} \cdot \sqrt{x} = x\sqrt{x}$$

and so on.

The following three examples illustrate the simplification, under this assumption, of radicals involving variables.

Example 1: $\sqrt{x^5} = \sqrt{x^4 \cdot x} = \sqrt{x^4}\sqrt{x} = x^2\sqrt{x}$

Example 2: $\sqrt{18x^3y^5} = \sqrt{9 \cdot 2 \cdot x^2 \cdot x \cdot y^4 \cdot y}$

$$= \sqrt{9} \cdot \sqrt{x^2} \cdot \sqrt{y^4} \cdot \sqrt{2xy} = 3xy^2\sqrt{2xy}$$

Example 3: $\sqrt{\dfrac{x^2}{y^2}} = \sqrt{\dfrac{x^2}{y^3} \cdot \dfrac{y}{y}} = \dfrac{\sqrt{x^2 y}}{\sqrt{y^4}} = \dfrac{\sqrt{x^2}\sqrt{y}}{\sqrt{y^4}} = \dfrac{x\sqrt{y}}{y^2}$

(Here, of course, $y \neq 0$, and we make such assumptions whenever variables appear in the denominator.)

EXERCISES 5.5

In exercises 1–54 simplify the given expression under the assumption that the replacement set for each variable is the set of nonnegative real numbers.

1. $\sqrt{4x^2}$

2. $\sqrt{16y^2}$

3. $\sqrt{9b^3}$

4. $\sqrt{25x^3}$

5. $\sqrt{24y^3}$

6. $\sqrt{32a^3}$

7. $\sqrt{36x^3y}$

8. $\sqrt{16xy^3}$

9. $\sqrt{18x^3y^3}$

10. $\sqrt{8b^2x}$

11. $\sqrt{3x} \cdot \sqrt{6x}$

12. $\sqrt{10a} \cdot \sqrt{5a}$

13. $3\sqrt{12x^3} \cdot 2\sqrt{3y}$

14. $4\sqrt{2x^2} \cdot 5\sqrt{6y^5}$

15. $2\sqrt{32a^2x} \cdot 3\sqrt{50a^3x^2}$

16. $4\sqrt{18x^5} \cdot 5\sqrt{12x^3}$

17. $\sqrt{12x^3y^5}$

18. $\sqrt{28z^3b^5}$

19. $\sqrt{x^6y^3}$

20. $\sqrt{x^3y^8}$

21. $\sqrt{8a^7b^2}$

22. $\sqrt{18x^9y^2}$

23. $\sqrt{54x^{10}y^8}$

24. $\sqrt{16a^4b^9}$

25. $\sqrt{50x^5y^6}$

26. $\sqrt{0.25x^2y}$

27. $\sqrt{0.49ab^2}$

28. $\sqrt{0.09x^4y^5}$

29. $\sqrt{0.27x^2y^2}$

30. $\sqrt{0.50x^{10}y^7}$

31. $\sqrt{\dfrac{4x^2}{25}}$

32. $\sqrt{\dfrac{9y^2}{16}}$

33. $\sqrt{\dfrac{3a^2}{2}}$

44. $\sqrt{\dfrac{8x^8}{y^6}}$

34. $\sqrt{\dfrac{2x^2}{3}}$

45. $\sqrt{\dfrac{18x^5}{25y^7}}$

35. $\sqrt{\dfrac{3a^5}{b^4}}$

46. $\sqrt{\dfrac{54x^7}{y^8}}$

36. $\sqrt{\dfrac{5x^3}{y^6}}$

47. $\sqrt{\dfrac{16x^8y^{10}}{8x^4y^5}}$

37. $\sqrt{\dfrac{x^5y^3}{3a}}$

48. $\sqrt{\dfrac{81x^9y^{25}}{27x^3y^{17}}}$

38. $5\sqrt{\dfrac{a^3b^5}{c^2d}}$

49. $\dfrac{2}{\sqrt{3x}}$

39. $\dfrac{5x}{2}\sqrt{\dfrac{5y^2}{2x^3}}$

50. $\dfrac{5}{\sqrt{2a}}$

40. $\dfrac{2y}{3}\sqrt{\dfrac{7y^5}{3x^3}}$

51. $\dfrac{5x}{\sqrt{75x^3}}$

41. $\sqrt{\dfrac{2}{7x^3}}$

52. $\sqrt{(x+y)^2}$

42. $\sqrt{\dfrac{3}{5a^5}}$

53. $\sqrt{(x-y)^4}$

43. $7x\sqrt{\dfrac{50x^5y^6}{48xy}}$

54. $\sqrt{(x^2+y^2)^2}$

5.6 Addition and Subtraction of Square Roots

The methods we use to simplify square roots are often helpful in simplifying sums and differences of square roots, as shown in the following examples.

Example 1: $\sqrt{3} + \sqrt{12} = 1 \cdot \sqrt{3} + 2\sqrt{3} = (1 + 2)\sqrt{3} = 3\sqrt{3}$
(Note the use of the distributive property.)

Example 2: $\sqrt{63} - \sqrt{28} = 3\sqrt{7} - 2\sqrt{7} = (3 - 2)\sqrt{7} = \sqrt{7}$

Example 3: $8\sqrt{\dfrac{1}{2}} - \dfrac{1}{2}\sqrt{8} = 8\sqrt{\dfrac{1}{2} \cdot \dfrac{2}{2}} - \dfrac{1}{2}\sqrt{4 \cdot 2}$

$$= \dfrac{8}{2}\sqrt{2} - \dfrac{2}{2}\sqrt{2} = (4 - 1)\sqrt{2} = 3\sqrt{2}$$

Example 4: $5\sqrt{xy^3} + y\sqrt{4xy} = 5y\sqrt{xy} + 2y\sqrt{xy} = (5y + 2y)\sqrt{xy}$

$$= 7y\sqrt{xy}$$

Example 5: $y^3\sqrt{\dfrac{a}{y^3}} + \sqrt{4ay} = y^3\sqrt{\dfrac{ay}{y^4}} + 2\sqrt{ay}$

$$= \dfrac{y^3}{y^2}\sqrt{ay} + 2\sqrt{ay} = (y + 2)\sqrt{ay}$$

EXERCISES 5.6

In exercises 1–35 simplify the given expression when possible.

Example (a): $2\sqrt{8} + 3\sqrt{2} = 2 \cdot \sqrt{4 \cdot 2} + 3\sqrt{2}$

$$= 2 \cdot 2\sqrt{2} + 3\sqrt{2}$$

$$= 4\sqrt{2} + 3\sqrt{2}$$

$$= 7\sqrt{2}$$

Example (b): $\sqrt{6} + \sqrt{18} = \sqrt{6} + 3\sqrt{2}$ (Not possible to simplify further.)

1. $3\sqrt{2} + 5\sqrt{2}$
2. $5\sqrt{3} - 3\sqrt{3}$
3. $4\sqrt{5} + 2\sqrt{5}$
4. $3\sqrt{2} - 7\sqrt{2}$
5. $2\sqrt{5} - 3\sqrt{5}$
6. $2\sqrt{3} - 4\sqrt{5}$
7. $\sqrt{12} - \sqrt{48}$
8. $\sqrt{75} + \sqrt{12}$
9. $\sqrt{50} + \sqrt{8}$
10. $\sqrt{27} + \sqrt{48}$
11. $\sqrt{45} - \sqrt{20}$
12. $\sqrt{98} - \sqrt{72}$
13. $\sqrt{80} + \sqrt{45}$
14. $\sqrt{128} + \sqrt{18}$
15. $\sqrt{108} - \sqrt{48}$
16. $4\sqrt{12} - 6\sqrt{27}$
17. $5\sqrt{8} - 3\sqrt{18}$
18. $-7\sqrt{98} + 5\sqrt{50}$

19. $6\sqrt{28} + 2\sqrt{63}$
20. $-3\sqrt{75} + 4\sqrt{27}$
21. $2\sqrt{40} - 3\sqrt{90}$
22. $9\sqrt{8} - \sqrt{72}$
23. $2\sqrt{98} - 5\sqrt{18}$
24. $\sqrt{80} - \sqrt{320}$
25. $\dfrac{1}{3}\sqrt{288} - \dfrac{2}{5}\sqrt{450}$
26. $2\sqrt{8} - 3\sqrt{2} + \sqrt{6}$
27. $3\sqrt{18} - 5\sqrt{2} + \sqrt{3}$
28. $3\sqrt{5} - 2\sqrt{3} + \sqrt{20} - \sqrt{12}$
29. $\sqrt{6} - \sqrt{20} - 3\sqrt{45} - \sqrt{180}$
30. $3\sqrt{12} - 5\sqrt{20} + 7\sqrt{28} - 4\sqrt{18}$
31. $-4\sqrt{20} - 2\sqrt{72} + \sqrt{45} - \sqrt{18}$
32. $2\sqrt{32} - 5\sqrt{72} + 2\sqrt{40} - 3\sqrt{90}$
33. $5\sqrt{50} - 3\sqrt{75} - 2\sqrt{125}$
34. $4\sqrt{63} - 5\sqrt{28} + 3\sqrt{20}$
35. $5\sqrt{75} + 7\sqrt{108} - 6\sqrt{245}$

In exercises 36–50 simplify the given expression under the assumption that the replacement set for each variable is the set of positive real numbers.

Example (a):
$$3\sqrt{4x} - \sqrt{x} = 3 \cdot 2\sqrt{x} - \sqrt{x}$$
$$= 6\sqrt{x} - 1 \cdot \sqrt{x}$$
$$= (6 - 1)\sqrt{x}$$
$$= 5\sqrt{x}$$

Example (b):
$$6\sqrt{9x^3} - 2\sqrt{4x^3} = 6 \cdot 3x\sqrt{x} - 2 \cdot 2x\sqrt{x}$$
$$= 18x\sqrt{x} - 4x\sqrt{x}$$
$$= (18x - 4x)\sqrt{x}$$
$$= 14x\sqrt{x}$$

Example (c):
$$\sqrt{\frac{a^2}{4} + \frac{a^2}{9}} = \sqrt{\frac{9a^2 + 4a^2}{36}}$$
$$= \sqrt{\frac{13a^2}{36}}$$
$$= \frac{a}{6}\sqrt{13}$$

36. $2\sqrt{9y} - \sqrt{y}$

37. $4\sqrt{16x^3} - 3\sqrt{9x^3}$

38. $2x\sqrt{8x} - \sqrt{2x^3}$

39. $2\sqrt{a} - \sqrt{4a} + \sqrt{8a}$

40. $\sqrt{72x^3} + \sqrt{8x^3} - \sqrt{18x^3}$

41. $4\sqrt{x^3} - 2\sqrt{x^5} + \sqrt{x}$

42. $\sqrt{32a^2x} - \sqrt{8b^2x} + \sqrt{50x}$

43. $\sqrt{2y^3} - 8\sqrt{8y^5} + y^2\sqrt{98y}$

44. $\sqrt{72b^2x} + \sqrt{98x} - \sqrt{18a^2x}$

45. $\sqrt{36x^3y} + \sqrt{16xy^3} + \sqrt{18xy} - \sqrt{8x^3y}$

46. $\sqrt{\dfrac{x^2}{4} + \dfrac{x^2}{9}}$

47. $\sqrt{\dfrac{a^2}{4} + \dfrac{a^2}{16}}$

48. $\sqrt{\dfrac{2y^2}{7} + \dfrac{3y^2}{2}}$

49. $\dfrac{1}{y}\sqrt{4xy} - \sqrt{\dfrac{x}{y}}$

50. $\dfrac{4}{a}\sqrt{9a^3b} - \sqrt{\dfrac{4a}{b}}$

In exercises 51-60 perform the indicated operations and express your answer in simplest form.

Example: $\sqrt{\dfrac{1}{2}} + 2\sqrt{2} = \sqrt{\dfrac{1 \cdot 2}{2 \cdot 2}} + 2\sqrt{2}$

$$= \dfrac{1}{2}\sqrt{2} + 2\sqrt{2}$$

$$= \dfrac{5}{2}\sqrt{2}$$

51. $\sqrt{\dfrac{1}{3}} + 3\sqrt{3}$

52. $\sqrt{\dfrac{3}{2}} - \dfrac{1}{\sqrt{6}}$

53. $8\sqrt{\dfrac{1}{2}} - \dfrac{1}{2}\sqrt{8}$

54. $6\sqrt{\dfrac{3}{4}} - \sqrt{\dfrac{4}{3}} - \sqrt{27}$

55. $5\sqrt{\dfrac{1}{2}} - \sqrt{\dfrac{1}{4}} + \sqrt{\dfrac{1}{8}}$

56. $\sqrt{\dfrac{16}{3}} + \dfrac{1}{\sqrt{3}} - \sqrt{\dfrac{1}{12}}$

57. $\sqrt{\dfrac{2}{3} + \dfrac{3}{2}} + 2 - \sqrt{\dfrac{3}{2}}$

58. $\sqrt{\dfrac{12}{5}} + \sqrt{\dfrac{20}{3}} + \sqrt{\dfrac{15}{4}}$

59. $12\sqrt{\dfrac{2}{3}} - 2\sqrt{1\dfrac{1}{2}} - \dfrac{1}{3}\sqrt{24}$

60. $\sqrt{\dfrac{25}{5}} + \sqrt{\dfrac{1}{5}} - \sqrt{\dfrac{9}{20}}$

In exercises 61-70 find the solution of the given equation, express it in simplest form, and check your solution.

Example (a): $2\sqrt{48} + 4x = 3\sqrt{27}$

$$4x = 3\sqrt{27} - 2\sqrt{48}$$

$$4x = 3 \cdot 3\sqrt{3} - 2 \cdot 4\sqrt{3}$$

$$4x = \sqrt{3}$$

$$x = \dfrac{1}{4}\sqrt{3}$$

Check: $2\sqrt{48} + 4 \cdot \frac{1}{4}\sqrt{3} \overset{?}{=} 3\sqrt{27}$

$$8\sqrt{3} + \sqrt{3} \overset{?}{=} 9\sqrt{3}$$

$$9\sqrt{3} = 9\sqrt{3} \quad \text{(checks)}$$

Example (b): $\quad \sqrt{2x} + 6 = 18$

$$\sqrt{2x} = 18 - 6$$

$$\sqrt{2x} = 12$$

$$(\sqrt{2x})^2 = (12)^2$$

$$2x = 144$$

$$x = 72$$

Check: $\quad \sqrt{2 \cdot 72} + 6 \overset{?}{=} 18$

$$\sqrt{144} + 6 \overset{?}{=} 18$$

$$12 + 6 \overset{?}{=} 18$$

$$18 = 18 \quad \text{(checks)}$$

61. $2\sqrt{24} + 3x = 3\sqrt{54}$
62. $5\sqrt{3} - 6x = 2\sqrt{3}$
63. $4\sqrt{5} + 5x = 7\sqrt{5} + 7x$
64. $\sqrt{\dfrac{16}{3}} = 6y + \sqrt{\dfrac{1}{12}} - \dfrac{1}{\sqrt{3}}$
65. $6\sqrt{3} + 2x = 3\sqrt{3} + 5x$

66. $\sqrt{3}\, x + 6 = 18$
67. $\sqrt{3}\, a - 7 = 2$
68. $5\sqrt{7}\, b + 2 = 7\sqrt{7}\, b + 10$
69. $4\sqrt{5}\, y + 8 = 3\sqrt{5}\, y + 8$
70. $\dfrac{2}{3}x + 2\sqrt{180} = 4\sqrt{125}$

5.7 Other Roots

We know that the equation $x^2 = a$ for $a > 0$ has two solutions, $\sqrt{a}$ and $-\sqrt{a}$. Consider now the equation $x^3 = a$. The solution of this equation is called the *cube root* of a. Thus, since $2^3 = 8$, 2 is the cube root of 8, and since $(-2)^3 = -8$, -2 is the cube root of -8. We write $\sqrt[3]{8} = 2$ and $\sqrt[3]{-8} = -2$. In contrast to square roots, every real number, whether positive, negative, or zero, has a single real number cube root although, as was the case for square roots, many of them are irrational numbers.

Similarly, we can consider *fourth roots, fifth roots,* and so on. If $x^4 = a$, then x is a fourth root of a, and if $x^5 = a$, then x is a fifth root of a. As was the case for square roots, any positive real number has two real number fourth roots. Thus, $2^4 = 16$ and also $(-2)^4 = 16$. But the symbol $\sqrt[4]{a}$ for $a \geqslant 0$ will always refer to the nonnegative root, so that $\sqrt[4]{16} = 2$. Furthermore, only nonnegative numbers have real numbers as fourth roots, because the equation $x^4 = a$ has no real number solution if $a < 0$: $x^4 = x^2 \cdot x^2$ is nonnegative for all real numbers x.

On the other hand, fifth roots are like cube roots in that every real number has an unique real number fifth root. We have, for example,

$$\sqrt[5]{32} = 2$$

since $2^5 = 32$ and for no real number $x \neq 2$ is it true that $x^5 = 32$. Likewise

$$\sqrt[5]{-32} = -2$$

since $(-2)^5 = -32$ and for no real number $x \neq -2$ is it true that $x^5 = -32$.

We call the symbol $\sqrt[n]{a}$ a *radical*, a the *radicand* of the radical, and n the *index* of the radical. Thus

$$\overset{\displaystyle \text{the } \textit{index}}{\sqrt[7]{55}} \quad \text{is a } \textit{radical}$$

the *radicand*

If $\sqrt[n]{a} = x$, then $x^n = a$ for $n = 2, 3, 4, \ldots$. So, if $\sqrt{a} = x$, then $x^2 = a$ and the index of $\sqrt{a}$ is 2. That is, $\sqrt{a} = \sqrt[2]{a}$. For square roots, however, it is customary to omit writing the index.

EXERCISES 5.7

In exercises 1–12 determine the index and the radicand of the given radical.

1. $\sqrt[3]{5}$ 4. $\sqrt[10]{100}$ 7. $\sqrt[27]{1372}$ 10. $\sqrt[6]{x^2 y}$

2. $\sqrt[5]{a}$ 5. $\sqrt[7]{xy}$ 8. $\sqrt{x}$ 11. $\sqrt[3]{\dfrac{1}{8}}$

3. $\sqrt{3}$ 6. $\sqrt[3]{15}$ 9. $\sqrt[4]{32}$ 12. $\sqrt[5]{\dfrac{1}{25}}$

In exercises 13–28 find a rational number equal to the given radical.

13. $\sqrt[3]{8}$ 15. $\sqrt[4]{16}$ 17. $\sqrt[3]{-27}$ 19. $\sqrt[4]{256}$

14. $\sqrt[3]{-8}$ 16. $\sqrt[5]{32}$ 18. $\sqrt[4]{81}$ 20. $\sqrt{\dfrac{16}{25}}$

21. $\sqrt[3]{\dfrac{8}{-27}}$ 23. $\sqrt[3]{\dfrac{-125}{64}}$ 25. $\sqrt[5]{1024}$ 27. $\sqrt[4]{\dfrac{81}{625}}$

22. $\sqrt[4]{\dfrac{16}{81}}$ 24. $\sqrt[4]{\dfrac{625}{256}}$ 26. $\sqrt[5]{-3125}$ 28. $\sqrt[5]{\dfrac{1024}{3125}}$

In exercises 29–44 perform the indicated computation to obtain a rational number as an answer.

Example (a): $\sqrt[3]{27} + \sqrt[4]{16} = 3 + 2 = 5$

Example (b): $\sqrt[3]{27} \cdot \sqrt[4]{16} = 3 \cdot 2 = 6$

29. $\sqrt[3]{8} + \sqrt[3]{27}$ 35. $\sqrt[5]{32} + \sqrt[3]{64}$ 41. $\sqrt[5]{-3125} \cdot \sqrt[4]{\dfrac{81}{16}}$

30. $\sqrt[3]{-8} + \sqrt[3]{-27}$ 36. $\sqrt[3]{8} \cdot \sqrt[3]{125}$ 42. $\dfrac{\sqrt[4]{256}}{\sqrt[3]{-64}}$

31. $\sqrt[4]{16} + \sqrt[4]{81}$ 37. $\sqrt[3]{-125} \cdot \sqrt[3]{-27}$ 43. $\dfrac{\sqrt[5]{1024}}{\sqrt[4]{256}}$

32. $\sqrt[4]{16} - \sqrt[4]{81}$ 38. $\sqrt[4]{625} \cdot \sqrt[3]{-8}$ 44. $\dfrac{\sqrt[3]{-125}}{\sqrt[4]{81}}$

33. $\sqrt[3]{125} + \sqrt[4]{256}$ 39. $\sqrt[4]{\dfrac{16}{81}} \cdot \sqrt[3]{\dfrac{125}{64}}$

34. $\sqrt[3]{-125} - \sqrt[4]{256}$ 40. $\sqrt[5]{\dfrac{-32}{243}} \cdot \sqrt[4]{\dfrac{256}{625}}$

REVIEW EXERCISES

Section 5.1

In exercises 1–20 simplify the given expression.

1. $P^5 \cdot P^6$
2. $x^{13} \cdot x^2$
3. $(2x^2 y)(3x^3 y^2)$
4. $(-2a^2 b^3)(-3a^3 b^2)$
5. $(5xyz)^2 (3x^2 yz^3)$
6. $y^a \cdot y^{2a}$

7. $(x^2)^{13}(13x)^2$
8. $(-2a^2 b^3)^3$
9. $(-x^3)^8 (-x^2)^4$
10. $\left(\dfrac{1}{2}y^2\right)\left(\dfrac{2}{3}x^2 y^3\right)$
11. $\left(\dfrac{-3}{4}x^2 y^2\right)\left(\dfrac{3}{2}x^2 y^3\right)^2$
12. $x^{3m} \cdot x^{4m}$

13. $(x^m)^4$

17. $\dfrac{8x^3y^4}{12x^2y^2}$

14. $\dfrac{y^8}{y^2}$

18. $\dfrac{120p^8q^6t^4}{-45p^4q^6z^2}$

15. $\dfrac{x^2y^3z^5}{xyz}$

19. $\dfrac{132x^5y^3z^7}{22x^4y^2}$

16. $\left(\dfrac{3a^2b}{4ab}\right)^2$

20. $\dfrac{x^{2n+3}}{x^2}$

Section 5.2

In exercises 21–40 simplify the given expression. Write all answers without negative exponents.

21. $4x^0$

31. $(2x^0y^2)^3\,(y-2)^2\,(x^3y^2)^0$

22. $2^{-3}\cdot 5$

32. $(x^{-3}y^{-2})^{-3}$

23. $\dfrac{x^0+4^{-2}}{y^0+2^{-3}}$

33. $\dfrac{x^{-16}}{x^{-17}}$

24. $(3^{-1}+3^{-2})^{-2}$

34. $\dfrac{(x^{-3})^2}{(x^4)^{-2}}$

25. $\dfrac{(a+b)^0}{9^{-1}}$

35. $\dfrac{6x^4y^3}{4x^{-3}y^{-3}}$

26. $10^3\cdot 10^0\cdot 10^{-2}$

36. $\dfrac{5a^{-2}b}{3ab^{-3}}$

27. $\dfrac{10^7\cdot 10^{-4}}{10^3}$

37. $\dfrac{(xy^3)^{-2}}{(x^2y)^{-3}}$

28. $(26^0+37^0)^{-5}$

38. $\dfrac{x^{-3}y}{x^{-5}y^{-2}}$

29. $(2x^3y^2)(3x^{-2}y^{-3})$

39. $\dfrac{(ab^2)^{-3}\,(a^2b^{-1})^2}{(a^3b^4)^{-1}}$

30. $(3y^{-2})^{-3}$

40. $\dfrac{(x+y)^{-1}}{(a-b)^{-1}}$

Section 5.3

In exercises 41–50 find a rational number equal to the given square root.

41. $\sqrt{36}$

43. $\sqrt{400}$

42. $\sqrt{121}$

44. $\sqrt{6400}$

45. $\sqrt{\dfrac{25}{4}}$ 48. $\sqrt{0.16}$

46. $\sqrt{\dfrac{1}{9}}$ 49. $\sqrt{0.0004}$

47. $\sqrt{\dfrac{36}{16}}$ 50. $\sqrt{4.1616}$

In exercises 51-55 find, to two decimal places, a rational number approximation to the irrational number given that, to three decimal places, $\sqrt{3} = 1.732$ and $\sqrt{5} = 2.236$.

51. $2\sqrt{3}$ 54. $\sqrt{3} + \sqrt{5}$

52. $3\sqrt{5}$

53. $\dfrac{3}{\sqrt{5}}$ 55. $\dfrac{\sqrt{3}}{\sqrt{5}}$

Section 5.4

In exercises 56-70 simplify the given expression.

56. $\sqrt{32}$ 61. $\sqrt{72} \cdot \sqrt{32}$ 66. $\sqrt{\dfrac{3}{2}}$

57. $\sqrt{12}$ 62. $3\sqrt{6} \cdot \sqrt{27}$ 67. $\sqrt{\dfrac{1}{3}}$

58. $\sqrt{75}$ 63. $2\sqrt{5} \cdot \sqrt{15}$ 68. $\sqrt{\dfrac{3}{8}}$

59. $\sqrt{27}$ 64. $2\sqrt{8} \cdot 3\sqrt{10}$ 69. $\sqrt{\dfrac{9}{2}}$

60. $\sqrt{800}$ 65. $5\sqrt{6} \cdot \sqrt{24}$ 70. $\sqrt{\dfrac{2}{27}}$

Section 5.5

In exercises 71-85 simplify the given expression under the assumption that the replacement set for each variable is the set of positive real numbers.

71. $\sqrt{12x^3}$ 74. $\sqrt{98xy^3}$ 77. $\sqrt{3a} \cdot \sqrt{6a^3}$

72. $\sqrt{16t^6}$ 75. $\sqrt{28x^2y}$ 78. $\sqrt{ab^3} \cdot \sqrt{ab}$

73. $\sqrt{32x^4}$ 76. $\sqrt{2x} \cdot \sqrt{8x}$ 79. $\sqrt{xy^2} \cdot \sqrt{x^3y^3}$

80. $\sqrt{\dfrac{3}{5x}}$

82. $\sqrt{\dfrac{a^3}{2}}$

84. $\sqrt{\dfrac{3a}{2}}$

81. $\sqrt{\dfrac{5}{12y}}$

83. $\sqrt{\dfrac{4c^3}{3}}$

85. $\sqrt{\dfrac{3}{8x^3}}$

Section 5.6

In exercises 86–95 simplify the given expression.

86. $\sqrt{32} + \sqrt{72}$

90. $2\sqrt{3} - \sqrt{75}$

94. $\sqrt{\dfrac{1}{3}} + \sqrt{27}$

87. $\sqrt{18} - \sqrt{50}$

91. $\sqrt{27} + 5\sqrt{3}$

95. $\sqrt{\dfrac{2}{5}} + \sqrt{40}$

88. $\sqrt{75} - \sqrt{27}$

92. $\sqrt{72} - \sqrt{32}$

89. $\sqrt{24} - 3\sqrt{6}$

93. $\sqrt{\dfrac{1}{2}} - \sqrt{\dfrac{1}{8}}$

In exercises 96–110 simplify the given expression under the assumption that the replacement set for each variable is the set of positive real numbers.

96. $\sqrt{8x^2} - x\sqrt{18}$

97. $3\sqrt{18x^2} + \sqrt{50x^2}$

98. $3a\sqrt{16a} - \sqrt{144a^3}$

99. $2x\sqrt{25xy^2} - 3y\sqrt{4x^3}$

100. $n\sqrt{8m^3n} - m\sqrt{32mn^3}$

101. $\sqrt{12x^2} + 4\sqrt{3x^2}$

102. $y\sqrt{4y} - 25\sqrt{25y^3} + \sqrt{81y^3}$

103. $\sqrt{18x^4} - 3\sqrt{8x^4} + 2\sqrt{50x^4}$

104. $5\sqrt{\dfrac{x}{2}} + \dfrac{1}{4}\sqrt{8x} - \sqrt{\dfrac{25x}{2}}$

105. $\sqrt{2a^2} - \sqrt{18a^2} - \dfrac{1}{2}a\sqrt{8}$

106. $\sqrt{32x} - \sqrt{72x} + 2\sqrt{2x}$

107. $13\sqrt{4x^2} - 2\sqrt{9x^2} - 4\sqrt{121x^2}$

108. $\dfrac{1}{x}\sqrt{16xy} - \sqrt{\dfrac{4y}{x}}$

109. $\dfrac{8}{ab}\sqrt{81ab^3} - \sqrt{\dfrac{4b}{a}}$

110. $\sqrt{\dfrac{x^2}{16} + \dfrac{x^2}{25}}$

Section 5.7

In exercises 111–140 simplify the given expression to obtain a rational number as the answer.

111. $\sqrt[3]{125}$

112. $\sqrt{625}$

113. $\sqrt[3]{-125}$

114. $\sqrt[4]{625}$

115. $\sqrt[5]{243}$

116. $\sqrt[4]{\dfrac{625}{1296}}$

117. $\sqrt[3]{\dfrac{64}{125}}$

118. $\sqrt[5]{\dfrac{3125}{7776}}$

119. $\sqrt[3]{\dfrac{-125}{216}}$

120. $\sqrt{9} + \sqrt[3]{27}$

121. $\sqrt[3]{-27} - \sqrt{9}$

122. $\sqrt[5]{32} - \sqrt[5]{-32}$

123. $\sqrt[4]{256} + \sqrt[3]{216}$

124. $\sqrt[5]{32} + \sqrt[5]{1024}$

125. $\sqrt[3]{\dfrac{8}{27}} + \sqrt{\dfrac{16}{25}}$

126. $\sqrt[4]{\dfrac{81}{256}} - \sqrt[3]{\dfrac{-64}{125}}$

127. $\sqrt[5]{\dfrac{1024}{32}} + \sqrt[3]{\dfrac{216}{343}}$

128. $\sqrt[4]{\dfrac{16}{81}} - \sqrt[3]{\dfrac{-8}{27}}$

129. $\sqrt{16} \cdot \sqrt[4]{16}$

130. $\sqrt[5]{32} \cdot \sqrt[3]{343}$

131. $\sqrt{49} \cdot \sqrt[3]{-343}$

132. $\sqrt{\dfrac{4}{9}} \cdot \sqrt[3]{\dfrac{8}{27}}$

133. $\sqrt[3]{\dfrac{-64}{125}} \cdot \sqrt[4]{\dfrac{81}{256}}$

134. $\sqrt[5]{\dfrac{1024}{3125}} \cdot \sqrt[3]{-\dfrac{27}{64}}$

135. $\dfrac{\sqrt[4]{81}}{\sqrt{36}}$

136. $\dfrac{\sqrt[5]{1024}}{\sqrt[4]{16}}$

137. $\dfrac{\sqrt[3]{-125}}{\sqrt[3]{-8}}$

138. $(\sqrt[3]{8} \cdot \sqrt[4]{16}) \cdot \sqrt[4]{256}$

139. $\sqrt{25} \, (\sqrt[4]{81} + \sqrt[3]{-125})$

140. $\sqrt[5]{3125} \, (\sqrt[3]{-27} - \sqrt[5]{243})$

CUMULATIVE REVIEW C

1. $2 + 3 \cdot 5 + 2 = ?$

2. $3 \, (5 + x) = ?$

3. $-7a \cdot 9ab = ?$

4. $\dfrac{4}{5} - \dfrac{5 - 9}{7} = ?$

5. $\left(\dfrac{-5}{-3}\right)\left(\dfrac{-6}{7}\right) = ?$

6. If r is a rational number, then $-r$ is called the ___?___ of r.

7. What decimal is equal to $\dfrac{3}{7}$?

8. What fraction in lowest terms is equal to $0.\overline{1}$?

In exercises 9–22 find the solution set of the given equation or inequality. For each inequality graph the solution set on a number line.

9. $7 - 3x = 22 - 8x$

10. $14x - 5 + 2x = -5x + 16x + 18x$

11. $-8 + 5x - 6 = x + 14$

12. $-3x + 7 = 2x + 7$

13. $2(x + 5) - 4 = x - 1$

14. $2(x - 4) = 7(2x - 1) - 13$

15. $(5x + 2) - 3 = (x + 4) + (4x - 3)$

16. $16 + 3x = 4x - 2(x - 8)$

17. $x - 5 = 5 - x$

18. $12 - 8x < 20 - 4x$

19. $12x + 4 < 8(2x - 1)$

20. $(3x + 4) + (2x + 7) < 3x + 5$

21. $x - (4 - x) > 3x + 5$

22. $x(2x + 8) < 2(x^2 + 8)$

In exercises 23–28 graph the given equations on the same coordinate plane.

23. $x + y = 5$ 25. $x = 2$ 27. $x + y = -2$

24. $2y - x = 2$ 26. $y = 2$ 28. $x - 2y = -10$

In exercises 29 and 30 solve the given system of equations by graphing.

29. $\begin{cases} 2x + y = 6 \\ y - x = 3 \end{cases}$ 30. $\begin{cases} 2x - 3y = 1 \\ x + 2y = 4 \end{cases}$

In exercises 31–34 find the solution set of the given system of equations.

31. $\begin{cases} x + y = 7 \\ 2x - y = -1 \end{cases}$ 33. $\begin{cases} 2x - 3y = -14 \\ 3x + 7y = 48 \end{cases}$

32. $\begin{cases} 4x + 5y = 7 \\ 3x + 4y = 6 \end{cases}$ 34. $\begin{cases} 6x + 2y = -3 \\ 5x - 3y = -6 \end{cases}$

35. A small cake and a large cake together make 15 servings. Ten of the small cakes and 6 of the large cakes serve exactly 110 people. How many servings are there in each type of cake?

36. How many cubic centimeters of an acid and water solution that is 50% acid must be mixed with 400 cubic centimeters of a 24% acid solution to make a 40% acid solution?

37. John can do a job in 6 days, Geraldine can do the same job in 8 days, and Joe can do the same job in 12 days. How long will it take them to do the job working together?

38. The sum of two numbers is 50. The second number is 2 more than the first. What are the two numbers?

39. A car left a service station traveling at a rate of 45 miles per hour. Two hours later a second car left the same station going in the same direction but traveling at a rate of 60 miles per hour. Assuming that the rates will remain constant, find how long it will take the second car to overtake the first car.

40. A total of $5000 is invested in two companies, A and B. Company A pays a 5% return on the amount invested, and company B pays 6%. If the total return is $262, how much is invested in each company?

In exercises 41–44 write the given expression with positive exponents and simplify.

41. $\dfrac{3^2 - 2^2}{3^{-1} + 2^{-1}}$

42. $\dfrac{6^{-1}x^4}{3x^{-2}}$

43. $\dfrac{24x^3 y^2 z}{8x^2 yz^3}$

44. $(a^2 m^3)^{-2}$

In exercises 45–50 simplify the given expression.

45. $\sqrt{27}$

46. $\sqrt{\dfrac{1}{8}}$

47. $\sqrt{75} + 2\sqrt{80}$

48. $\sqrt{3} \cdot \sqrt{6}$

49. $\sqrt[4]{81}$

50. $\sqrt{112x^4 y^{-2}}$ $(x, y \geq 0)$

6

POLYNOMIALS

Expressions such as $4x^2$, $\frac{1}{2}x^5$, $\sqrt{7}\,x$, and, in general, ax^n where a is a real number and n is a natural number are called *monomials* in x over the real numbers. Similarly, $5y$, $\frac{2}{3}y^2$, and πy^5 are monomials in y over the real numbers. The real number a is called the *coefficient* and the natural number n is called the *degree* of the monomial. Note that since $x = x^1$, the degree of ax is 1. Thus

$$4x^2 \overset{\text{degree } 2}{\underset{\text{coefficient } 4}{}} \, , \quad \frac{1}{2}x^5 \overset{\text{degree } 5}{\underset{\text{coefficient } \frac{1}{2}}{}} \, , \quad \sqrt{7}\,x = \sqrt{7}\,x^1 \overset{\text{degree } 1}{\underset{\text{coefficient } \sqrt{7}}{}}$$

A *constant* such as 4, $\frac{5}{3}$, or $\sqrt{5}$ is also called a monomial and is said to have degree zero if the constant is not zero. (Note that this is in agreement with the fact that $ax^0 = a$.) Finally, the zero monomial, 0, is not assigned any degree.

A *polynomial* in x over the real numbers is either a monomial or the sum of monomials. Thus

$$0, \quad \sqrt{2}, \quad \frac{1}{2}x^2, \quad 2x^3 + \sqrt{5}\,x, \quad -3x^4 + \frac{7}{3}x^2 + 2x + 3, \quad \text{and} \quad x^{12}$$

are all polynomials in x over the real numbers.

If the coefficients of a polynomial are all rational numbers, then the polynomial can be considered as a polynomial over the rational numbers. Thus

$$\frac{2}{3}y^6 + \frac{3}{4}y^5 + 6y^4 + \frac{7}{8}y^3 + 3y + \frac{7}{5}$$

is a polynomial in y over the rational numbers as well as over the real numbers.

Likewise, if the coefficients are all integers, the polynomial can be considered as a polynomial over the integers. Thus, for example, the polynomial $2x^2 - 5x + 3$ is a polynomial over the integers as well as a polynomial over the rational numbers and a polynomial over the real numbers.

The nonzero monomials of a polynomial are called the *terms* of the polynomial. Thus $2x^2$ is a polynomial with the single term $2x^2$ whereas

$$5x^3 + 6x^2 - \frac{1}{2}$$

has the three terms $5x^3$, $6x^2$, and $-\frac{1}{2}$ . .

The *degree of a polynomial* is equal to the highest degree of the terms of the polynomial. Thus

$$2x^2 - 5x + 3 \ (= 2x^2 + (-5)x + 3)$$

has degree 2 and the terms $2x^2$, $-5x$, and 3

$$5x^{10} - 3x^2 + 5 \ (= 5x^{10} + (-3)x^2 + 5)$$

has degree 10 and the terms $5x^{10}$, $-3x^2$, and 5

$$3 \text{ has degree 0 and the single term 3}$$

$$5x \text{ has degree 1 and the single term } 5x$$

$$0 \text{ has no degree and the single term 0}$$

It is customary, although not essential, to write a polynomial with the monomial of highest degree first, the monomial of the second highest degree next, and so on. For example, we would ordinarily write

$$\frac{1}{2} - 3y^2 + y + \sqrt{7}\,y^3 \quad \text{as} \quad \sqrt{7}\,y^3 - 3y^2 + y + \frac{1}{2}$$

Other kinds of polynomials can be formed by using more than one variable. For example, each of

$$2x^2 + 3xy + 6y^2, \quad yx^3 + 8x^2y + 12xy^2 - 3y - 88, \quad \text{and} \quad x^2y^2$$

are polynomials in x and y over the integers.

The degree of a monomial in more than one variable is the sum of the exponents of the variables. Thus

$$-4x^2v^3 \quad \text{has degree } 2 + 3 = 5$$

$$6u^2vw \quad \text{has degree } 2 + 1 + 1 = 4$$

$$\sqrt{7}r^3t^5 \quad \text{has degree } 3 + 5 = 8$$

Then the degree of a polynomial in more than one variable is equal to the highest degree of the terms of the polynomial. Thus

$$-3xy \quad \text{has degree } 2$$

$$-6x^3y + 2xy^2 + \sqrt{7}y^3 \quad \text{has degree 4 (from } x^3y)$$

$$2s^6 + \sqrt{3}r^3s^3 + 4rst \quad \text{has degree 6 (from both } s^6 \text{ and } r^3s^3)$$

These last two polynomials have each been written in descending order of powers of one of the variables (x and s, respectively). We can rewrite each of them in descending powers of y and r, respectively, as $\sqrt{7}y^3 + 2xy^2 - 6x^3y$ and $\sqrt{3}r^3s^3 + 4rst + 2s^6$.

We have called a polynomial of one term a monomial. A polynomial with two terms, such as $2x^2 - y^2$, is called a *binomial*, and a polynomial with three terms, such as $x^2 + 5x + 6$, is called a *trinomial.*

EXERCISES 6.1

In exercises 1-15 give the degree, if any, of the given monomial.

1.	$3x^2$	6.	0	11.	$2xy$
2.	$\frac{1}{2}y^3$	7.	$\frac{7}{5}x$	12.	$\sqrt{5}x^2y^3$
3.	$\sqrt{3}x^6$	8.	π	13.	$7x^{13}y^2$
4.	5	9.	$75y^{13}$	14.	$\frac{3}{2}a^2b^4$
5.	y	10.	$\sqrt{10}$	15.	$5xyz^2$

In exercises 16-30 give the degree, if any, of the given polynomial.

16.	$5x^2 + 3$	21.	$3x^3 + 0x^7 + 5$
17.	$x^2 + 4x + 3x^4 + 2$	22.	$5x^4 + 6x + 1$
18.	3	23.	$\frac{1}{2}x^3 - x$
19.	$5x^3$	24.	0
20.	$7x^6 - 2x^3 + 3x + \sqrt{3}$	25.	$x^2 - x + 5$

26. $2x + 1 - x^2$

27. $3x^7 + \sqrt{2}$

28. $5x^8 + 3x^7 + 3x^2 + 5$

29. $\sqrt{3}$

30. $7.6x^2 - 0.32x + 1.68$

In exercises 31-50 indicate which expressions are polynomials over the integers.

31. $27x^3 + 4x - 1$

32. $13\sqrt{x} + 2$

33. $\dfrac{3}{2}x^2 + 6x - \dfrac{5}{3}$

34. $2x^2 + 6x - 5$

35. $3\sqrt{3}\,x - 2$

36. $5x^7 + 3x^4 - 3x + 2$

37. $3x^2 + 2x + 1$

38. $\sqrt{5}x^2 - 2x + 3$

39. 0

40. 5

41. $5x$

42. $15x^2$

43. $3x^2 - 2 + \dfrac{1}{5x}$

44. $\dfrac{x^2}{3} + \dfrac{x}{2} - 5$

45. $35x^{41} + 7x^{25} - 3x^2 + 1$

46. $\sqrt{2}x^2 - 3x + \sqrt{5}$

47. $\dfrac{3}{2x^2} + \dfrac{1}{6x} + 5$

48. $x^2 + 3x - 4$

49. $\dfrac{x^2 + 1}{x}$

50. $3x^{-2} + 5$

In exercises 51-70 let A be the set of polynomials over the integers, B the set of polynomials over the rational numbers, and C the set of polynomials over the real numbers. List the set or sets to which each polynomial belongs. (Note that a polynomial may belong to more than one set.)

Example (a): $3x^2 - 4$

Solution: A, B, C

Example (b): $5x^2 - \dfrac{1}{2}x + \sqrt{7}$

Solution: C

51. $x^2 + 5$

52. $2x^4 + \dfrac{1}{3}$

53. $x^2 + 3x + \pi$

54. $\dfrac{1}{2}x^2 + 5x - 6$

55. 5

56. $x^3 + 3x + 3$

57. $\sqrt{3}x^2 + 2x + 3$

58. $3y^2 + 3$

59. $5z^2 - 3z + 1$

60. $\dfrac{1}{2}w^2 + 2w$

61. $2xy + 5$

62. $3x^2 - \dfrac{1}{3}xy + \dfrac{5}{2}y^2$

63. $15r^3 - \sqrt{7}x - 5$

64. $32a^2b^2 - 16ab + 1$

65. $y^2 - x^2 - \pi$

66. $\sqrt[3]{16}\, x^3 + 2$

67. $4y^2 - 9x^2$

68. $\frac{1}{2}a^2 - \frac{3}{8}ab + b^2$

69. $\frac{1}{2}y^{2n} - 2y^n x^n - \frac{5}{2}x^n$

70. $\sqrt{6}\, w^3 - \sqrt{5}\, t^2$

6.2 Addition of Polynomials

Two monomials are called *like* monomials if they contain the same variables each to the same degree. Thus

$$5x^3 \qquad \text{and} \qquad -2x^3$$

$$-\frac{1}{2}x^3y^2 \qquad \text{and} \qquad 5x^3y^2$$

and

$$\sqrt{7}\, rst \qquad \text{and} \qquad -6rst$$

are all pairs of like monomials. Sums or differences of like monomials can be expressed as single monomials by use of the distributive property as shown by the following examples.

Example 1: $5x^2 + 4x^2 = (5 + 4)x^2 = 9x^2$

Example 2: $3xy + 2xy + \sqrt{3}xy = (3 + 2 + \sqrt{3})xy = (5 + \sqrt{3})xy$

Example 3: $5x^3y^2 - 2x^3y^2 = (5 - 2)x^3y^2 = 3x^3y^2$

Example 4: $ax^4 + x^4 = ax^4 + 1 \cdot x^4 = (a + 1)x^4$

Notice that, in the last example, we regarded x^4 as a polynomial in x with coefficient 1.

The procedures for addition and subtraction of monomials suggest procedures for adding polynomials as shown by the next three examples.

Example 5: $(3x^2 + 2x + 1) + (4x^2 - 3x - 4)$

$$= (3x^2 + 4x^2) + (2x - 3x) + (1 - 4)$$

$$= (3 + 4)x^2 + (2 - 3)x + (-3) = 7x^2 - x - 3$$

(Note that the commutative and associative properties of addition as well as the distributive property are used extensively in obtaining the answer.)

Example 6: $(3x^2 + 2x + 3) + (4x^3 - 2x^2 + 3x - 2)$

$= 4x^3 + (3 - 2)x^2 + (2 + 3)x + (3 - 2)$

$= 4x^3 + x^2 + 5x + 1$

If several polynomials are involved in an addition problem, it may help to avoid errors if you arrange the polynomials in columns as shown in the next example.

Example 7: Add $3x^2 + 4x + 1, - 5x^3 + 3x + 2, 6x^4 + x,$ and

$-3x^4 + 2x^3 - 5x^2 - 4.$

$$
\begin{array}{r}
3x^2 + 4x + 1 \\
- 5x^3 \qquad\quad + 3x + 2 \\
6x^4 \qquad\qquad\qquad + x \\
-3x^4 + 2x^3 - 5x^2 \qquad\quad - 4 \\
\hline
3x^4 - 3x^3 - 2x^2 + 8x - 1
\end{array}
$$

To subtract a polynomial Q from a polynomial P we note that $P - Q = P + (-Q)$ where $Q + (-Q) = 0$. Our next two examples illustrate subtraction of polynomials.

Example 8: $(6x^2 + x - 3) - (- 3x^2 - 5x + 4)$

$= (6x^2 + x - 3) + (3x^2 + 5x - 4) = 9x^2 + 6x - 7$

Example 9: $(- 3x^2y + 5xy - 3) - (5x^2y + 5xy - 2)$

$= (-3x^2y + 5xy - 3) + (-5x^2y - 5xy + 2) = - 8x^2y - 1$

Our last example illustrates how addition or subtraction of polynomials may be involved in finding solution sets of equations.

Example 10: Find the solution set of the equation $(3x - 5) + (2x - 7) = 2$

Solution: Adding the two polynomials we have

$$(3x + 2x) + [-5 + (-7)] = 2$$

$$5x - 12 = 2$$

$$5x = 14$$

$$x = \frac{14}{5}$$

Solution set = $\{ \frac{14}{5} \}$.

EXERCISES 6.2

In exercises 1–25 express the given polynomial as a monomial.

Example (a): $2x^2 + 3x^2 = (2 + 3)x^2 = 5x^2$

Example (b): $3y^3 - 5y^3 = (3 - 5)y^3 = -2y^3$

1. $8a + 4a$
2. $11n - 6n$
3. $231x^6 + 5x^6$
4. $54y^5 + 100y^5$
5. $5a - 12a$
6. $-7m - 8m$
7. $-10a^2 + 4a^2$
8. $-bc + 6bc$
9. $6x^3 + 10x^3$
10. $4x^2 - 3x^2$
11. $\sqrt{5}a^4 - \sqrt{5}a^4$
12. $34y^7 - 7y^7 - 5y^7$
13. $-3d - 10d - 4d$

14. $-7s - s + 2s$
15. $23x - 11x + 3x - 4x$
16. $14s^9 - 6s^9 - 2s^9$
17. $3x^2y - 2x^2y$
18. $5ab - 6ab$
19. $3abc - 2abc + 5abc$
20. $-18m^2n^3 - 27m^2n^3$
21. $36a^3bc^2 - 19a^3bc^2$
22. $16x^2y + 13x^2y - 18x^2y$
23. $\dfrac{1}{2}a^2b + \dfrac{3}{4}a^2b - \dfrac{1}{4}a^2b$
24. $1.7m^3n^2 - 0.5m^3n^2 + 1.8m^3n^2$
25. $\dfrac{4}{3}bc^2 - \dfrac{5}{3}bc^2 + \dfrac{7}{3}bc^2$

In exercises 26–50 simplify the given polynomial.

Example (a): $6x - 3 - 5x = 6x - 5x - 3$
$$= (6 - 5)x - 3$$
$$= x - 3$$

Example (b): $13y^5 - 12y^3 - 3y^5 + 5y^3 = 13y^5 - 3y^5 - 12y^3 + 5y^3$
$$= (13 - 3)y^5 + (-12 + 5)y^3$$
$$= 10y^5 - 7y^3$$

26. $3a + 5 + 6a$
27. $4xy + 6 - 2xy$
28. $5a + 6b + 7a + 2b$
29. $3x^2 + 4x + 2x^2 + x$
30. $-20xy^2 + 8 + 21xy^2$

31. $5ab - 4ab + 10$

32. $7x + 4 + 6x - 5$

33. $-3x + 12 + 7x + 3$

34. $4y^2 + 3y + 6y + 2$

35. $3x + 4y - w + 6x - 3y + 7w$

36. $13y^5 - 12y^3 - 3y^5$

37. $11 + 6a + 4 - 15a$

38. $16m^2 + 5m - 4m^2 - 6m$

39. $15ab - 14b^2 - 13ab + 4b^2$

40. $8x + 5 - 2x - 3$

41. $4y^2 + 3y + 2y + 6y^2$

42. $13y^2 + 4y - 3 + 2y^2 + 6y - 7$

43. $4\sqrt{5}xy + 6\sqrt{3}yz - 8\sqrt{5}xy + 32\sqrt{3}yz$

44. $7st + 9s^2t + 5st^2 - 8s^2t^2 - 5st - 9s^2t - 5st^2$

45. $-10a^2b + 41a^2b - 18ab^2 - 2a^2b + 5ab^2$

46. $3y^4 - y^5 + 5 + 2y^5 + 1 + 4y^4 - y^7$

47. $\dfrac{2}{3}a + \dfrac{5}{6}a^2 - \dfrac{1}{2}a + \dfrac{2}{3}a^2$

48. $\dfrac{3}{4}xy^2 + \dfrac{5}{6}xy - \dfrac{9}{12}xy^2 - \dfrac{10}{12}xy$

49. $5\dfrac{3}{4}mn + 1\dfrac{1}{4}mn - \dfrac{3}{4}m^2n^2 + \dfrac{3}{4}mn$

50. $0.76x^2y^2 + 0.23x^2y^2 - 1.62xy + 0.01x^2y^2$

In exercises 51–75 find the sum or difference.

51. $(4x^2 + 2x - 1) + (5x^2 - 3x - 1)$

52. $(9a + 6b - 4c) + (-15a - 13b - 11c)$

53. $(2a^2 - 5ab - b^2) + (7a^2 + 3ab - 9b^2)$

54. $(4x - 3x^2 - 11 + 5x^3) + (12x^2 - 7 - 8x^3 - 15x)$

55. $(-5a^2 + 2a - 3) + (6a^2 - 2a + 3)$

56. $(6p - 7q + r) - (8r - 8q - 8p)$

57. $(4a - 6b + 3c) - (5a + 2b - 9c)$

58. $(6rs - 2r + 3s + 4) - (7rs + 2r - 3s + 1)$

59. $(2x^2 - 5xy - y^2) - (-4x^2 - 6xy + 8y^2)$

60. $(3x - 8y) - (5x - 8y)$

61. $(\sqrt{2}x^2 + 3x + \sqrt{12}) + (\sqrt{18}x^2 + 6x + \sqrt{27})$

62. $\sqrt{98}x^3 - (16x^3 - \sqrt{75})$

63. $(5x^2y - 3xy + 2xy^2) - (4xy + 3x^2y - 5xy^2)$

64. $(3x - 8y) + (7y - 6z) + (5z - 2x)$
65. $(x^3 - 3xy^2 - 2x^2 y) + (3x^2 y - 5x^3 - 4xy^2)$
66. $2x - (4x^2 - 6x + 2)$
67. $(4m^2 - 3mn + n^2) - (3m^2 - n^2)$
68. $(-6x + 4y - 2z) + (4x + 9y - 6z) - (7x + 2y + 8z)$
69. $(7a^2 + 2b^2) + (3a^2 - 4c) - (3a^2 + 7c)$
70. $(3y + 4) - (2y + 11) - (7y - 6) + (3y - 2)$
71. $- (6x + 3y - 2z) - (9x + 5y - 12z) + (7x + 16y + 4z)$
72. $(3ab - 2a + 8b - 5) - (2ab + 14a - 3b + 2)$
73. $(4x - 5) - [(6x + 7) - (3x + 4)] - (2x + 6)$
74. $9a - (6b + 4) - (8a - 2) - [(2b + 7) - (18a - 3)]$
75. $(13x^2 + 3 - 4x + 8x^3) + (-9x + 5 + 16x^3 + x^2) + (-15 - 6x^2 - 7x^3 + 11x)$
    $- (-10x^3 - 12x + 14x^2 - 17)$

In exercises 76–100 find the solution set of the given equation or inequality.

76. $(3y - 5) + (2y - 7) = -2$
77. $(5x + 1) + (3x - 6) = -1$
78. $(3a + 3) + (5a - 5) = 7a$
79. $(4y + 5) + (y + 5) = 35$
80. $(7x - 8) + (x - 3) < 61$
81. $(4b + 2) + (3b - 4) \geqslant 21 + 5b - 17$
82. $(2y + 4) - (7y - 3) = 3$
83. $(-7a + 3) - (-3a - 5) = a + 2$
84. $(-8y - 4) - (-5y - 2) \leqslant 3y - 1$
85. $(6r - 4) - (-2r - 8) > (18 - 3r) - (8 - 9r)$
86. $(2y + 23) - (6 - y) = (4y + 13) - (3y - 14)$
87. $(-5x - 2) + (-3x - 4) = 2(3x - 1)$
88. $(-7x - 1) + (-2x - 5) \geqslant 3(2x + 1)$
89. $(60 + 3x) - (5x - 2) < (7x - 15) + 2 (x - 11)$
90. $(8y + 7) - 2(2y - 1) + 7 = 5(3 - y) + (-9y + 5)$
91. $(9y - 4) + (2y - 3) + (-y + 7) = 0$
92. $(2x + 5) - (4x - 5) + (7x - 2) + 7x < 3x + 5$
93. $(3x - 2) + x = 2(x + 3) - (-3x + 8)$
94. $(x - 1) - (x + 2) - (x - 3) = x$
95. $x + (2x + 3) \leqslant (x + 3) + (x + 4)$
96. $(4x + 5) - 3(2x - 5) \geqslant 9 - (8x - 1) + x$
97. $(-5a^2 + 2a - 3) - (-5a^2 + 7a + 3) = 0$

98. $(-7m^2 - 3m + 4) - (-7m^2 + 5m - 3) \leqslant 0$
99. $(3\sqrt{2}x - 8) + (5\sqrt{2}x + 7) = 0$
100. $6(\sqrt{80}x - 2\sqrt{5}) = 14(\sqrt{125}x - 14\sqrt{5})$

6.3 Multiplication Involving Monomials

Multiplication involving monomials depends first of all upon the commutative and associative properties of multiplication and on the properties of exponents as given in Chapter 5. We have, for all real numbers a and b and natural numbers m and n,

$$(ax^n)(bx^m) = (ab)(x^n x^m) = abx^{n+m}$$

Thus

$$(6x^3)(3x^2) = (6 \cdot 3)(x^3 \cdot x^2) = 18x^{3+2} = 18x^5$$

$$\left(\frac{1}{8}rs\right)(3r^2 s^3 t)(4s^2 t) = \left(\frac{1}{8} \cdot 3 \cdot 4\right)(r \cdot r^2)(s \cdot s^3 \cdot s^2)(t \cdot t)$$

$$= \frac{3}{2}r^{1+2}s^{1+3+2}t^{1+1}$$

$$= \frac{3}{2}r^3 s^6 t^2$$

To multiply any polynomial by a monomial we simply apply the distributive property, as illustrated by the following examples:

$$6(x + y) = 6x + 6y$$
$$x^2(x + 2) = (x^2 \cdot x) + (x^2 \cdot 2) = x^3 + 2x^2$$
$$3r^2 s(r^2 + 3rs + s^2) = (3r^2 s)r^2 + (3r^2 s)(3rs) + (3r^2 s)s^2$$
$$= 3r^4 s + 9r^3 s^2 + 3r^2 s^3$$

EXERCISES 6.3

In exercises 1–40 express the given product as a monomial.

Example: $(6x^2 y^2)(-3xy^2) = (6 \cdot -3)(x^2 \cdot x)(y^2 \cdot y^2) = -18x^3 y^4$

1. $(4x^2 y)(2x^3)$
2. $(5yx)(3cx)$
3. $(5y)(-yx)$
4. $(-6x^2 y)(2xy)$
5. $(4ab)(5ab^2)$
6. $(-2m^2)(-3m^4 n)$

7. $(-9rs^3)(-3ws^2)$
8. $(6abc)(4cdx)$
9. $(-xy^2)(-7x^3)$
10. $(5abc)(2ab^2)$
11. $(-3xyz^2)(-8xz^2)$
12. $(-18a^2b^3)(3abc)$
13. $(-7x^2y^3z^2)(-x^3y^2z^3)$
14. $\left(-\dfrac{1}{2}s^2w\right)\left(\dfrac{1}{5}w^2\right)$
15. $\left(-\dfrac{3}{5}t^3\right)\left(-\dfrac{2}{3}t^2\right)$
16. $\left(-\dfrac{1}{3}\right)\left(\dfrac{3}{8}ab\right)$
17. $(0.6a^8b)(0.3a^2b^3)$
18. $(-1.4x^2y)(-0.5xyz^2)$
19. $(0.07m^2n)(-0.01m^2y)$
20. $(3.7xy^2z)(-0.15x^2yz^2)$
21. $(5\sqrt{3}x)(7\sqrt{3}x)$

22. $(\sqrt{7}x)(\sqrt{2}z)$
23. $(4\sqrt{5}xy)(\sqrt{5}x^2y^3)$
24. $(8\sqrt{3}ab)(\sqrt{27}b^3)$
25. $(-x)(-x)(-x)$
26. $(2xy^2)(3x^2y^3)(-4xy)$
27. $(-5ab)(-6bc)(-cd)$
28. $(-7xy^2)(x^5z)(-3x^2yz)$
29. $(-3x)(-3x)(-3x)(-3x)$
30. $(xy^2)(-2x)(3yz)(-4x^2y)$
31. $\left(\dfrac{1}{3}x^2y\right)\left(\dfrac{1}{6}xy^2\right)(12x^2y^2)$
32. $\left(\dfrac{2}{5}a\right)\left(\dfrac{1}{3}ab\right)(5a^2b^2)$
33. $\left(-\dfrac{2}{3}m^2\right)(6mn)\left(-\dfrac{2}{3}n^2\right)$
34. $\left(\dfrac{3}{4}xy\right)\left(-\dfrac{7}{8}x^2y\right)\left(\dfrac{8}{7}x^2y^2\right)$
35. $\left(1\dfrac{2}{3}ab^2\right)\left(1\dfrac{1}{6}a^3\right)\left(-\dfrac{9}{5}b^3\right)$

In exercises 36–40 consider all exponents to be positive integers.

36. $(2x^ay^a)(3x^ay^a)$
37. $(2a^x)(3a^{2x})$
38. $(-4x^ky)(-3x^ky^a)$

39. $\left(\dfrac{1}{6}x^b\right)(18x^by)$
40. $(-3t^2)(-7t^a)$

In exercises 41–65 express the given product as a polynomial.

Example: $9(2x - y) = (9 \cdot 2x) - (9 \cdot y)$
$$= 18x - 9y$$

41. $2(x + 5)$
42. $8(y - 6)$
43. $-2(5a + 6)$
44. $-4(x^2 - 3)$

45. $10(3y - 7x)$
46. $-9(m^2 - 2m)$
47. $x^3(3x^2 + 5)$
48. $-a(a^2 + 3)$
49 $-x^2(4x^4 - 1)$
50. $m^2(2m^6 - 5m^4 - m^2)$
51. $x(x^3 + 3x^2 + x)$
52. $ab(a^2b - a^2b^2 + 2ab^3)$
53. $x^2y^2(-x^2y + 2xy - 3xy^2)$
54. $-a(a - 10a^2)$
55. $-x(-x^2 - x)$
56. $-mn(3m^2n - 4mn^2)$
57. $\frac{2}{3}(9 - 6x + 12x^2)$
58. $\sqrt{5}a(2\sqrt{5}a - 4\sqrt{5})$
59. $\sqrt{3}y(\sqrt{27}y^3 - \sqrt{12}y^2)$
60. $\frac{3}{4}x^2y^2(8x^2 + 4x^2y - 16xy^2 - 20y^3)$

In exercises 61–65 consider all exponents to be positive integers.

61. $x^a(x^a - y^a)$
62. $a^n(a^n + b^n)$
63. $x^2(x^b + y^b)$

64. $x^{2a}(x^a + x^{2a})$
65. $a^bx(ax + bx^2)$

6.4 Multiplication of Polynomials

To multiply any two polynomials we continue to make extensive use of the distributive property, as illustrated by the following examples.

Example 1:

$$(2x + 3)(x - 1) = (2x + 3)[x + (-1)]$$
$$= (2x + 3)x + (2x + 3)(-1)$$
$$= (2x)x + 3x + (2x)(-1) + (3)(-1)$$
$$= 2x^2 + 3x - 2x - 3$$
$$= 2x^2 + x - 3$$

Example 2: $(2x + 3)^2 = (2x + 3)(2x + 3)$

$$= (2x + 3)2x + (2x + 3) \cdot 3$$

$$= 4x^2 + 6x + 6x + 9$$

$$= 4x^2 + 12x + 9$$

Study Examples 1 and 2 carefully to find a way to determine the product of two binomials quickly without any intermediate steps. The following diagrams will assist you in your search.

$$(2x + 3)(x - 1) = 2x^2 + x - 3$$

$$(2x + 3)(2x + 3) = 4x^2 + 12x + 9$$

Once you grasp the basic principles of multiplying polynomials, there is no need to write down all the steps leading to the answer. Various mechanical procedures are possible, as shown in the next two examples. Remember, however, that accuracy and understanding are more important than speed.

Example 3: Multiply $x^3 + 3x^2 - 2x + 1$ by $2x^3 - 3x - 4$.

$$
\begin{array}{r}
x^3 + 3x^2 - 2x + 1 \\
2x^3 - 3x - 4 \\
\hline
2x^6 + 6x^5 - 4x^4 + 2x^3 \\
- 3x^4 - 9x^3 + 6x^2 - 3x \\
- 4x^3 - 12x^2 + 8x - 4 \\
\hline
2x^6 + 6x^5 - 7x^4 - 11x^3 - 6x^2 + 5x - 4
\end{array}
$$

Example 4: Multiply $x^2 + 3xy - y^2$ by $x + 2y$.

$$
\begin{array}{l}
x^2 + 3xy \;\;-\;\; y^2 \\
\underline{\qquad\quad x \;\;+\; 2y} \\
x^3 + 3x^2y - \;\; xy^2 \\
\underline{\qquad 2x^2y + 6xy^2 - 2y^3} \\
x^3 + 5x^2y + 5xy^2 - 2y^3
\end{array}
$$

EXERCISES 6.4

In exercises 1–56 express the given product as a polynomial.

Example (a):
$$(x + 1)(x + 2) = (x + 1)x + (x + 1) \cdot 2$$
$$= (x^2 + x) + (2x + 2)$$
$$= x^2 + 3x + 2$$

Example (b):
$$(a + b)(a^2 - ab + b^2) = (a + b)a^2 - (a + b)ab + (a + b)b^2$$
$$= a^3 + a^2b - a^2b - ab^2 + ab^2 + b^3$$
$$= a^3 + b^3$$

or

$$
\begin{array}{l}
a^2 - ab \;\;+\;\; b^2 \\
\underline{\qquad\quad a \;\;+\; b} \\
a^3 - a^2b + ab^2 \\
\underline{\qquad\quad a^2b - ab^2 + b^3} \\
a^3 + \;\; 0 \;\; + \;\; 0 \;\; + b^3 = a^3 + b^3
\end{array}
$$

1. $(x + 4)(x + 3)$
2. $(n + 9)(n + 8)$
3. $(a + 1)(a + 1)$
4. $(x + 3)(x + 6)$
5. $(a + 12)(a + 3)$

6. $(n + 2)(n + 8)$
7. $(y + 2)(y + 5)$
8. $(m + 7)(m + 3)$
9. $(s + 5)(s + 4)$
10. $\left(x + \dfrac{1}{2}\right)\left(x + \dfrac{3}{4}\right)$

11. $(x - 4)(x - 3)$

12. $(n - 9)(n - 8)$

13. $(a - 1)(a - 1)$

14. $(x - 3)(x - 6)$

15. $(a - 12)(a - 3)$

16. $(x - 4)(x + 3)$

17. $(n - 9)(n + 8)$

18. $(x - 3)(x + 6)$

19. $(a - 12)(a + 3)$

20. $(x + 4)(x - 3)$

21. $(n + 9)(n - 8)$

22. $(x + 3)(x - 6)$

23. $(a + 12)(a - 3)$

24. $(2x + 3)(2x + 3)$

25. $(4c + 5)(3c + 2)$

26. $(2a + 4)(3a + 4)$

27. $(5y + 7)(2y - 3)$

28. $(2n + 7)(2n + 5)$

29. $(2x - 3)(2x + 3)$

30. $(4c - 5)(3c + 2)$

31. $(2a - 4)(3a + 4)$

32. $(5y - 7)(2y + 3)$

33. $(2n - 7)(2n + 5)$

34. $(2a + 3)(2a - 3)$

35. $(4c + 5)(3c - 2)$

36. $(2a + 4)(3a - 4)$

37. $(2y - 3)(5y + 7)$

38. $(2n + 7)(2n - 5)$

39. $(4c - 5)(3c - 2)$

40. $(2a - 4)(3a - 4)$

41. $(5y - 7)(2y - 3)$

42. $(3y + 1)(4y^2 - 3y + 2)$

43. $(y - 1)(y^2 - 2y + 1)$

44. $(a - 8)(3a^2 + 2a - 7)$

45. $(x + 1)(x^2 + 2x + 1)$

46. $(n - 2)(n^2 + 4n - 4)$

47. $(y + 2)(y^2 + 8y - 1)$

48. $(4x - 5)(2x^2 - x + 3)$
49. $(5y + 2x)(7xy - y^2 + 6x^3)$
50. $(7b - 3a)(2b^2 - 4ba - a^2)$
51. $(3x - 4)(2x^2 - 5x + 3)$
52. $(-n - 8)(3n^2 + 2n - 7)$
53. $(a - 1)(a^5 + 3a^4 - 3a^3 + a^2 - 4a + 3)$
54. $(x^2 - x + 1)(x^3 - x^2 + x - 1)$
55. $(4a^2 - 3a - 2)(a^2 + 5a + 4)$
56. $(a + b + c)(a - b + c)$

In exercises 57-76 perform the indicated multiplications and then simplify the resulting sum.

Example: $3x(x + 2y) + (x + y)(2x - 3y)$

$$= 3x^2 + 6xy + 2x^2 + 2xy - 3xy - 3y^2$$

$$= 5x^2 + 5xy - 3y^2$$

57. $6x^2 + x(3x - 5)$
58. $ax + 2x(a + x)$
59. $6(2x + 5) + 3$
60. $2a(5a + 4) - 9a$
61. $-3n(-3n + 8) - 6n$
62. $a(b - c) + c(a - b)$
63. $12x^2 + 1 + 2x(5 - 4x)$
64. $x^2 + 4x - 2(x + 3) + x(2x - 4)$
65. $-3a(a - 3) + (a - 5)(a + 5)$
66. $-5b(b + c) + (b + c)(b - 2c)$
67. $2t(t + 2) + (t - 3)(t - 5)$
68. $(x + y)(x + y) + (x - y)(x + y)$
69. $(a - 3b)(a + 3b) + (5a - 4b)(2a + 5b)$
70. $(a - b)(a - b) - (a + b)(a - b)$
71. $(x^2 - y^2)(x^2 + y^2) - (x^2 - y^2)(x^2 - y^2)$
72. $(y + 1)(y + 2) - (y - 4)(y + 5)$
73. $(x - 1)(x + 1) - (x + 1)(x + 1)x$
74. $n(n - 8) - 6(n + 1)(n - 1) + \frac{1}{4}(16n^2 - 12n)$
75. $(3x^2 + 1)(y^2 - 3y + 1) - (y^2 + 2)(3x^2 + 2x - 4)$
76. $3x - [2y - (z - x)] - [2z - (y - x)]$

6.5 Equations and Inequalities Involving Multiplication of Polynomials

The following two examples illustrate how multiplication of polynomials may be involved in finding the solution sets of equations and inequalities.

Example 1: Find the solution set of the equation

$$(x + 1)(x + 2) - (x + 2)(x + 3) = 0$$

Solution: Multiplying the polynomials, we have

$$(x^2 + 3x + 2) - (x^2 + 5x + 6) = 0$$
$$-2x - 4 = 0$$
$$-2x = 4$$
$$x = -2$$

Solution set = { -2 }.

Example 2: Find the solution set of the inequality

$$(x - 3)(2x + 3) < (2x - 1)(x + 4)$$

Solution: Multiplying the polynomials, we have

$$2x^2 - 3x - 9 < 2x^2 + 7x - 4$$
$$-10x < 5$$
$$x > -\frac{1}{2}$$

Solution set = $\{ x : x > -\frac{1}{2} \}$.

EXERCISES 6.5

In these exercises find the solution set of the given equation or inequality.

1. $6(2y + 3) - 6 - 4(2y + 3) = -3y$
2. $4n - 3(3n - 1) + 12 < 15(n + 9)$
3. $6(2a - 5) + 2(4a + 3) > 9(8a - 7)$
4. $22 - 5(3 - 2x) \leqslant 2x - (5x + 32)$
5. $7(2n + 6) - 13 - (3n - 1) = 2(-5n - 6)$
6. $2(3y + 4) - 6(y - 4) = 5 + (3y - 2) - (4y + 6)$

7. $8x + x(x - 6) = 57 - (17 - x^2)$
8. $3a^2 - a(3a + 4) \leqslant 5(6 - 2a)$
9. $4(n^2 + 2n + 1) - 14 > 2n(2n + 9)$
10. $6x(x - 3) + 20 \leqslant 8 - 3x(5 - 2x)$
11. $5(4y - 10) + 2y^2 \geqslant 4(30 + y^2) - y(2y - 3)$
12. $2(x + 3) - 3(x - 2) = 5(x + 6) + 4(3 - 2x)$
13. $4a + 5(2a - 1) - 23 > 37 - 3(6a - 5)$
14. $(x - 2) \, (x - 3) + 25 = (x + 1) \, (x + 1) - 12$
15. $(y + 2) \, (y + 3) - (y + 1) \, (y + 4) = 0$
16. $(2x + 1) \, (x + 3) - (x + 5) \, (2x + 3) = 0$
17. $(n + 2) \, (2n + 3) > (2n + 1) \, (n + 3)$
18. $(2a - 1) \, (2a + 2) > (2a + 2) \, (2a - 3)$
19. $(2y - 3) \, (y - 3) - (y - 1) \, (2y - 5) > 0$
20. $(x - 1) \, (3x - 2) - (3x - 5) \, (x - 2) = 0$
21. $(x + 2) \, (x + 2) = x^2 + 8$
22. $(n + 3) \, (n - 3) = (n + 7) \, (n - 2)$
23. $(y - 4) \, (y - 1) - y^2 = 9$
24. $(3x + 1) \, (2x - 5) = 6x^2$
25. $(4n - 1) \, (3n + 7) < 12n^2$

6.6 Monomial and Binomial Factors

In Sections 6.3 and 6.4 you learned how to multiply two polynomials to form a polynomial of equal or higher degree. The opposite process is to write a polynomial as a product of polynomials. Thus

$$x^2 + 7x + 12 = (x + 3) \, (x + 4)$$

This process is called *factoring,* and $x + 3$ and $x + 4$ are called *factors* of $x^2 + 7x + 12$.

Recall that a natural number other than 1 is said to be a prime number if its only natural number factors are 1 and itself. Thus 2, 3, 5, 7, and 11 are prime numbers, whereas 4 is not a prime number since it has the factor 2 in addition to the factors 1 and 4. Clearly, however, if we do not restrict ourselves to natural number factors, a prime number such as 3 has many other factors, such as $\frac{1}{2}, \sqrt{3}, \pi$, etc., and we can write 3 not only as $1 \cdot 3$ and $3 \cdot 1$ but also as $\frac{1}{2} \cdot 6, \sqrt{3} \cdot \sqrt{3}, \pi \cdot \frac{3}{\pi}$, etc. Thus the number of factors a number has depends upon the kind of numbers we allow as factors.

The same situation applies to polynomials. If we allow rational numbers as factors, we can write

$$x + 2 = \frac{1}{2}(2x + 4) = \frac{1}{3}(3x + 6) = \frac{1}{4}(4x + 8), \text{ etc.}$$

For most purposes, however, our interest lies in polynomials over the integers and their factorization into polynomials over the integers. Hence we make the following definition: A polynomial $P \neq 0$ over the integers is said to be *prime*, or *irreducible*, if it cannot be written as a product of factors (other than $1, -1, P,$ or $-P$) that are polynomials over the integers.

Thus we consider $x + 2$ as irreducible even though

$$x + 2 = \frac{1}{2}(2x + 4) = \frac{1}{3}(3x + 6) = \frac{1}{4}(4x + 8), \text{ etc.}$$

since $\frac{1}{2}, \frac{1}{3}, \frac{1}{4}$, etc., are not polynomials over the *integers*. Some examples of prime factorization over the integers are shown below.

Polynomial	Prime Factorization
$3a + 3b$	$3(a + b)$
$5xy + 5xz$	$5x(y + z)$
$x^2 + 7x + 12$	$(x + 3)(x + 4)$
$6x^3 + 12x^2 - 18x$	$6x(x + 3)(x - 1)$

Note that in the last example the prime factorization is given as $6x(x + 3)(x - 1)$ rather than as $2 \cdot 3x(x + 3)(x + 1)$. By convention, we do not give the prime factorization of a numerical coefficient.

Unfortunately, there is no general method available for identifying an irreducible polynomial or of finding the prime factors of one that is not irreducible. The special situations that we will examine in this chapter, however, cover most of the situations met in practice.

The first thing to look for in factoring a polynomial is a factor common to all terms. For example, $5ax^2 + 15ax + 25a$ has the factor $5a$ common to all terms so that, by the distributive property, we can write

$$5ax^2 + 15ax + 25a = 5a(x^2 + 3x + 5)$$

Notice that we could also write

$$5ax^2 + 15ax + 25a = 5(ax^2 + 3ax + 5a)$$

but then $ax^2 + 3ax + 5a$ would certainly not be irreducible, since each term has the common factor a. Generally, when we ask for the factorization of a polynomial we mean that we should, if possible, find the irreducible factors.

Here are more examples.

Example 1. $8x^4 - 4x^3 - 2x^2 = 2x^2(4x^2 - 2x - 1)$

Example 2: $2ar^2 + 8ars - 16art = 2ar(r + 4s - 8t)$

The common factor may be a binomial, as illustrated in the next two examples.

Example 3: $6a(x + y) + 5b (x + y) = (6a + 5b) (x + y)$

Example 4: $x^2(x - 2) + 3(x - 2) = (x^2 + 3) (x - 2)$

Polynomials with four terms can sometimes be factored by finding a common binomial factor, as illustrated by our concluding examples.

Example 5: $ax - ay + bx - by = a(x - y) + b(x - y)$
$$= (a + b) (x - y)$$

Example 6: $2x^3 + 2x^2 - x - 1 = 2x^2 (x + 1) - 1(x + 1)$
$$= (2x^2 - 1) (x + 1)$$

(Notice that $-x - 1$ is written as $- 1(x + 1)$ rather than as $-(x + 1)$ in order to emphasize that we get $2x^2 - 1$ and not $2x^2$ as the first factor.)

EXERCISES 6.6

In exercises 1–50 factor the given polynomial by finding a factor common to all terms and applying the distributive property.

Example (a): $2x - 2y = 2(x - y)$

Example (b): $12x^2 + 18x = 6x(2x + 3)$

Example (c): $2xy^2 + 6xy + 8x = 2x(y^2 + 3y + 4)$

1. $3a + 3b$
2. $cn + cb$
3. $5a + 15$
4. $8x - 2$
5. $4xy - 8xz$
6. $3x + 12y$
7. $3a + 9b$
8. $3xy + 9yz$
9. $4x^2 + 15xy$
10. $12x^2 - 6x$
11. $9y + 27x$
12. $6s - 6$
13. $3t^2 + t$
14. $p + prt$
15. $18ab + 12ac$
16. $5x^2y^2 + 10x^2y$
17. $12x^2 + 18x^3$
18. $a^2b - 16ab^2$
19. $7a^2 - 49ab$
20. $24x^2 - 16xy^2$
21. $2\pi rh + 2\pi r^2$
22. $7a^3 - 21a^2$
23. $\frac{1}{4}ha + \frac{1}{4}hb$
24. $38a^2x^2 - 57ax$
25. $5x + 5y - 10z$

26. $ac + ad - ae$
27. $rs^2 - 2rs + r^2s$
28. $3x^2 - 6xy + 9y^2$
29. $36a^3 + 24a^2b - 12a^4b^2$
30. $nb^4 + nb^3 - nb^2$
31. $x^2yz - xy^2 + xyz^2$
32. $15x^3 + 10ax^3 - 15z^2x$
33. $7c^4 - 28c^3 - 49c^2$
34. $9x^3y + 36x^2y^2 + 6xy^3$
35. $a^2b^2 - 3a^3b^3 - 6a^4b^4$
36. $12b^3 + 16b^2 + 8b$
37. $30d^2n + 6d^3 - 24d^2p$
38. $\frac{2}{3}ar - \frac{2}{3}as + \frac{2}{3}at$
39. $0.4w + 0.6w^2 + 0.2w^3$
40. $12x^3y^5 + 24xy^3 - 42x^2y^6$
41. $14a^5b^4 + 21a^4b^3 - 49a^3b^2$
42. $81m^4n + 54m^3n^2 + 9m^2n^3$
43. $48x^4y^2 - 144x^3y^2 + 108x^2y^4$
44. $70a^4x^5 - 126a^2x^4 - 112a^5x^6$
45. $x^4y^2 + x^3y^3 + x^2y^4 + xy^5$
46. $4a^4b - 6a^3b + 12a^2b - 10ab$
47. $9ax^2 - 12a^2x^2 - 60x + 3a$
48. $-27a^5x - 36a^4x^2 + 126x^2$
49. $-9x^3 + 36x^2y^2 - 6xy^3$
50. $-5y^2 - 10y - 5$

In exercises 51–74 factor out the common binomial factor.

Example (a): $a(x + y) - b(x + y) = (a - b)(x + y)$

Example (b): $xy + ay + xz + az = (xy + ay) + (xz + az)$
$$= (x + a)y + (x + a)z$$
$$= (x + a)(y + z)$$

51. $r(s + t) + n(s + t)$
52. $m(2x + y) + n(2x + y)$

53. $2a(x - y) - b(x - y)$
54. $5x(2a - b) - 3y(2a - b)$

55. $2a(x - 3y) + c(x - 3y)$

56. $4(a^2 + b^2) - x(a^2 + b^2)$

57. $a(2a - 3b) - 3b(2a - 3b)$

58. $(x - a) + (x - a)^2$

59. $x^2(a - b) + y^2 (a - b) + z^2 (a - b)$

60. $a(a + b) - 3b(a + b) + 5(a + b)$

61. $ab + an + bm + mn$

62. $ax - ay + bx - by$

63. $rx + ry + sx + sy$

64. $12ac + 4bc - 3ad - bd$

65. $xy - xz - ty + tz$

66. $x^2y + ax + bxy + ab$

67. $a^2 + ab + ac + bc$

68. $x^4 + a^2x^2 + b^2x^2 + a^2b^2$

69. $3mx - nx - 3my + ny$

70. $a^3 + a^2 + a + 1$

71. $4x^3 - 5x^2 - 4x + 5$

72. $2 + 3y - 8y^2 - 12y^3$

73. $3x^3 + 6x^2 + x + 2$

74. $a - b + ac - bc$

6.7 Factoring Special Binomials

We rarely need to factor any binomials except those with a common monomial factor and those of the form $a^2 - b^2$, $a^3 - b^3$, or $a^3 + b^3$. A binomial of the form $a^2 - b^2$ is often called "the difference of squares." Similarly, $a^3 - b^3$ is called "the difference of cubes" and $a^3 + b^3$ "the sum of cubes." We will consider only the difference of squares here, leaving the sum and difference of cubes for a later course.

It is easy to verify that

$$a^2 - b^2 = (a + b)(a - b)$$

and this equality enables us to obtain very easily the following factorizations.

Example 1: $4a^2 - 9b^2 = (2a)^2 - (3b)^2 = (2a + 3b)(2a - 3b)$

Example 2: $36x^2y^2 - 121s^2t^2 = (6xy)^2 - (11st)^2$

$$= (6xy + 11st)(6xy - 11st)$$

Example 3: $x^4 - 1 = (x^2 + 1)(x^2 - 1) = (x^2 + 1)(x + 1)(x - 1)$

Notice that the first factorization in Example 3 gives us a polynomial, $x^2 - 1$, which can be factored further. What about the polynomial $x^2 + 1$? Can it be factored?

Example 4: $6x^2 - 24y^2 = 6(x^2 - 4y^2) = 6(x + 2y)(x - 2y)$

Note that $6x^2 - 24y^2$ is not a difference of squares but that, after factoring out the common factor of 6, we obtain a difference of squares.

Always look first for common factors!

EXERCISES 6.7

In exercises 1–42 factor completely the given polynomial. If any common factors occur they should be factored out first.

Example (a): $a^2 - 4 = (a + 2)(a - 2)$

Example (b): $8x^4 - 8 = 8(x^4 - 1) = 8(x^2 + 1)(x^2 - 1)$

$$= 8(x^2 + 1)(x + 1)(x - 1)$$

1. $a^2 - b^2$	15. $a^4 - b^2$	29. $16x^4 - x^2$
2. $a^2 - 1$	16. $(2a - b)^2 - 4$	30. $8m^4 - 8$
3. $x^2 - 4$	17. $4 - (x + y)^2$	31. $25d^2 - 9c^2$
4. $x^2y^2 - 9$	18. $9 - (a - b)^2$	32. $16y^2 - 9$
5. $a^2x^2 - b^2y^2$	19. $(x + y)^2 - (a + b)^2$	33. $9y^2 - 4$
6. $9a^2 - 16$	20. $(x - y)^2 - (a - b)^2$	34. $100m^2 - 121n^2$
7. $9a^2b^2 - 25c^2$	21. $3x^2 - 3y^2$	35. $a^2 - 0.01b^2$
8. $x^6 - y^2$	22. $8a^2 - 2y^2$	36. $t^{12} - 9$
9. $4a^2 - 9x^2$	23. $18ax^2 - 8a$	37. $64a^2 - 36b^2$
10. $16s^2 - 25t^2$	24. $a^4 - 1$	38. $289ax^2 - 576ay^2$
11. $(x + y)^2 - 1$	25. $x^4 - y^4$	39. $t^2 - t^4$
12. $(a + b)^2 - 4$	26. $2b^4 - 2$	40. $18a^2 - 32b^2$
13. $81x^2 - 64y^2$	27. $6a^4 - 24b^4$	41. $0.04x^2 - 0.09y^2$
14. $25x^2y^2z^2 - 4$	28. $b^8 - 1$	42. $x^{2n} - y^{4n}$

6.8 Factoring Trinomials of the Form x² + bx + c

In Section 6.4 you learned to multiply two binomials. Given the binomials $x + 3$ and $x + 4$, we can find their product:

$$(x + 3)(x + 4) = x^2 + 3x + 4x + 12$$
$$= x^2 + (3 + 4)x + 12$$
$$= x^2 + 7x + 12$$

In this section we will reverse the process and consider how to factor such trinomials as $x^2 + 7x + 12$. Thus, given $x^2 + 7x + 12$, we will write

$$x^2 + 7x + 12 = (x + 3)(x + 4)$$

Given the binomials $x + n$ and $x + s$, we can find their product:

$$(x + n)(x + s) = x^2 + nx + sx + (n \cdot s)$$
$$= x^2 + (n + s)x + (n \cdot s)$$

Therefore the trinomial $x^2 + (n + s)x + (n \cdot s)$ has $x + n$ and $x + s$ as factors and

$$x^2 + (n + s)x + n \cdot s = (x + n)(x + s)$$

Using this idea to factor $x^2 + 7x + 12$ we think

$$x^2 + 7x + 12 = x^2 + (n + s)x + (n \cdot s) = (x + n)(x + s)$$
$$\text{where} \quad n + s = 7$$
$$\text{and} \quad n \cdot s = 12$$

Now the possible factorizations of 12 are

$$12 \cdot 1, \quad 6 \cdot 2, \quad \text{and} \quad 4 \cdot 3$$

But

$$12 + 1 \neq 7 \quad \text{and} \quad 6 + 2 \neq 7$$

whereas

$$4 + 3 = 7$$

We conclude that

$$x^2 + 7x + 12 = x^2 + (4 + 3)x + 4 \cdot 3 = (x + 4)(x + 3)$$

Here are some other examples of the use of this procedure for factoring trinomials.

Example 1: $x^2 + 5x + 6 = (x + 2)(x + 3)$

Here we found integers n and s such that $n + s = 5$ and $n \cdot s = 6$: $n = 2$ and $s = 3$.

Example 2: $x^2 - 5x - 24 = (x + 3)(x - 8)$

Here we found integers n and s such that $n + s = -5$ and $n \cdot s = -24$: $n = 3$ and $s = -8$.

Example 3: $x^2 - 4x + 4 = (x - 2)(x - 2) = (x - 2)^2$

Here we found integers n and s such that $n + s = -4$ and $n \cdot s = 4$: $n = s = -2$. Since $x^2 - 4x + 4 = (x - 2)^2$, we call $x^2 - 4x + 4$ a *trinomial perfect square*.

Example 4: $x^2 - 7xy + 12y^2 = (x - 4y)(x - 3y)$

Here we found integers n and s such that $n + s = -7$ and $n \cdot s = 12$: $n = -4$ and $s = -3$.

Example 5: $x^2 + 7x + 3 = ?$

Here we need to find integers n and s such that $n + s = 7$ and $n \cdot s = 3$. But if $n \cdot s = 3$ and n and s are integers, the only possibilities are $n = 1$ and $s = 3$, $n = 3$ and $s = 1$, $n = -1$ and $s = -3$, or $n = -3$ and $s = -1$. Since for none of these pairs of values of n and s is it true that $n + s = 7$, we conclude that $x^2 + 7x + 3$ is irreducible over the integers.

Example 6: $2x^2 + 4x + 2 = ?$

This trinomial is not of the form $x^2 + bx + c$. But, factoring out the common factor of 2, we have

$$2x^2 + 4x + 2 = 2(x^2 + 2x + 1) = 2(x + 1)^2$$

EXERCISES 6.8

In exercises 1–50 factor the given trinomial over the integers.

Example (a): $x^2 - 5x + 6 = (x - 2)(x - 3)$

Example (b): $5x^2 + 5x - 30 = 5(x^2 + x - 6) = 5(x + 3)(x - 2)$

1. $x^2 + 3x + 2$
2. $x^2 + 6x + 8$

3. $y^2 + 5y + 6$
4. $w^2 + 4w + 3$

5.	$x^2 + 7x + 10$	28.	$x^2 + 4x - 12$
6.	$x^2 + 8x + 15$	29.	$x^2 + 3x - 40$
7.	$b^2 + 12b + 27$	30.	$a^2 + 5a - 24$
8.	$c^2 + 11c + 24$	31.	$x^2 - 2x - 15$
9.	$x^2 + 15x + 54$	32.	$a^2 - 2a - 24$
10.	$y^2 + 15y + 56$	33.	$b^2 - b - 42$
11.	$x^2 - 4x + 3$	34.	$y^2 - y - 20$
12.	$y^2 - 5y + 6$	35.	$x^2 - 2x - 48$
13.	$a^2 - 4a + 3$	36.	$y^2 - 3y - 4$
14.	$w^2 - 6w + 8$	37.	$s^2 - 7s - 8$
15.	$x^2 - 7x + 12$	38.	$b^2 - 12b - 28$
16.	$y^2 - 9y + 20$	39.	$x^2 - 2x - 63$
17.	$a^2 - 10a + 21$	40.	$y^2 - 2y - 80$
18.	$x^2 - 8x + 15$	41.	$2x^2 + 8x - 24$
19.	$a^2 - 17a + 72$	42.	$4x^2 - 24x + 32$
20.	$b^2 - 14b + 48$	43.	$3x^2 + 9x - 30$
21.	$x^2 + 2x - 15$	44.	$7x^2 + 21x - 28$
22.	$y^2 + 2y - 8$	45.	$x^4 - 12x^2 - 28$
23.	$a^2 + a - 2$	46.	$z^4 + 4z^2 - 96$
24.	$b^2 + b - 6$	47.	$12 - x - x^2$
25.	$x^2 + 3x - 10$	48.	$30 + y - y^2$
26.	$y^2 + 3y - 4$	49.	$72 + b^3 - b^6$
27.	$w^2 + 2w - 15$	50.	$2r^4 + 12r^2 - 80$

In exercises 51–60 replace each ? by a positive number in such a way that each trinomial thus formed will be a trinomial perfect square.

Example (a): $x^2 + \boxed{?} \, x + 9$

If $x^2 + (n + s)x + (n \cdot s) = (x + n)(x + s)$ is to be a perfect square, then $n = s$. Thus, in $x^2 + \boxed{?} \, x + 9$, $n = s = 3$ since $3^2 = 9$ (rejecting $n = s = -3$ since $-3 < 0$). Then $n + s = 6$ and $x^2 + \boxed{?} \, x + 9 = x^2 + 6x + 9 = (x + 3)(x + 3) = (x + 3)^2$.

Example (b): $t^2 + 10t + \boxed{?}$

$t^2 + (n + s)t + (n \cdot s) = (t + n)(t + s)$. If $n = s$, then $n + s = 2n$ and $n \cdot s = n^2$. So $2n = 10$, $n = 5$, and $n^2 = 25$ and we have $t^2 + 10t + \boxed{?} = t^2 + 10t + 25 = (t + 5)(t + 5) = (t + 5)^2$.

51.	$x^2 + \boxed{?} \, x + 4$	53.	$a^2 + \boxed{?} \, a + 81$
52.	$y^2 + \boxed{?} \, y + 16$	54.	$x^2 + 4x + \boxed{?}$

55. $t^2 + 6t + \boxed{?}$

56. $r^2 - 10r + \boxed{?}$

57. $n^2 + \boxed{?} \, n + 49$

58. $b^2 + \boxed{?} \, b + 25$

59. $x^2 - 14x + \boxed{?}$

60. $a^2 + 24a + \boxed{?}$

In exercises 61–78 factor completely over the integers the given trinomial or state that it is irreducible.

Example (a): Factor $ax^2 + 12ax + 36a$.

Solution: $a(x^2 + 12x + 36) = a(x + 6)^2$

Example (b): Factor $x^3y - 3x^2y^2 + 2xy^3$.

Solution: $xy(x^2 - 3xy + 2y^2) = xy(x - 2y)(x - y)$

Example (c): Factor $x^2 + 3x + 3$.

Solution: Irreducible; not factorable over the integers.

61. $y^2 + 2yx - 8x^2$

62. $a^2 + 7ab + 12b^2$

63. $x^2 + 9xy + 18y^2$

64. $m^2 + 5mn + 6n^2$

65. $x^2 + 2xy + y^2$

66. $a^2 - 6ab + 9b^2$

67. $c^2 + 3cd - 28d^2$

68. $x^2 + 3xy - 9y^2$

69. $2x^2 + 8x + 8$

70. $3y^2 - 30y + 75$

71. $y^3x^2 + 9y^2xz - 10yz^2$

72. $3xa^2 + 36x^2a - 135x^3$

73. $36xy - 3xy^2 - 3xy^3$

74. $4x^2 - 8xy - 320y^2$

75. $x^2 + 4ax + 10a^2$

76. $x^2t^2 - 11xyt^2 + 24y^2t^2$

77. $2w^2 - 28wx - 144x^2$

78. $bx^2 - 6bxy + 9by^2$

6.9 Factoring Other Trinomials

When the coefficient a in $ax^2 + bx + c$ is not equal to 1, the process of finding binomial factors is still basically the same — but it can be considerably more complicated. For example, when we attempt to factor $2x^2 - 3x - 5$ we need to consider not only the factors of 5 but also the factors of 2. Study carefully the following examples and then try the exercises. Practice won't necessarily bring perfection, but it will help!

Example 1: $2x^2 - 3x - 5 = (2x - 5)(x + 1)$

Here

$$2 \cdot 1 = 2, \quad (-5) \cdot 1 = -5, \quad \text{and} \quad (2 \cdot 1) + (-5 \cdot 1) = -3$$

It is easy, of course, to find a pair of integers whose product is 2 and a pair of integers whose product is -5. The problem is to decide which pairs will produce the correct coefficient of -3 for x. Experience will help you get the correct choices right away, but be prepared to reject incorrect choices. Thus the choices

$$(2x + 1)(x - 5), (2x - 1)(x + 5), \quad \text{and} \quad (2x + 5)(x - 1)$$

all give the correct coefficient, 2, for x^2 and the correct constant term, -5, but they fail to give the correct coefficient, -3, of x. (Check this by multiplying each of the above pairs of binomials.) Notice that since the constant term is negative we must have one "+" and one "−" in our binomial factors.

Example 2:   Factor $6x^2 - 17x + 12$.

Since $6 = 6 \cdot 1 = 3 \cdot 2$ and $12 = 12 \cdot 1 = 6 \cdot 2 = 4 \cdot 3$, the following unsuccessful attempts might be made:

$$(6x - 3)(x - 4), (6x - 2)(x - 6), (6x - 1)(x - 12),$$
$$(6x - 4)(x - 3), (6x - 6)(x - 2), (6x - 12)(x - 1),$$
$$(3x - 3)(2x - 4), (3x - 2)(2x - 6), (3x - 1)(2x - 12)$$
$$(3x - 4)(2x - 3), (3x - 6)(2x - 2), (3x - 12)(x - 1)$$

Notice, however, that in none of these attempts did we have a "−" in one of our binomial factors and a "+" in the other. This is because the product of the constant terms of the binomials must be 12 and this could not be obtained from, for example, $(6x - 3)(x + 4)$. Also, none of the attempts have a "+" in both binomial factors since we certainly cannot get $-17x$ if both of the constant terms are positive.

Many of these choices of factors, however, can be quickly rejected. For example, in $(6x - 3)(x - 4)$ the fact that $6 \cdot 4 = 24 > 17$ tells us right away that this is not a factorization of $6x^2 - 17x + 12$.

Sooner or later (and sooner with experience!) we arrive at the correct choice of factors and obtain

$$6x^2 - 17x + 12 = (2x - 3)(3x - 4)$$

Example 3:   Factor $4x^2 + 8xy + 3y^2$.

Here it should be clear that both binomials must have a "+". Thus the possible attempts are:

$$(4x + 3y) (x + y), (4x + y) (x + 3y), \text{ and } (2x + 3y) (2x + y)$$

Since $(2x + 3y) (2x + y) = 4x^2 + 8xy + 3y^2$, we have

$$4x^2 + 8xy + 3y^2 = (2x + 3y) (2x + y)$$

for our factorization.

Example 4: Factor $4x^2 + 12xy + 9y^2$

This is another trinomial perfect square and we have

$$4x^2 + 12xy + 9y^2 = (2x + 3y)^2$$

Note that $2\sqrt{4} \cdot \sqrt{9} = 12$ where 4 is the coefficient of x^2, 9 the coefficient of y^2, and 12 the coefficient of xy. Such relationships can help you recognize a trinomial perfect square.

Example 5: Factor $2x^2 + 3x + 6$.

Both binomial factors must have a "+" (why?) and hence the possibilities for factorization are:

$$(2x + 6) (x + 1), (2x + 1) (x + 6), (2x + 2) (x + 3), \text{ and } (2x + 3) (x + 2)$$

Since none of these products yield $2x^2 + 3x + 6$, we conclude that this trinomial is irreducible over the integers.

EXERCISES 6.9

In exercises 1–50 factor the given trinomial over the integers or show that the trinomial is irreducible over the integers.

1.	$2x^2 + 5x + 3$	10.	$2a^2 - 5a + 2$	19.	$2a^2 + 13a - 7$
2.	$2x^2 + 7x + 3$	11.	$2a^2 + 5a + 2$	20.	$2x^2 + x - 6$
3.	$2x^2 - 5x + 3$	12.	$2a^2 + 3a - 2$	21.	$3t^2 - 5t + 2$
4.	$2x^2 - 7x + 3$	13.	$3y^2 - 2y - 5$	22.	$10b^2 + 7b - 12$
5.	$2x^2 + x - 3$	14.	$4x^2 + 11x - 6$	23.	$6x^2 + 11x + 4$
6.	$2x^2 - 5x - 3$	15.	$4x^2 + 6x + 2$	24.	$7y^2 + 6y + 13$
7.	$2x^2 - x - 3$	16.	$10t^2 - t - 3$	25.	$12x^2 + 8x - 15$
8.	$2x^2 + 5x - 3$	17.	$3x^2 + 17x + 10$	26.	$6x^2 - 17x + 12$
9.	$2a^2 - 3a + 2$	18.	$3y^2 + 7y + 2$	27.	$3x^2 - 4xy - 4y^2$

28. $5x^2 + 34xy + 7y^2$ 36. $2a^2 - 13a + 15$ 44. $8 + 2x - 15x^2$
29. $3b^2 + 20b - 7$ 37. $2t^2 + 11t + 15$ 45. $12t^2 + 8t - 15$
30. $2x^2 - 21x + 40$ 38. $6x^2 + 7x - 3$ 46. $18w^2 + 9w - 20$
31. $27t^2 + 3t - 2$ 39. $6y^2 + 13y + 15$ 47. $8c^2 - 2cd + 15d^2$
32. $6a^2 - ab - 12b^2$ 40. $2a^2 + 11a - 21$ 48. $5x^2 + 28xy - 12y^2$
33. $4s^2 - 8st - 5t^2$ 41. $2 - 6x - 8x^2$ 49. $21a^2 + 13ab - 20b^2$
34. $6x^2 - 11x + 4$ 42. $4 + 4y - 15y^2$ 50. $45m^2 - 8mn - 4n^2$
35. $6y^2 + 5y - 4$ 43. $7 - 19a - 6a^2$

6.10 Summary of Factoring

Factoring polynomials over the integers takes much insight, ingenuity, and perseverance. (Good luck will also help!) There is no easy road to success, but the following steps should help you attack the factoring of polynomials.

1. First, check to see if there is a common monomial factor. If there is, use the distributive property to factor it out before trying any other method.

2. Next, determine how many terms are in the polynomial, i.e., if it is a binomial, a trinomial, or a polynomial of more than three terms.

3. If it is a binomial (two terms), by methods we have discussed you can only factor it if it is a difference of two squares:
$$a^2 - b^2 = (a + b)(a - b)$$

4. If it is a trinomial, check to see if it is a trinomial perfect square:
$$a^2 \pm 2ab + b^2 = (a \pm b)^2$$

5. If it is not a trinomial perfect square, then try to factor by trial and error.

6. If there are more than three terms, the polynomial may fall into one of the above types if you group the terms.

Example 1: Factor $3x^2 - 27$.

Solution: $3x^2 - 27 = 3(x^2 - 9)$ Monomial factor

$\qquad\qquad\quad = 3(x - 3)(x + 3)$ Difference of two squares

Example 2: Factor $4at^4 - 64a$.

Solution: $4at^4 - 64a = 4a(t^4 - 16)$ Monomial factor

$$= 4a(t^2 + 4)(t^2 - 4) \qquad \text{Difference of two squares}$$

$$= 4a(t^2 + 4)(t - 2)(t + 2) \qquad \text{Difference of two squares}$$

Example 3: Factor $3y^2 - 18y + 27$.

Solution: $3y^2 - 18y + 27 = 3(y^2 - 6y + 9)$ Monomial factor

$$= 3(y - 3)^2 \qquad \text{Trinomial perfect square}$$

Example 4: Factor $2ab^2 + 4ab - 30a$.

Solution: $2ab^2 + 4ab - 30a = 2a(b^2 + 2b - 15)$ Monomial factor

$$= 2a(b + 5)(b - 3) \qquad \text{Factoring a trinomial}$$

Example 5: Factor $a^2x^2 - a^2y^2 - b^2x^2 + b^2y^2$.

Solution: $a^2x^2 - a^2y^2 - b^2x^2 + b^2y^2$

$$= (a^2x^2 - a^2y^2) - (b^2x^2 - b^2y^2) \qquad \text{Grouping (be careful of signs)}$$

$$= a^2(x^2 - y^2) - b^2(x^2 - y^2) \qquad \text{Monomial factor}$$

$$= (a^2 - b^2)(x^2 - y^2) \qquad \text{Factoring out } (x^2 - y^2) \text{ as a common factor}$$

$$= (a - b)(a + b)(x - y)(x + y) \qquad \text{Difference of two squares}$$

Example 6: Factor $x^2 + x + 1$.

Solution: Since the only possible factorization is $(x + 1)(x + 1)$ (why?) and $(x + 1)(x + 1) = x^2 + 2x + 1 \neq x^2 + x + 1$, we conclude that $x^2 + x + 1$ is irreducible over the integers.

EXERCISES 6.10

Find the prime factors of the given polynomial over the integers or show that it is irreducible over the integers.

1. $8a^2 - 8b^2$
2. $ax^2 - ay^2$
3. $2x^2 - 72$
4. $9x^3 - 16xy^2$
5. $abx^2 + 6abx + 8ab$
6. $3y^2 + 24y - 45$
7. $30 - 4y - 2y^2$
8. $6x^3 + 34x^2 + 20x$
9. $m^3 + 4m^2 + 4m$
10. $y^3 + 3y^2 - 4y$
11. $x^4 - x^2a^2$
12. $2a^2 - 18b^2c^2$
13. $6t^2 - 15t + 9$
14. $10a^2 - 15a + 5$
15. $24a^2b^2 - 3ab^3 - 36b^4$
16. $x^4 - 16$
17. $7a^3 - 63a$
18. $a^2x^3 - 64x^3y^2$
19. $ax^2 - 10ax + 25a$
20. $8y^2 - 4y - 24$
21. $4ax^2 + 38ax + 70a$
22. $3r^2 - 48$
23. $x^4y^4 - 1$
24. $ab^4 - ac^4$
25. $x^8 - y^8$

26. $x^4 - 5x^2y^2 + 4y^4$
27. $2ax^2 - 16ax + 24a$
28. $x^2 + 11xy + 28y^2$
29. $8a^2 + 112a + 392$
30. $10t^2 + 12t + 3$
31. $m^2n + 6mn + 9n$
32. $12s^2 - 28s - 24$
33. $3y^4 - 3$
34. $8x^2 + 8$
35. $xyz - x^7y^5z^3$
36. $x^2 - 4y^2 - x + 2y$
37. $m^2n^2 - m^2 - n^2 + 1$
38. $x^2 - 4y^2 + x - 2y$
39. $axy + bcxy - az - bcz$
40. $t^3 - 2t^2 - 4t + 8$
41. $xy^2 - xz^2 + by^2 - bz^2$
42. $acx + adx + bc + bd$
43. $3y^4 - 2y^3 + 3y^2 - 2y$
44. $a^2 - (b + c)^2$
45. $(a + b)^2 - c^2$
46. $(2x - 1)^2 - (2x - 3)^2$
47. $x^2(2x + 1) - 4(4x^2 - 1)$
48. $4(a - b)^3 - (a - b)$
49. $(2x - 3)^2 - (x - 1)^2$
50. $(x^2 - y^2) - (x - y)^2$

REVIEW EXERCISES

Section 6.1

Use the expressions below to do exercises 1–7.

(a) $\frac{3}{4}x + \frac{1}{2}y$

(b) $3y^2 + 5y + 1$

(c) $\frac{1}{5}a^4b^3c^3 - c^8 + 14a^7c^6$

(d)   25

(e) $\frac{a}{3}$

(f) $5 - x^2$

(g) $\frac{x^3}{3} + \frac{7x^2}{2} + \frac{x}{5} + 6$

(h)   0

(i) $\frac{29}{37}$

(j) $2x^2 - xy$

(k) $\frac{a}{b} + 5a(2 - b) + 5(2 - b) - \frac{3}{8}$

(l) $\frac{4}{2}x^6 - \frac{8}{4}x^4 + \frac{12}{x}$

(m) $\sqrt{7}x$

(n) $\sqrt{5}$

1.   Which of these expressions are polynomials?
2.   Which of the polynomials are monomials?
3.   Which of the polynomials are binomials?
4.   Which of the polynomials are trinomials?
5.   Which of the polynomials are polynomials over the integers?
6.   Which of the polynomials are polynomials over the rational numbers?
7.   What is the degree of each polynomial?

Section 6.2

In exercises 8–20 express the given polynomial as a monomial.

8. $4ab + 3ab$

9. $3xy - 2xy$

10. $5t + 3t + 2t$

11. $3mn - 5mn + 7mn$

12. $\frac{1}{2}y + \frac{1}{3}y$

13. $\frac{1}{2}x - \frac{3}{4}x$

14. $7y + 0.7y - 0.07y$

15. $\frac{1}{3}xy - \frac{1}{6}xy - \frac{1}{30}xy$

16. $9ab^2 - 6ab^2 - 3ab^2$

17. $-4abx - 6abx$

18. $x^3 - x^3$

19. $-9x^2y + 9x^2y$

20. $\frac{1}{2}x^2y^3 - 7x^2y^3 + \frac{1}{4}x^2y^3$

In exercises 21–30 simplify the given polynomial.

21. $2x^2 - 5 + 2x^2 + 3$

22. $9a^2 - 4b^2 - 2a^2 + 7b^2$

23. $-2x - 9 + x - 2$

24. $4x + 4y - 3 + 2x + 8y - 5$

25. $9x^2 - 4x + 5x^2 + 6 + 4x - 3$

26. $4x^2 - 5x + 3x + 2x^2 - 4$

27. $-3a - 2c + 5a + 4b$
28. $3ab - 4bc + 4bc - 3ab$

29. $x^2 + xy + xy + y^2$
30. $3a + 5b + 2a - 4c - 5b - 4c$

In exercises 31–50 simplify the given sum or difference.

31. $(6x - 2y) + (3x + 5y)$
32. $(2a - 3b + 4c) + (7a - 5b - 3c)$
33. $(a^2 + 2ab) + (-3ab - b^2)$
34. $(a^2 - 6ax) + (6ax - x^2)$
35. $(x^2 + y^2) - (2x^2 - 3y^2)$
36. $(4cd - d^2) - (cd + 2d^2)$
37. $(5m^3 - n^2) - (-2m^3 + 4n^2)$
38. $(4a + 3b + 6c) - (2a + b + 2c)$
39. $(4a^2 - 2ab - 9b^2) + (-a^2 + 4ab - b^2)$
40. $(3x^3 + 2x^2 - 4x) + (2x^3 + 8x^2 - 5x)$
41. $(3r^3 - 1) - (2r^3 - 4r^2 - 5r)$
42. $(2c^2 - cd + 3d^2) - (c^2 - 2d^2)$
43. $(x^2 - x + 2) - (2x^2 + x - 2)$
44. $(b^3 + 3b^2 - 4b) + (-3b^2 + 4b - 27)$
45. $(5y^2 - 3y + 2) + (4y^2 - 7y)$
46. $(x - a + 2b) - (x - a + 2b)$
47. $0 - (4y^2 - 2y^2)$
48. $(t^2 - t + 2) - (2t^2 + t - 2)$
49. $(2b - 2a - c) + (-6b - 3c)$
50. $(6a^2b^2 + 3ab - 5) + (-5a^2b^2 - 4ab)$

Section 6.3

In exercises 51–71 express the given product as a monomial or as the sum of monomials.

51. $(x^2)(x^3)$
52. $(7a^6)(2a^5)$
53. $(-5b^5)(-5b^5)$
54. $(10y^6)(-4y)$
55. $(4a^2b)(2a^3)$
56. $(2y^2)(3y^4z)$

57. $(4xyz^2)(-7xz^2)$
58. $(-6a^2x)(2ab)$
59. $(-9ab^3)(-3bc^2)$
60. $(5a^2y)(9a^5y^2)$
61. $-x(-y)$
62. $(-4a^2b)(5ab^2)$

63. $(-4m^4)\,(4m^4)$
64. $(xy^2)\,(-2x)\,(3xy)\,(-4x^2y)$
65. $(-7ab^2)\,(a^4b)\,(-3ab^3)$
66. $x(x + 3)$
67. $-a(a^2 - 3)$

68. $6x(4x^2 + 3x)$
69. $-3a(2b^2 - 5b)$
70. $\frac{1}{2}(4y^2 - 8y + 6)$
71. $-6xy^2(x^3 - x^2y^2 - 3y^4)$

Section 6.4

In exercises 72–90 express the given product as a polynomial.

72. $(a + 6)\,(a + 2)$
73. $(x + y)\,(x + y)$
74. $(a - b)\,(a - b)$
75. $(5xy - 3)\,(3xy - 8)$
76. $(x + y)\,(x - y)$
77. $(t^2 - 3)\,(t^2 + 5)$
78. $(2x + 5)\,(3x + 4)$
79. $(6x - 9)\,(-2x - 5)$

80. $(8 + 2b)\,(5 + 7b)$
81. $(2a - 4)\,(9 + 5a)$
82. $(2x^2y - 8)\,(-x^2y + 7)$
83. $(5a + 3b)\,(4a - 3b)$
84. $(a^2 + 4a + 4)\,(a - 2)$
85. $(y + 2)\,(1 + 8y - y^2)$
86. $(x + y + z)\,(x - y - z)$

87. $(x^2 + x + 3)\,(x^2 - x - 2)$
88. $(4a^2 - 3a + 2)\,(a^2 + 5a + 4)$
89. $(x^3 - x^2 + x - 1)\,(x^2 - x + 1)$
90. $(y^4 + 2y^3 - 3y^2 + 7y + 5)\,(3y - 4)$

In exercises 91–100 perform the multiplications and then simplify the resulting sum.

91. $2(x + 3) + 7$
92. $5(3a + 5) - 3a$
93. $4(2x - 3y) - 7y - 3x$
94. $4t - 3t\,(5t - 4) + 20t^2$
95. $(y + 2)\,(y + 3) + (y - 2)\,(y - 1)$
96. $(a + 2)\,(a + 2) + (a + 2)\,(a + 4)$
97. $8x - (2x + 3)\,(2x + 5) - 3x(x + 2)$
98. $3mn - (m - n)\,(m - 2n) - 5m(m + n)$

99. $(5x - 3y)(2x - 7y) - (3x + 2y)(3x - 5y)$
100. $y(x - y) - x(x + y) + x^2 - (x - y)(x + y)$

Section 6.5

In exercises 101-120 find the solution set of the given equation or inequality.

101. $3(x - 1) - 2(3 - x) -4x = 1$
102. $5(2y + 1) - 14 = 3(3y - 2)$
103. $x(x + 1) - 2(4 - x) > x^2 + 1$
104. $3(2 - 5a) + a(3 - a) = 6 - a^2$
105. $y(y + 5) - 4(3 - y) = y^2 + 6$
106. $3(x + 1) - 4(1 - x) < 13$
107. $2(1 - 3b^2) = 3(2 - b) - 6b^2$
108. $5(x + 2) = -(3 - 2x)$
109. $2(x - 2)(x - 3) \geqslant (2x - 3)(x + 4)$
110. $(x + 2)(2x - 1) = x(2x + 5)$
111. $(2a + 1)(2a - 3) = (4a + 3)(a - 1)$
112. $(x - 7)(3x - 5) < (2x + 3)(x - 11) + x^2 + 5$
113. $2(x + 1)(x - 1) + 3(x + 2)(x - 3) = 5x^2$
114. $4(w + 1)(w + 1) + 6(w + 2)(w + 3) = 2(5w^2 + 1)$
115. $(x + 7)(2x - 1) \leqslant (2x + 3)(x - 15) - 12$
116. $2 - (a + 1)^2 = a - (a - 2)^2$
117. $2(3y + 7)^2 = (y + 1)(2y + 1) - (7 - y - 16y^2)$
118. $(x + 2)(x + 1) - (x - 2)(x - 1) < 18$
119. $2(x - 15)(13 - 3x) - (2 - 3x)(5 + 2x) = 0$
120. $(x + 3)^2 + 25 = (x - 3)^2 + 9$

Section 6.6

In exercises 121-140 factor the given polynomial by finding a common monomial factor and using the distributive property.

121. $4x - 8y$
122. $10a - 5b$
123. $m^2 - 2mn$
124. $a^3 + 6a^2$
125. $3a^2 - 12a$
126. $18a^2xy + 20x^2y$

127. $x^2y + 3x^3y^2 - 5xy^2$

128. $2a^3 + a^2 - 3a$

129. $4mn - 8m + 6n$

130. $6x^6 - 36x^8 - 18x^2$

131. $2x^3y + 4x^2y^2 + 16xy^3$

132. $\frac{1}{8}m^2n - \frac{3}{8}mn^2 - \frac{7}{8}mn$

133. $2x^2 - 4xy + 12y^2$

134. $a^6 - 2a^4 + 3a^2$

135. $\frac{1}{4}t^2 - \frac{3}{8}t^2r + \frac{7}{16}tr^2$

136. $-4xy - 2yz - 2$

137. $24a^6 - 12a^4 + 42a^3$

138. $\frac{hB}{2} + \frac{hG}{2}$

139. $vt + \frac{1}{2}gt^2$

140. $2a^5 - 4a^4 + 8a^3 + 10a^2$

In exercises 141-150 factor out a common binomial.

141. $r(s + t) + n(s + t)$

142. $2a(x - y) - b(x - y)$

143. $4(a^2 + b^2) - x(a^2 + b^2)$

144. $a(x - y) + c(x - y)$

145. $3z(a^2 + b^2) - (a^2 + b^2)$

146. $ax + bx + 3a + 3b$

147. $x^2 - ax + bx - ab$

148. $ax - 3by + ayx - 3b$

149. $6x + 6y - 4x - 4y$

150. $ab^2 - b^2c - ad + cd$

Section 6.7

In exercises 151-165 factor completely the given polynomial. If any common factors occur they should be factored out first.

151. $a^2 - 4$

152. $1 - x^2$

153. $4b^2 - 1$

154. $9t^2 - 4r^2$

155. $49x^2 - 16y^2$

156. $x^2 - \frac{1}{4}$

157. $\frac{1}{16} - m^2$

158. $n^4 - 1$

159. $x^6 - 1$

160. $3 - 27x^4$

161. $2a^2 - 8b^2$

162. $12m^2 - 27n^2$

163. $a^5b^2 - a^3$

164. $\frac{3}{4}x^2 - \frac{27}{16}y^2$

165. $5a^2 - 20b^4$

Section 6.8

In exercises 166-175 factor the given trinomial completely.

166. $x^2 + 5x + 6$
167. $b^2 - 12b + 36$
168. $a^2 - 2a + 3$
169. $x^2 + 2x + 1$
170. $t^2 - 7t + 12$

171. $m^2 - 10m + 16$
172. $x^2 + 4x - 5$
173. $a^2 + 6ax + 9x^2$
174. $x^2 - 2xy - 3y^2$
175. $x^2 + 2xy - 3y^2$

Section 6.9

In exercises 176–189 factor the given trinomial completely. If any common factors occur they should be factored out first.

176. $3a^2 + 24a + 48$
177. $5a^2 + 10a - 15$
178. $2ax^2 + 6ax - 20a$
179. $5x - 36 + x^2$
180. $x^2 - 7xy + 10y^2$
181. $2x^2 - 3x - 9$
182. $9a^2 - 12a + 4$

183. $2m^2 + 9mn - 5n^2$
184. $10x^2 - 19xy + 6y^2$
185. $6a^2 - 7ab - 3b^2$
186. $6n^2 - n - 7$
187. $2x^2 - \frac{5}{2}x - \frac{3}{4}$
188. $2a^4 - 12a^2 + 16$
189. $6x^2 + 28x - 10$

Section 6.10

In exercises 190–219 factor the given polynomial completely.

190. $x^2 + 10x + 9$
191. $m^2 - n^2$
192. $mx + nx + 5m + 5n$
193. $3x^2 - 12x - 15$
194. $x^2 + 10x + 25$
195. $2a^2 + 2a + 1$
196. $9x^2 - 1$
197. $9a^2b^2 + 6abk + k^2$
198. $6a^2 + 3ab - 2ba - b^2$
199. $6x^2 - 24$
200. $(2x - y)^2 - 1$
201. $3x^2 - ax + 3xy - ay$
202. $x^2 + 5x + 3$

203. $3x^4y^4 - 3$
204. $by^4 - 64b^5$
205. $ab^2 + 7ab - 30a$
206. $a^4 + 2a^2k^2 - 35k^4$
207. $y^4 + y^2 - 42$
208. $9x^2 - 29xy + 20y^2$
209. $a^3 + 3a^2 + 6a + 18$
210. $ky^2 - 6k^2y + 5k^3$
211. $4a^2x + 9ax - 2x$
212. $a^2b - bc^2 + a^2c - c^3$
213. $x^2 - 3x - ax + 3a$
214. $16s^4 + 40s^2t^2 + 25t^4$
215. $36x^2 + 168x + 196$

216. $a^2k + 10ak + 9k$

217. $5t^2 + 33t + 48$

218. $10kr + 21js + 6jr + 35ks$

219. $3x^4y + 33x^2y + 90y$

CUMULATIVE REVIEW D

1. $-12 - (-8) -11 = ?$
2. What must be added to -5 to make a sum of -2?
3. True or false: It is possible to find two negative numbers such that their difference is zero. If your answer was "true," give an example.
4. $\dfrac{3}{4} - \dfrac{6-9}{5} = ?$
5. $\dfrac{3}{7} - \dfrac{3}{14} - \dfrac{2}{7} = ?$
6. The fraction $\frac{7}{3}$ written as a repeating decimal is ___?___ .
7. What fraction in lowest terms is equal to 0.32?

In exercises 8–20 find the solution set of the given equation or inequality. The domain of the variable is the set of real numbers.

8. $5 - 2x \geqslant 12$
9. $4x - 12 = 6x + 16$
10. $7x - 12 + 2x = -8x + 14 + 4x$
11. $6(2x - 3) -4(5 - x) = 26$
12. $6 + x = 5x - 3(x + 2)$
13. $3x - (2x + 6) = 4x - 3$
14. $(2x + 4) - (3 - x) = 4x + 2$
15. $3 - 4x > 11$
16. $2x + 3 > 3(x - 1)$
17. $6 - 10x < 12 - 13x$
18. $x + 3(x + 1) < 2x + 17$
19. $16 < 5x + 2(x + 8)$
20. In the formula $f = \dfrac{pq}{p + q}$ solve for q.

In exercises 21 and 22 graph the given equation.

21. $3x - 2y = 5$

22. $2x + 3y = 5$

In exercises 23 and 24 find the solution set of the given system of equations.

23. $\begin{cases} 3x - 4y = 17 \\ 2x + 3y = 0 \end{cases}$

24. $\begin{cases} 7x - 9y = 15 \\ 8y - 5x = -17 \end{cases}$

25. Enough tea at $1.60 a kilogram is mixed with 20 kilograms of tea that sells for $1.28 a kilogram so that the mixture may sell at $1.40 a kilogram. How many kilograms of the $1.60 tea were used?

26. The larger of two integers is 2 more than the smaller. If 3 times the smaller added to twice the larger is 89, what are the integers?

27. A car is stolen at 4:00 P.M. and driven away at 50 miles per hour. A police car leaves the scene of the crime 10 minutes later at 70 miles per hour to give chase. At what time will the police car overtake the stolen car?

28. A total of $5000 is invested in two companies, A and B. Company A pays 10% interest and company B pays 12%. How much is invested in each company if the total return is $524?

29. One pipe can be used alone to fill a tank in 5 minutes. A second pipe can be used alone to fill the tank in 7 minutes. A third pipe can be used alone to fill the tank in 10 minutes. How long will it take, to the nearest minute, to fill the tank if all three pipes are used at the same time?

In exercises 30–35 simplify the given expression so that no negative exponents appear and no irrational numbers are in the denominator.

30. $x^2(x^{-3} + x^{-1})$

31. $\sqrt{128}$

32. $\sqrt{\dfrac{5}{8}}$

33. $\sqrt{98} + 3\sqrt{50} - 5\sqrt{8}$

34. $\sqrt{6} \cdot 5\sqrt{10}$

35. $\dfrac{\sqrt{3}}{\sqrt{2}}$

In exercises 36–41 perform the indicated operation.

36. $-3x^2(-2x - 4a + 2)$

37. $(p - q + 2) - (2p - 2q) + (3p - 4r)$

38. $3a(4a - 7) - 1$

39. $(x + 4)(x + 3)$

40. $(2a - 3)^2$

41. $(x + 2)(x^2 - 2x + 4)$

In exercises 42–50 factor the given polynomial completely.

42. $y^2 + 9y + 8$

43. $14xy^5 - 20xz$

44. $2x^2 + 5x + 3$

45. $64 - 16x + x^2$

46. $16a^4 - 81b^2$

47. $x^2 + 5x - bx - 5b$

48. $2y^2 - 6xy - 9xy + 18x^2$

49. $4a^2 + 20ab + 25b^2$

50. $2x^2 - 26x + 84$

7

RATIONAL ALGEBRAIC EXPRESSIONS

7.1 Definitions and Basic Principles

As we have seen, any two polynomials over the real numbers can be added, subtracted, or multiplied to form another polynomial over the real numbers. However, it is not always possible to divide one polynomial by another and get another polynomial as the quotient. Here, of course, we use the same definition for division of polynomials as we did for real numbers. Just as $a \div b = c$ if and only if $a = b \cdot c$ for real numbers a, b, and c, so $P \div G = Q$ if and only if $P = G \cdot Q$ for polynomials P, G, and Q. For example:

$$\overset{a}{\underset{\downarrow}{12}} \div \overset{b}{\underset{\downarrow}{3}} = \overset{c}{\underset{\downarrow}{4}} \qquad \text{because} \qquad \overset{a}{\underset{\downarrow}{12}} = \overset{b}{\underset{\downarrow}{3}} \cdot \overset{c}{\underset{\downarrow}{4}}$$

and

$$(\overset{P}{\underset{\downarrow}{x^2 - 5x + 6}}) \div (\overset{G}{\underset{\downarrow}{x - 3}}) = \overset{Q}{\underset{\downarrow}{x - 2}} \text{ because } \overset{P}{\underset{\downarrow}{x^2 - 5x + 6}} = (\overset{G}{\underset{\downarrow}{x - 3}})(\overset{Q}{\underset{\downarrow}{x - 2}})$$

It is easy to see, then, that under this definition

$$(x - 1) \div (x + 1)$$

is not a polynomial since there is no polynomial Q such that

$$x - 1 = (x + 1) \cdot Q$$

Just as the set of integers can be extended to the set of rational numbers, so the set of polynomials can be extended to the set of *rational algebraic expressions*

$$\frac{P}{Q}$$

where P and Q are polynomials and $Q \neq 0$. Thus

$$\frac{x^2 + x + 1}{x - 1} \,, \quad \frac{x + y}{2x - y} \,, \quad \sqrt{3}x \,, \quad \frac{2x}{x - 1} \,, \frac{t^2 - 1}{\sqrt{5t + 2}} \,, \frac{\frac{1}{2}x}{\pi y}$$

are examples of rational algebraic expressions over the real numbers.

Note that just as we can consider integers as special kinds of fractions with denominator 1, so we can consider polynomials such as $\sqrt{3}x$ as special kinds of rational algebraic expressions with denominator 1. Note also that the *coefficients* of the variables need not be rational numbers: $\sqrt{3}, \sqrt{5}, \pi$, etc., are all permissible coefficients when we are considering rational algebraic expressions over the real numbers.

When the numerical coefficients in a rational algebraic expression are real numbers, we say that we have a rational algebraic expression over the real numbers. If the coefficients are rational numbers, we have a rational algebraic expression over the rational numbers, and if the restriction is to integers, we have a rational algebraic expression over the integers. A rational algebraic expression over the rational numbers can always be written as a rational algebraic expression over the integers. Thus,

$$\frac{1}{2}x = \frac{x}{2} \quad \text{and} \quad \frac{\frac{1}{3}x - \frac{1}{2}y}{\frac{1}{6}x + 2y} = \frac{6\left(\frac{1}{3}x - \frac{1}{2}y\right)}{6\left(\frac{1}{6}x + 2y\right)} = \frac{2x - 3y}{x + 12y}$$

Just as it is often useful to reduce a fraction to lowest terms, so it is often useful to reduce a rational algebraic expression over the integers to *lowest terms* — meaning that the numerator and the denominator are polynomials over the integers having no common factors other than 1 and –1.

To reduce a rational algebraic expression to lowest terms we use the following extension of our basic principle for fractions (Section 2.3):

If $P, Q,$ and R, with Q and $R \neq 0$, are any polynomials, then

$$\frac{P}{Q} = \frac{R \cdot P}{R \cdot Q}$$

Thus

$$\frac{x + 1}{x + 2} = \frac{(x + 3)(x + 1)}{(x + 3)(x + 2)}$$

so that, given $\frac{x^2 + 4x + 3}{x^2 + 5x + 6}$, we can write

$$\frac{x^2 + 4x + 3}{x^2 + 5x + 6} = \frac{(x + 3)\,(x + 1)}{(x + 3)\,(x + 2)} = \frac{x + 1}{x + 2}$$

This basic principle, together with an ability to factor polynomials, enables us to perform such reductions to lowest terms as shown in the following examples.

Example 1: $\dfrac{ax + ay}{ax - ay} = \dfrac{a(x + y)}{a(x - y)} = \dfrac{x + y}{x - y}$

Example 2: $\dfrac{ax + a}{x^2 + 3x + 2} = \dfrac{a(x + 1)}{(x + 2)\,(x + 1)} = \dfrac{a}{x + 2}$

Example 3: $\dfrac{3y^2 - 48}{y^2 + 8y + 16} = \dfrac{3(y^2 - 16)}{(y + 4)\,(y + 4)} = \dfrac{3(y + 4)\,(y - 4)}{(y + 4)\,(y + 4)} = \dfrac{3(y - 4)}{y + 4}$

Example 4: $\dfrac{4 - y^2}{y^2 - 4y + 4} = \dfrac{(2 - y)\,(2 + y)}{(y - 2)\,(y - 2)}$

To proceed further in Example 4 we note that $2 - y = (-1)\,(y - 2)$ and so we have

$$\frac{(2 - y)\,(2 + y)}{(y - 2)\,(y - 2)} = \frac{(-1)\,(y - 2)\,(y + 2)}{(y - 2)\,(y - 2)} = \frac{(-1)\,(y + 2)}{y - 2} = -\frac{y + 2}{y - 2}$$

Example 5: $\dfrac{x - y}{y - x} = \dfrac{(-1)\,(y - x)}{y - x} = -1$

Just as rational numbers are often called fractions, so it is common practice to call rational algebraic expressions *algebraic fractions.*

EXERCISES 7.1

In exercises 1–15 show how the right-hand expression can be obtained from the left-hand expression.

Example (a): $\dfrac{a}{a - b} = \dfrac{-a}{b - a}$.

Solution: $\dfrac{a}{a - b} = \dfrac{(-1)\,(a)}{(-1)\,(a - b)} = \dfrac{-a}{-a + b} = \dfrac{-a}{b - a}$

Example (b): $\dfrac{x-y}{a-b} = \dfrac{y-x}{b-a}$.

Solution: $\dfrac{x-y}{a-b} = \dfrac{-1(x-y)}{-1(a-b)} = \dfrac{-x+y}{-a+b} = \dfrac{y-x}{b-a}$

Example (c): $\dfrac{x-y}{a-b} = -\dfrac{x-y}{b-a}$.

Solution: $\dfrac{x-y}{a-b} = -\dfrac{x-y}{-(a-b)} = -\dfrac{x-y}{-a+b} = -\dfrac{x-y}{b-a}$

1. $\dfrac{y}{x-y} = -\dfrac{y}{y-x}$

2. $\dfrac{2}{x-1} = \dfrac{-2}{1-x}$

3. $\dfrac{5}{3x-2} = \dfrac{-5}{2-3x}$

4. $\dfrac{x-4}{7} = -\dfrac{4-x}{7}$

5. $\dfrac{a-b}{x-y} = \dfrac{b-a}{y-x}$

6. $\dfrac{b-a}{y-x} = -\dfrac{a-b}{y-x}$

7. $\dfrac{x-1}{2-x} = -\dfrac{x-1}{x-2}$

8. $\dfrac{3x-5}{1-x} = -\dfrac{3x-5}{x-1}$

9. $\dfrac{x^2-xy-y^2}{a-b} = \dfrac{y^2+yx-x^2}{b-a}$

10. $\dfrac{-a-b}{x-y} = \dfrac{a+b}{y-x}$

11. $\dfrac{(a-b)\,(a+b)}{x-y} = \dfrac{(b-a)\,(b+a)}{y-x}$

12. $\dfrac{x+y}{y-x} = -\dfrac{x+y}{x-y}$

13. $\dfrac{-6m^2+2m-4}{m+2} = -\dfrac{6m^2-2m+4}{m+2}$

14. $\dfrac{-(x-y)}{(y-x)\,(x+y)} = \dfrac{1}{x+y}$

15. $\dfrac{(x+y)\,(2x-y)}{(a+b)\,(2a-b)} = \dfrac{(y+x)\,(y-2x)}{(b+a)\,(b-2a)}$

In exercises 16–63 reduce the given fraction to lowest terms.

16. $\dfrac{21}{36}$

17. $\dfrac{24}{42}$

18. $\dfrac{42}{64}$

19. $\dfrac{5x}{9x}$

20. $\dfrac{11y^2}{15y^2}$

21. $\dfrac{5abc}{11abc}$

22. $\dfrac{rsy}{rty}$

23. $\dfrac{15a^2b}{9ab}$

24. $\dfrac{10mn^2}{15mn}$

25. $\dfrac{7x^2y}{14xy^2}$

26. $\dfrac{10x^3y^2}{12x^2y^3}$

27. $\dfrac{32x^3y^2z}{56x^2y^2z^2}$

28. $\dfrac{2(x + 5)}{5(x - 5)}$

29. $\dfrac{2a(c + d)}{3a^2(c + d)}$

30. $\dfrac{5m - 5n}{3m + 3n}$

31. $\dfrac{a + b}{7a + 7b}$

32. $\dfrac{3m + 3n}{5m + 5n}$

33. $\dfrac{4x - 4}{4x + 4}$

34. $\dfrac{x^2 - x}{x^2 + x}$

35. $\dfrac{4a + 12}{6a^2 + 18a}$

36. $\dfrac{3x + 3y}{x^2 - y^2}$

37. $\dfrac{4m - 4n}{m^2 - n^2}$

38. $\dfrac{br - bs}{br + bs}$

39. $\dfrac{p^2 - q^2}{(p - q)^2}$

40. $\dfrac{2x + 6y}{x - 3y}$

41. $\dfrac{4r - 6s}{4r^2 + 9s^2}$

42. $\dfrac{(x - 3)^2}{5x - 15}$

43. $\dfrac{2(a - b)^2}{3a - 3b}$

44. $\dfrac{3ay + a}{ay - 3a}$

45. $\dfrac{x^2 - 2x + 1}{x^2 - 1}$

46. $\dfrac{x^2 - 2x}{x^2 - 5x + 6}$

47. $\dfrac{c^2 - 9}{c^2 - 6c + 9}$

48. $\dfrac{m^2 - 5m}{m^2 - 4m - 5}$

49. $\dfrac{a^2 + 3ab + 2b^2}{a^2 - 4b^2}$

50. $\dfrac{9x^2 - 25}{3x^2 - 11x + 10}$

51. $\dfrac{x^2 - 6x + 9}{x^2 - 4x + 3}$

52. $\dfrac{y^2 + 3y - 18}{y^2 - 8y + 15}$

53. $\dfrac{ab - ac - ad}{rb - rc - rd}$

54. $\dfrac{ax^2 + 7ax - 18a}{ax^2 - 3ax + 2a}$

55. $\dfrac{2b^2 + 7b + 3}{b^2 + 2b - 3}$

56. $\dfrac{6c^2 + c - 1}{6c^2 + 5c + 1}$

57. $\dfrac{3x^2 - 8x + 5}{3x^2 + x - 10}$

58. $\dfrac{2y^2 - 2y - 24}{4y^2 - 20y + 16}$

59. $\dfrac{6 - 2x}{x^2 - 9}$

60. $\dfrac{2s - 2r}{4r^2 - 4s^2}$

61. $\dfrac{(y - x)^2}{x^2 - y^2}$

62. $\dfrac{2m^2 - 2}{4 + 4m}$

63. $\dfrac{3 - 3t^2}{4 + 4t}$

7.2 Multiplication and Division

Multiplication and division of rational algebraic expressions are defined in the same way that multiplication and division of rational numbers are defined.

Thus if P_1, P_2, P_3, and P_4 are polynomials, we define

$$\frac{P_1}{P_2} \cdot \frac{P_3}{P_4} = \frac{P_1 \cdot P_3}{P_2 \cdot P_4} \quad \text{and} \quad \frac{P_1}{P_2} \div \frac{P_3}{P_4} = \frac{P_1}{P_2} \cdot \frac{P_4}{P_3}$$

The following examples illustrate multiplication and division of rational algebraic expressions.

Example 1: $\dfrac{x+2}{x+3} \cdot \dfrac{x+1}{x-1} = \dfrac{(x+2)\,(x+1)}{(x+3)\,(x-1)} = \dfrac{x^2+3x+2}{x^2+2x-3}$

Example 2: $\dfrac{r+s}{r-s} \cdot \dfrac{2r-s}{r-s} = \dfrac{(r+s)\,(2r-s)}{(r-s)^2} = \dfrac{2r^2+rs-s^2}{r^2-2rs+s^2}$

Example 3: $\dfrac{a-b}{c+d} \div \dfrac{c-d}{a-b} = \dfrac{a-b}{c+d} \cdot \dfrac{a-b}{c-d} = \dfrac{(a-b)^2}{(c+d)\,(c-d)}$

$$= \dfrac{a^2-2ab+b^2}{c^2-d^2}$$

When there are common factors in numerators and denominators, and we wish to express the answer in lowest terms, the procedure illustrated in the following examples is useful.

Example 4: $\dfrac{3x+15}{6x-18} \cdot \dfrac{2x+6}{x^2+5x} = \dfrac{3(x+5)\,(2)\,(x+3)}{6(x-3) \cdot x \cdot (x+5)}$

$$= \dfrac{6(x+5)\,(x+3)}{6(x+5) \cdot x(x-3)} = \dfrac{x+3}{x(x-3)}$$

We can, of course, also write the answer to Example 4 as $\dfrac{x+3}{x^2-3x}$. Frequently, however, it is more useful to leave the numerator and denominator in factored form.

Example 5: $\dfrac{x^2+x}{3y-3} \div \dfrac{2x}{3y} = \dfrac{x(x+1)}{3(y-1)} \cdot \dfrac{3y}{2x} = \dfrac{3x \cdot y\,(x+1)}{3x \cdot 2(y-1)}$

$$= \dfrac{y(x+1)}{2(y-1)}$$

Example 6: $\dfrac{x + y}{x - y} \div (x + y) = \dfrac{x + y}{x - y} \div \dfrac{x + y}{1}$

$$= \dfrac{x + y}{x - y} \cdot \dfrac{1}{x + y} = \dfrac{(x + y) \cdot 1}{(x + y)(x - y)}$$

$$= \dfrac{1}{x - y}$$

EXERCISES 7.2

In exercises 1–50 find the given product and express it in lowest terms.

1. $\dfrac{2}{3} \cdot \dfrac{5}{7}$

2. $\dfrac{3}{8} \cdot \dfrac{4}{15}$

3. $\dfrac{3a}{5b} \cdot \dfrac{b}{2a}$

4. $\dfrac{4x}{5y} \cdot \dfrac{5y}{8x}$

5. $\dfrac{10m}{9n} \cdot \dfrac{3n}{5m}$

6. $\dfrac{4ax}{15by} \cdot \dfrac{3ay}{8bx}$

7. $\dfrac{5c}{8d} \cdot \dfrac{12d}{10c}$

8. $\dfrac{4x^2}{6y} \cdot \dfrac{2y}{8x}$

9. $\dfrac{9a^2 b}{bc} \cdot \dfrac{c^2}{3ab^2}$

10. $\dfrac{18xy}{25ab} \cdot \dfrac{10ax}{27by}$

11. $\dfrac{2r^2}{3t^2} \cdot \dfrac{9t}{8r^2}$

12. $\dfrac{100ab}{63c^2} \cdot \dfrac{7ac}{16b}$

13. $\dfrac{28xy^2}{91z} \cdot \dfrac{65z^3 y}{4x}$

14. $\dfrac{21ab^2 c}{13x^2 yz^4} \cdot \dfrac{39x^3 yz^2}{28ab^3 c^3}$

15. $\dfrac{4r^2 h}{3} \cdot \dfrac{1}{r^2 h}$

16. $\dfrac{4x}{3} \cdot \dfrac{9}{7x - 3}$

17. $\dfrac{2r}{3s} \cdot \dfrac{s - 1}{r + 2}$

18. $\dfrac{t + w}{t - u} \cdot \dfrac{w}{t}$

19. $\dfrac{3 + a}{4 - b} \cdot \dfrac{a - 3}{4 + b}$

20. $\dfrac{b + 3}{y} \cdot \dfrac{y}{b - 3}$

21. $\dfrac{a^2 - b^2}{10a^2} \cdot \dfrac{5a^3}{(a - b)^2}$

22. $\dfrac{6m - 12}{3m + 9} \cdot \dfrac{5m + 15}{4m - 8}$

23. $\dfrac{x + 1}{x} \cdot \dfrac{x^2}{x^2 + x}$

24. $\dfrac{a}{b + c} \cdot \dfrac{b^2 - c^2}{2a}$

25. $\dfrac{rm - rx}{5m + 5x} \cdot \dfrac{tm + tx}{wm - wx}$

26. $\dfrac{a - 3}{2a - 1} \cdot \dfrac{4a^2 - 1}{a^2 - 9}$

27. $\dfrac{bc}{4c - 4d} \cdot \dfrac{2c - 2d}{ac}$

28. $\dfrac{4a^2 - b^2}{a^2 - b^2} \cdot \dfrac{a + b}{2a + b}$

29. $\dfrac{x^2 + 2x}{2x - 1} \cdot \dfrac{10x^2 - 5x}{12x^3 + 24x^2}$

30. $\dfrac{3a + 3b}{2a - 2b} \cdot \dfrac{a^2 - b^2}{(a + b)^2}$

31. $\dfrac{k^2 - 4}{(k - 2)^2} \cdot \dfrac{7k - 14}{3k + 6}$

32. $\dfrac{8 - 4x}{4x + 4} \cdot \dfrac{x^2 - 1}{6 - 3x}$

33. $\dfrac{5t + 15}{8t + 4} \cdot \dfrac{4t + 2}{3t + 9}$

34. $\dfrac{x^2 - 8x + 15}{x^2 + 9x + 14} \cdot \dfrac{x^2 - 6x - 16}{x^2 + 4x - 21}$

35. $\dfrac{a^2 + 4a}{2a^2 + 9a + 4} \cdot \dfrac{2a^2 + 5a + 2}{a^2 - 4}$

36. $\dfrac{y^2 + 3y - 10}{3y^2 + 22y + 35} \cdot \dfrac{6y^2 + 11y - 7}{3y^2 + y - 14}$

37. $\dfrac{7t^2 - 7t}{t^2 - 2t + 1} \cdot \dfrac{t^2 - 4t + 3}{42t^2}$

38. $\dfrac{a^2 - 2a - 3}{a^2 - 9} \cdot \dfrac{a^2 + 6a + 9}{a^2 - 1}$

39. $\dfrac{4x + 12}{6x^2 + 24x + 24} \cdot \dfrac{3x^2 - 12}{7x + 21}$

40. $\dfrac{6y - 4x}{4w^2 - 9z^2} \cdot \dfrac{2w + 3z}{2x - 3y}$

41. $\dfrac{x^2 - 6xy - 27y^2}{x^2 - 12xy - 45y^2} \cdot \dfrac{x^2 - 14xy - 15y^2}{x^2 - 4xy - 45y^2}$

42. $\dfrac{y^2 + 9y + 18}{y^2 - 7y + 10} \cdot \dfrac{y^2 - y - 20}{y^2 + 7y + 12}$

43. $\dfrac{3a^2 - 20a - 7}{2a^2 - 3a - 5} \cdot \dfrac{4a^2 - 8a - 5}{3a^2 + 4a + 1}$

44. $\dfrac{b^2 - 9}{b^2 + 6b + 9} \cdot \dfrac{2b^2 + 18}{b^4 - 81}$

45. $\dfrac{x^2 + xy}{x + y} \cdot \dfrac{x^2 + xy + y^2}{x^2 - xy} \cdot \dfrac{x - y}{x^3 + x^2y + xy^2}$

46. $\dfrac{1 - a^2}{4a^2} \cdot \dfrac{3ab}{2 - 2a} \cdot \dfrac{8}{ab + a}$

47. $\dfrac{2b + 10}{3b - 9} \cdot \dfrac{12b - 36}{5b + 5} \cdot \dfrac{ab + a}{4b + 20}$

48. $\dfrac{xy - 2y^2}{x^2 - xy} \cdot \dfrac{(x - y)^2}{x^2 + xy} \cdot \dfrac{x^2 - y^2}{x^2 - 3xy + 2y^2}$

49. $\dfrac{x^2 + x - 6}{x^2 - 6x + 8} \cdot \dfrac{x^2 + x - 2}{x^2 - 3x + 2} \cdot \dfrac{x^2 - x - 12}{x^2 + 6x + 9}$

50. $\dfrac{a^2 + ab}{b} \cdot \dfrac{a^2 - 2ab + b^2}{a^4 - b^4} \cdot \dfrac{a^2 + b^2}{3a^2b - 3ab^2}$

In exercises 51–80 find the quotient and express it in lowest terms.

51. $\dfrac{2}{5} \div \dfrac{3}{4}$

52. $\dfrac{5}{3} \div \dfrac{5}{7}$

53. $\dfrac{2x}{3y} \div \dfrac{4x}{9y}$

54. $\dfrac{a^2}{b^2} \div \dfrac{b}{a}$

55. $\dfrac{x}{y} \div \dfrac{y}{x}$

56. $\dfrac{a+1}{b+1} \div \dfrac{b}{a}$

57. $\dfrac{3x}{4y} \div \dfrac{3x-1}{4y+1}$

58. $4ab \div \dfrac{3a}{2b}$

59. $\dfrac{x+1}{x-1} \div \dfrac{x^2-1}{x+1}$

60. $\dfrac{12abc}{25x^3} \div \dfrac{16b^2 x}{15ax^2}$

61. $\dfrac{86m^2 n^3 p}{17a^2 b^2 c^3} \div \dfrac{129mnp^2}{34ab^2 c}$

62. $\dfrac{18a^2 x^3}{25by} \div \dfrac{81a^3 x^2}{125b^2 y}$

63. $\dfrac{2a-6}{a+2} \div \dfrac{3a-9}{4a+8}$

64. $\dfrac{(x+1)^2}{(x-1)^2} \div \dfrac{5x+5}{4x-4}$

65. $\dfrac{3y-9}{4y+8} \div \dfrac{2y-6}{y+2}$

66. $\dfrac{5}{12-6b} \div \dfrac{10-5b}{6-3b}$

67. $\dfrac{ax-a}{bx-3b} \div \dfrac{cx-c}{dx-3d}$

68. $\dfrac{(x+y)^2}{x-y} \div \dfrac{6x+6y}{3x-3y}$

69. $\dfrac{m^2-mn}{4m+8n} \div \dfrac{m^2-n^2}{2m+4n}$

70. $\dfrac{x-4}{x^2-4} \div \dfrac{x^2-16}{x-2}$

71. $\dfrac{4t+12}{t^2-9} \div \dfrac{2t-6}{(t-3)^2}$

72. $\dfrac{ab+3b}{a^2-a} \div \dfrac{a+3}{a}$

73. $\dfrac{y^2-3y-10}{y^2+2y-35} \div \dfrac{y^2+9y+14}{y^2+4y-21}$

74. $\dfrac{2x^2 + x - 10}{x^2 + 5x - 14} \div \dfrac{2x^2 + 11x + 15}{x^2 + 10x + 21}$

75. $\dfrac{x^2 + 6x + 9}{x^2 - 2x - 15} \div \dfrac{x^2 - 2x - 3}{x^2 - 4x - 5}$

76. $\dfrac{2y^2 - 3y - 2}{4y^2 + y - 14} \div \dfrac{2y^2 + 5y + 2}{16y^2 - 49}$

77. $\dfrac{a^2 - x^2}{ax - 2x^2} \div \dfrac{a^2 - 2ax + x^2}{2a^2 - 8x^2}$

78. $\dfrac{y^3 + y^2 - 20y}{by + 7b} \div \dfrac{y^2 + 5y}{y^2 + 3y - 28}$

79. $\dfrac{6x^2 + 11x + 4}{2x^2 + 17x - 9} \div \dfrac{6x^2 - x - 12}{2x^2 + 15x - 27}$

80. $\dfrac{12a^2 - 28a + 15}{10a^2 - 19a + 6} \div \dfrac{6a^2 - 29a + 20}{10a^2 - 29a + 10}$

In exercises 81–90 perform the indicated operations and simplify your answer.

81. $\left(\dfrac{3y + 9}{2y + 14} \cdot \dfrac{6y - 30}{7y + 21}\right) \div \dfrac{3y - 15}{7y + 49}$

82. $\left(\dfrac{4a + 36}{6a + 10} \div \dfrac{10a + 5}{18a + 30}\right) \cdot \dfrac{20a + 10}{2a + 18}$

83. $\left(\dfrac{x^2 + xy}{3xy + 2y^2} \cdot \dfrac{9x + 6y}{4x^2 - 2xy}\right) \div \dfrac{xy + y^2}{2xy - y^2}$

84. $\dfrac{6 - 3a}{5 - 10a} \cdot \left(\dfrac{4 - 2a}{6a + 3} \div \dfrac{2a^2 - 8}{4a^2 - 1}\right)$

85. $\dfrac{x - 2x^2}{4x^2 - 1} \div \left(\dfrac{2 + 2x - 4x^2}{6x - 6x^2} \cdot \dfrac{3x}{1 + 4x + 4x^2}\right)$

86. $\left(\dfrac{x^2 - x - 6}{(x - 2)^2} \cdot \dfrac{x^2(5 - x)\,(x + 7)}{x^2 + 9x + 14}\right) \div \dfrac{(3 - x)\,(x - 5)}{x^3\,(x - 2)^2}$

87. $\left(\dfrac{(x - y)^2 - z^2}{(x - z)^2 - y^2} \cdot \dfrac{y}{x^2 + xy + xz}\right) \div \dfrac{xy - y^2 + yz}{(x + y)^2 - z^2}$

88. $\dfrac{2y^2 - 19y + 42}{30 + y - y^2} \div \left(\dfrac{3 - 2y}{2y^2 + 7y - 15} \cdot \dfrac{y + 2}{2y^2 - 3y - 14}\right)$

89. $\dfrac{9x^2 + 6x + 1}{2 + 4x - 6x^2} \cdot \left(\dfrac{x - 3x^2}{9x^2 - 1} \div \dfrac{24x}{8x - 8x^2}\right)$

90. $\left(\dfrac{2a - 3}{a^2 - 8a + 7} \cdot \dfrac{a^2 - 2a + 1}{1 - a^2}\right) \div \dfrac{3a - 2}{7 + 6a - a^2}$

7.3 Addition and Subtraction

Again, the procedure for adding and subtracting rational algebraic expressions is similar to the procedure for adding and subtracting rational numbers. And, just as the least common denominator or LCD (Section 2.3) is useful in adding and subtracting fractions, so the least common denominator is useful in adding and subtracting rational algebraic expressions.

Before defining the LCD of rational algebraic expressions, we will consider the more general concept of the *least common multiple (LCM)* of *polynomials.* First of all, a *common* multiple of two or more given polynomials is a polynomial or the product of polynomials which has each of the given polynomials as a factor. The least common multiple is the common multiple with the smallest possible number of prime factors. For example, from

$$10a^2 = 2 \cdot 5 \cdot a \cdot a \qquad \text{and} \qquad 4ab = 2 \cdot 2 \cdot a \cdot b$$

we get

$$2 \cdot 2 \cdot 5 \cdot a \cdot a \cdot b$$

so that $20a^2 b$ is the LCM of $10a^2$ and $4ab$.

We can use this result to write $\frac{3}{10a^2}$ and $\frac{5}{4ab}$ with the least common denominator (the LCM of the denominators) as $20a^2 b$. Thus

$$\frac{3}{10a^2} = \frac{3 \cdot 2b}{10a^2 \cdot 2b} = \frac{6b}{20a^2 b} \qquad \text{and} \qquad \frac{5}{4ab} = \frac{5 \cdot 5a}{4ab \cdot 5a} = \frac{25a}{20a^2 b}$$

Hence we have

$$\frac{3}{10a^2} + \frac{5}{4ab} = \frac{6b}{20a^2 b} + \frac{25a}{20a^2 b} = \frac{6b + 25a}{20a^2 b}$$

Similarly, to perform the subtraction

$$\frac{1}{x^2 + x} - \frac{3}{x^2 - 1}$$

we first factor the denominators to get

$$\frac{1}{x^2 + x} - \frac{3}{x^2 - 1} = \frac{1}{x(x + 1)} - \frac{3}{(x + 1)(x - 1)}$$

Here the LCD is $x(x + 1)(x - 1)$, and so we write

$$\frac{1}{x(x + 1)} = \frac{1 \cdot (x - 1)}{x(x + 1)(x - 1)}$$

and

$$\frac{3}{(x + 1)(x - 1)} = \frac{3x}{x(x + 1)(x - 1)}$$

Thus

$$\frac{1}{x(x + 1)} - \frac{3}{(x + 1)(x - 1)} = \frac{1 \cdot (x - 1)}{x(x + 1)(x - 1)} - \frac{3x}{x(x + 1)(x - 1)}$$

$$= \frac{1 \cdot (x - 1) - 3x}{x(x + 1)(x - 1)} = \frac{x - 1 - 3x}{x(x + 1)(x - 1)}$$

$$= \frac{-1 - 2x}{x(x + 1)(x - 1)}$$

$$= -\frac{2x + 1}{x(x + 1)(x - 1)}$$

Notice that in this problem we did not actually carry out the multiplications indicated in the denominator of the answer; the factored form is usually more useful. This is illustrated in the following example:

$$\frac{7}{x^2 + x - 12} + \frac{2}{x^2 - 8x + 15} = \frac{7}{(x + 4)(x - 3)} + \frac{2}{(x - 5)(x - 3)}$$

Here the LCD is $(x + 4)(x - 3)(x - 5)$ and we have

$$\frac{7}{(x + 4)(x - 3)} \cdot \frac{x - 5}{x - 5} + \frac{2}{(x - 5)(x - 3)} \cdot \frac{x + 4}{x + 4}$$

$$= \frac{7(x - 5) + 2(x + 4)}{(x + 4)(x - 3)(x - 5)}$$

$$= \frac{7x - 35 + 2x + 8}{(x + 4)(x - 3)(x - 5)}$$

$$= \frac{9x - 27}{(x + 4)(x - 3)(x - 5)}$$

$$= \frac{9(x - 3)}{(x + 4)(x - 3)(x - 5)} = \frac{9}{(x + 4)(x - 5)}$$

Here the fact that we did not perform the multiplications in the denominator made it easy for us to reduce the answer to lowest terms.

As another example, consider

$$\frac{x-1}{x+1} + x + 3 = \frac{x-1}{x+1} + \frac{x+3}{1} \cdot \frac{x+1}{x+1}$$

$$= \frac{x-1}{x+1} + \frac{(x+3)(x+1)}{x+1}$$

$$= \frac{(x-1) + (x^2+4x+3)}{x+1}$$

$$= \frac{x^2+5x+2}{x+1}$$

As a final example, consider

$$\frac{3}{y+3} - \frac{2}{3-y} - \frac{4y}{y^2-9}$$

Factoring $y^2 - 9$ gives us

$$\frac{3}{y+3} - \frac{2}{3-y} - \frac{4y}{(y+3)(y-3)}$$

At this point we might be tempted to say that the LCD is

$$(y+3)(3-y)(y-3)$$

If we notice, however, that $-(3-y) = y - 3$, we can write

$$\frac{2}{3-y} \qquad \text{as} \qquad \frac{(-1)(2)}{(-1)(3-y)} = \frac{-2}{y-3}$$

and thus have

$$\frac{3}{y+3} - \frac{2}{3-y} - \frac{4y}{(y+3)(y-3)}$$

$$= \frac{3}{y+3} - \frac{-2}{y-3} - \frac{4y}{(y+3)(y-3)}$$

with LCD $= (y + 3)(y - 3)$. Thus we have

$$\frac{3}{y + 3} - \frac{-2}{y - 3} - \frac{4y}{(y + 3)(y - 3)}$$

$$= \frac{3(y - 3)}{(y + 3)(y - 3)} - \frac{(-2)(y + 3)}{(y - 3)(y + 3)} - \frac{4y}{(y + 3)(y - 3)}$$

$$= \frac{3(y - 3) - (-2)(y + 3) - 4y}{(y + 3)(y - 3)}$$

$$= \frac{3y - 9 + 2y + 6 - 4y}{(y + 3)(y - 3)}$$

$$= \frac{y - 3}{(y + 3)(y - 3)} = \frac{1}{y + 3}$$

EXERCISES 7.3

In exercises 1–20 express the given sum or difference as a rational algebraic expression in lowest terms.

1. $\dfrac{7}{9} + \dfrac{5}{9}$

2. $\dfrac{9}{13} - \dfrac{5}{13}$

3. $\dfrac{5}{12} + \dfrac{4}{12} - \dfrac{1}{12}$

4. $\dfrac{5}{x} + \dfrac{3}{x}$

5. $\dfrac{11}{w} - \dfrac{7}{w}$

6. $\dfrac{2}{b} + \dfrac{5}{b} - \dfrac{3}{b}$

7. $\dfrac{m}{r} - \dfrac{n}{r}$

8. $\dfrac{2c}{d} + \dfrac{3a}{d}$

9. $\dfrac{r}{x} + \dfrac{25}{x} - \dfrac{5t}{x}$

10. $\dfrac{5}{x + y} + \dfrac{3}{x + y}$

11. $\dfrac{a}{a + b} + \dfrac{b}{a + b}$

12. $\dfrac{3}{5(x + 3)} + \dfrac{x}{5(x + 3)}$

13. $\dfrac{m}{r(m - n)} - \dfrac{n}{r(m - n)}$

14. $\dfrac{x^2}{x + y} - \dfrac{y^2}{x + y}$

15. $\dfrac{5x}{x^2 - x - 2} - \dfrac{4x - 1}{x^2 - x - 2}$

16. $\dfrac{x^2 - 4x}{x^2 - x - 6} - \dfrac{x - 6}{x^2 - x - 6}$

17. $\dfrac{x^2 + x - 7}{x^2 + 3x - 10} + \dfrac{x^2 + 5x - 13}{x^2 + 3x - 10}$

18. $\dfrac{c^2 + 7c + 3}{c^2 - 4c - 45} - \dfrac{2c + 3}{c^2 - 4c - 45}$

19. $\dfrac{a^2}{a^2 + 2a - 15} - \dfrac{25}{a^2 + 2a - 15}$

20. $\dfrac{x^2 - xy - y^2}{x^2 + 2xy + y^2} + \dfrac{x^2 + xy - y^2}{x^2 + 2xy + y^2}$

In exercises 21–34 find the least common multiple (LCM) of the given polynomials.

Example (a): ab, a^2, b^2; LCM $= a^2b^2$

Example (b): $5(a - b), (a - b)(a + b)$; LCM $= 5(a - b)(a + b)$

21. $3x, x, 4x$
22. xy, xz
23. rs, st, rt
24. $2x^2, 3xy, 6y^2$
25. $2r^2, 6r, 3$
26. $m - 3, m + 3$
27. $2x - 2y, 3x - 3y$
28. $r^2 - s^2, r - s$
29. $x^2 - 16, 2x + 8$
30. $t^2 - 2t - 8, t^2 - 4$
31. $2, a^2 - 9, 3a^2 - 3a - 18$
32. $x^2 + 2x - 8, x^2 + 5x + 4$
33. $x^2 - x - 12, x^2 + 3x - 10$
34. $6y^2 - 11y + 3, 12y - 4, 9y^2 - 1$

In exercises 35–104 express the given sum or difference as a rational algebraic expression in lowest terms.

35. $\dfrac{3}{8} + \dfrac{5}{4}$

36. $\dfrac{x}{2} - \dfrac{x}{4}$

37. $\dfrac{4a}{5} + \dfrac{3a}{2}$

38. $\dfrac{7x}{2} - \dfrac{x}{3}$

39. $\dfrac{7}{6y} + \dfrac{5}{2y}$

40. $\dfrac{3}{2x} + \dfrac{8}{3x}$

41. $\dfrac{5y}{4x} + \dfrac{7y}{6x}$

42. $\dfrac{7b}{4a} - \dfrac{7b}{8a}$

43. $\dfrac{7a}{4x} - \dfrac{3a}{5x}$

44. $\dfrac{15a}{8t} - \dfrac{3a}{10t}$

45. $\dfrac{9x^2}{16} + \dfrac{5y^2}{18}$

46. $\dfrac{9x}{8y} - \dfrac{11x}{16y}$

47. $\dfrac{a^2}{b^4} - \dfrac{ax}{b^3}$

48. $\dfrac{1}{a} + \dfrac{1}{b}$

49. $\dfrac{x-y}{4} - \dfrac{x+y}{8}$

50. $\dfrac{a+3}{3} - \dfrac{a+2}{5}$

51. $\dfrac{x-2}{6} - \dfrac{x+3}{2}$

52. $\dfrac{a+b}{3} + \dfrac{2(a+b)}{9} - \dfrac{a-b}{6}$

53. $\dfrac{x+3}{x} + \dfrac{x+2}{2}$

54. $\dfrac{ab+1}{a^2} + \dfrac{b+a}{a}$

55. $\dfrac{y-2x}{12y^2x} - \dfrac{2y+x}{10yx^2}$

56. $\dfrac{x-y}{xy} - \dfrac{a-y}{ay}$

57. $\dfrac{x+y}{2xy^2} + \dfrac{x^2-2y^2}{3x^2y^2} - \dfrac{2x-y}{4x^2y}$

58. $\dfrac{m^2-n^2}{mn} + \dfrac{m(m-3)}{4m} - \dfrac{m(n-2)}{6m}$

59. $\dfrac{a+4b}{5a^3b^3} - \dfrac{5a+b}{6a^2b^3} + \dfrac{3a-2b}{2a^2b^2}$

60. $\dfrac{3}{y+4} + \dfrac{6}{y+5}$

61. $\dfrac{9}{x-2} + \dfrac{2}{x-3}$

62. $\dfrac{10}{t-3} - \dfrac{2}{t-4}$

63. $\dfrac{2x}{x-2} + \dfrac{x}{x-1}$

64. $\dfrac{4a}{a-3} + \dfrac{a+3}{2a}$

65. $\dfrac{x}{x-y} - \dfrac{y}{x+y}$

66. $\dfrac{4}{a+7} - \dfrac{3a}{a-2}$

67. $\dfrac{y+5}{y-6} + \dfrac{3y}{y+5}$

68. $\dfrac{4m}{m-3} + \dfrac{m+3}{2m}$

69. $\dfrac{2x}{x-1} - \dfrac{x+3}{x+2}$

70. $\dfrac{x}{x-y} + \dfrac{x+y}{xy}$

71. $\dfrac{3}{a+b} + 4$

72. $\dfrac{x+y}{x-y} + 1$

73. $a+b + \dfrac{ab}{a+b}$

74. $x + \dfrac{x^2}{x-5}$

75. $a + 3b - \dfrac{a^2+b^2}{a+3b}$

76. $\dfrac{a}{5} + \dfrac{4b}{5(a+b)}$

77. $\dfrac{2x}{x+y} - \dfrac{8y}{3(x+y)}$

78. $\dfrac{5}{t-5} - \dfrac{1}{(t-5)^2}$

79. $\dfrac{5}{2(x+1)} + \dfrac{2}{3(x+1)}$

80. $\dfrac{x}{(x-4)^2} - \dfrac{1}{x-4}$

81. $\dfrac{y^2}{(x-y)^2} - \dfrac{x+y}{x-y}$

82. $\dfrac{5a}{8(a-b)} - \dfrac{a}{2(a+b)}$

83. $\dfrac{3m}{10(m+n)} - \dfrac{3n}{16(m-n)}$

84. $\dfrac{4}{(x+4)(x-2)} + \dfrac{2}{(x+4)(x+1)}$

85. $\dfrac{a-2}{2(a+3)(a-3)} + \dfrac{a+3}{3(a-3)(a+2)}$

86. $\dfrac{6}{x^2-9} - \dfrac{2}{x+3}$

87. $\dfrac{2x}{x-9} - \dfrac{20}{x^2-81}$

88. $\dfrac{3}{x+3} + \dfrac{3}{x^2+6x+9}$

89. $\dfrac{8a+1}{a^2+7a+10} + \dfrac{5}{a+2}$

90. $\dfrac{3x}{x^2-16x+60} + \dfrac{3}{x+8}$

91. $\dfrac{3y}{y-1} + \dfrac{4y^2+8}{y^2-1} - \dfrac{6y}{y+1}$

92. $\dfrac{k-4}{k-6} - \dfrac{3k}{k^2-3k-18}$

93. $\dfrac{x-2}{x-1} - \dfrac{3-3x}{x^2-2x+1}$

94. $\dfrac{2x+3y}{2x+2y} + \dfrac{x}{3x+3y}$

95. $\dfrac{5}{3a+15} + \dfrac{4}{a^2-25}$

96. $\dfrac{2x+6}{x^2-x-12} - \dfrac{3(x+5)}{x^2+3x-10}$

97. $\dfrac{a+2}{2a^2+3a-9} - \dfrac{a-2}{3a^2-11a+6}$

98. $\dfrac{3}{6y^2-11y+3} + \dfrac{3y+1}{12y-4} - \dfrac{2y-3}{9y^2-1}$

99. $\dfrac{2}{a-b} - \dfrac{1}{b-a}$

100. $\dfrac{4}{x-y} + \dfrac{2}{y-x}$

101. $\dfrac{x}{2x-8} + \dfrac{4x}{16-x^2}$

102. $\dfrac{2}{x + 3} + \dfrac{4}{3 - x} + \dfrac{3x + 15}{x^2 - 9}$

103. $\dfrac{a}{64 - a^2} + \dfrac{2}{3a - 24} - \dfrac{3}{2a + 16}$

104. $\dfrac{1}{x^2 + 5x + 6} - \dfrac{4}{4 - x^2}$

7.4 Rational Algebraic Expressions as Quotients of Polynomials

Just as $\frac{a}{b} = a \div b$ for real numbers a and b, we can think of $\frac{P_1}{P_2}$ as $P_1 \div P_2$ for polynomials P_1 and P_2. Then, just as it may sometimes happen that, for a and b integers, $a \div b$ is an integer, so it may sometimes happen that $P_1 \div P_2$ is a polynomial. Thus

$$(x^2 + 5x + 6) \div (x + 2) = \frac{(x + 2)\,(x + 3)}{x + 2} = x + 3$$

On the other hand, just as we can have a nonzero remainder when one integer is divided by another, so we can have a nonzero remainder when one polynomial is divided by another. Thus

$$7 \div 2 = \frac{7}{2} = 3 + \frac{1}{2} \quad \text{and} \quad (x^3 + 2) \div x = \frac{x^3 + 2}{x} = \frac{x^3}{x} + \frac{2}{x} = x^2 + \frac{2}{x}$$

We can also write these results as

$$7 = (3 \cdot 2) + 1 \quad \text{and} \quad x^3 + 2 = (x^2 \cdot x) + 2$$

In the division of 7 by 2 we call 7 the *dividend,* 2 the *divisor,* 3 the *quotient,* and 1 the *remainder.* Likewise, in the division of $x^3 + 2$ by x we call $x^3 + 2$ the dividend, x the divisor, x^2 the quotient, and 2 the remainder.

When the divisor is a polynomial of more than one term, a "long division" process is often used. Below we show an arithmetic long division and a similar division of polynomials. In the arithmetic example we see that if we first subtract ten 11s and then five 11s from 168 we

$$
\begin{array}{r}
15 \\
11\,)\overline{168} \\
110 \leftarrow 10 \times 11 = 110 \\
\overline{58} \\
55 \leftarrow 5 \times 11 = 55 \\
\overline{3}
\end{array}
\qquad
\begin{array}{r}
x\ + 5 \\
x + 1\,)\overline{x^2 + 6x + 8} \\
x^2 +\ x \leftarrow x(x + 1) = x^2 + x \\
\overline{5x + 8} \\
5x + 5 \leftarrow 5(x + 1) = 5x + 5 \\
\overline{3}
\end{array}
$$

get a remainder of 3; that is, $168 - (15 \cdot 11) = 3$ so that $168 = (15 \cdot 11) + 3$ and

$$\frac{168}{11} = 168 \div 11 = 15 + \frac{3}{11}$$

The quotient is 15 and the remainder is 3.

Similarly, for the polynomial division we first subtract x $(x + 1)$s and then $5 (x + 1)$s to get a quotient of $x + 5$ and a remainder of 3. That is,

$$(x^2 + 6x + 8) - (x + 5)(x + 1) = 3$$

so that

$$x^2 + 6x + 8 = (x + 5)(x + 1) + 3$$

and

$$\frac{x^2 + 6x + 8}{x + 1} = (x^2 + 6x + 8) \div (x + 1) = x + 5 + \frac{3}{x + 1}$$

The quotient is $x + 5$ and the remainder is 3.

Here are further examples of this process.

Example 1: Divide $x^3 - 3x^2 + x - 1$ by $x^2 - x + 1$

$$x \text{ is suggested since } x^3 \div x^2 = x \text{ and}$$
$$\text{thus } x^2 \cdot x = x^3$$
$$x(x^2 - x + 1) = x^3 - x^2 + x$$

Subtract

$$-2 \text{ is suggested since } -2x^2 \div x^2 = -2$$
$$\text{and thus } x^2 \cdot -2 = -2x^2$$
$$-2(x^2 - x + 1) = -2x^2 + 2x - 2$$

Subtract

The quotient is $x - 2$ and the remainder is $-2x + 1$. (Note that the division process stops when the remainder is 0 or when the degree of the remainder is less than the degree of the divisor.)

We can express the result of this division in the form

$$(x^3 - 3x^2 + x - 1) \div (x^2 - x + 1) = x - 2 + \frac{-2x + 1}{x^2 - x + 1}$$

or

$$\frac{x^3 - 3x^2 + x - 1}{x^2 - x + 1} = x - 2 + \frac{-2x + 1}{x^2 - x + 1}$$

or

$$x^3 - 3x^2 + x - 1 = (x - 2)(x^2 - x + 1) + (-2x + 1)$$

Check: $(x - 2)(x^2 - x + 1) + (-2x + 1)$

$$= (x^3 - 3x^2 + 3x - 2) + (-2x + 1) = x^3 - 3x^2 + x - 1$$

Example 2: Divide $y^3 - 2y + 1$ by $y - 3$.

When a term (here, y^2) is "missing" from the polynomial of the dividend, it is helpful to insert the term before dividing by using 0 as the coefficient. That is, we write

$$y^3 - 2y + 1 \qquad \text{as} \qquad y^3 + 0y^2 - 2y + 1$$

$$
\begin{array}{r}
y^2 + 3y + 7 \\
y - 3 \overline{)\, y^3 + 0y^2 - 2y + 1} \\
\underline{y^3 - 3y^2} \\
3y^2 - 2y + 1 \\
\underline{3y^2 - 9y} \\
7y + 1 \\
\underline{7y - 21} \\
22
\end{array}
$$

$-y^2$
$\begin{cases} y^2 \text{ is suggested since } y^3 \div y = y^2 \\ \text{and thus } y \cdot y^2 = y^3 \\ y^2(y - 3) = y^3 - 3y^2 \end{cases}$

$-3y$
$\begin{cases} 3y \text{ is suggested since } 3y^2 \div y = 3y \\ \text{and thus } y \cdot 3y = 3y^2 \\ 3y(y - 3) = 3y^2 - 9y \end{cases}$

7
$\begin{cases} 7 \text{ is suggested since } 7y \div y = 7 \text{ and} \\ \text{thus } y \cdot 7 = 7y \\ 7(y - 3) = 7y - 21 \end{cases}$

The quotient is $y^2 + 3y + 7$ and the remainder is 22. We have

$$\frac{y^3 - 2y + 1}{y - 3} = (y^3 - 2y + 1) \div (y - 3) = y^2 + 3y + 7 + \frac{22}{y - 3}$$

and

$$y^3 - 2y + 1 = (y^2 + 3y + 7)(y - 3) + 22$$

Check: $(y^2 + 3y + 7)(y - 3) + 22 = (y^3 - 2y - 21) + 22$

$$= y^3 - 2y + 1$$

Example 3: Divide $6x^2 - 17x + 12$ by $2x - 3$.

$$
\begin{array}{r}
3x - 4 \\
2x - 3 \overline{)\, 6x^2 - 17x + 12} \\
\underline{6x^2 - 9x} \quad \longleftarrow \; 3x(2x - 3) \\
-8x + 12 \\
\underline{-8x + 12} \longleftarrow -4(2x - 3) \\
0
\end{array}
$$

So

$$\frac{6x^2 - 17x + 12}{2x - 3} = (6x^2 - 17x + 12) \div (2x - 3) = 3x - 4$$

and

$$6x^2 - 17x + 12 = (3x - 4)(2x - 3)$$

The quotient is $3x - 4$ and the remainder is 0.

Check: $(3x - 4)(2x - 3) = 6x^2 - 17x + 12$

Example 4: Divide $x^2 - 2x + 2$ by $2x + 1$.

$$
\begin{array}{r}
\frac{1}{2}x - \frac{5}{4} \\
2x + 1 \overline{) x^2 - 2x + 2 } \\
x^2 + \frac{1}{2}x \longleftarrow \frac{1}{2}x(2x + 1) \\
\overline{} \\
-\frac{5}{2}x + 2 \\
-\frac{5}{2}x - \frac{5}{4} \longleftarrow -\frac{5}{4}(2x + 1) \\
\overline{} \\
\frac{13}{4}
\end{array}
$$

Thus

$$\frac{x^2 - 2x + 2}{2x + 1} = (x^2 - 2x + 2) \div (2x + 1)$$

$$= \frac{1}{2}x - \frac{5}{4} + \frac{\frac{13}{4}}{2x + 1}$$

and

$$x^2 - 2x + 2 = \left(\frac{1}{2}x - \frac{5}{4}\right)(2x + 1) + \frac{13}{4}$$

The quotient is $\frac{1}{2}x - \frac{5}{4}$ and the remainder is $\frac{13}{4}$.

Check: $\left(\frac{1}{2}x - \frac{5}{4}\right)(2x + 1) + \frac{13}{4} = \left(x^2 - 2x - \frac{5}{4}\right) + \frac{13}{4} = x^2 - 2x + 2$

Example 5: Divide $6x^3 + 4xy^2 - 5x^2y - 15y^3$ by $2x - 3y$.

When more than one variable is involved in a division problem we must arrange the dividend in descending powers of one of the variables. Thus, if we choose x as this variable, we begin by rewriting $6x^3 + 4xy^2 - 5x^2y - 15y^3$ as $6x^3 - 5x^2y + 4xy^2 - 15y^3$. Then we have

$$
\begin{array}{r}
3x^2 + 2xy + 5y^2 \\
2x - 3y \overline{\smash{\big)}\ 6x^3 - 5x^2y + 4xy^2 - 15y^3} \\
6x^3 - 9x^2y \quad\quad\quad\quad\quad \longleftarrow 3x^2(2x - 3y) \\
4x^2y + 4xy^2 - 15y^3 \\
4x^2y - 6xy^2 \quad\quad \longleftarrow 2xy(2x - 3y) \\
10xy^2 - 15y^3 \\
10xy^2 - 15y^3 \longleftarrow 5y^2(2x - 3y) \\
0
\end{array}
$$

So

$$
\frac{6x^3 - 5x^2y + 4xy^2 - 15y^3}{2x - 3y}
$$

$$
= (6x^3 - 5x^2y + 4xy^2 - 15y^3) \div (2x - 3y)
$$

$$
= 3x^2 + 2xy + 5y^2
$$

and

$$
6x^3 - 5x^2y + 4xy^2 - 15y^3 = (3x^2 + 2xy + 5y^2)(2x - 3y)
$$

The quotient is $3x^2 + 2xy + 5y^2$ and the remainder is 0.

Check: $(3x^2 + 2xy + 5y^2)(2x - 3y) = 6x^3 - 5x^2y + 4xy^2 - 15y^3$

EXERCISES 7.4

In exercises 1–30 divide the first polynomial by the second. In each of these problems the remainder will be 0. Check your answers by multiplication.

1. $(x^2 + 4x + 3) \div (x + 1)$
2. $(a^2 + 20a + 96) \div (a + 8)$
3. $(b^2 + 21b + 80) \div (b + 16)$
4. $(x^2 + 26x - 87) \div (x - 3)$
5. $(d^2 + 3d - 28) \div (d - 4)$
6. $(z^2 + 8z - 9) \div (z - 1)$
7. $(x^2 - 3x - 108) \div (x - 12)$
8. $(t^2 + 6t - 135) \div (t + 15)$
9. $(y^2 + 13y - 48) \div (y + 16)$

10. $(a^2 - 24a + 143) \div (a - 11)$

11. $(9 + 6x + x^2) \div (3 + x)$

12. $(21 - 4y - y^2) \div (7 + y)$

13. $(2x^2 + 13x + 6) \div (x + 6)$

14. $(10a^6 + a^3 - 24) \div (5a^3 + 8)$

15. $(6y^4 + 9y^2 - 42) \div (3y^2 - 6)$

16. $(16x^2 + 28x + 10) \div (4x + 5)$

17. $(35b^2 + 33b + 4) \div (7b + 1)$

18. $(4x^2 + 12xy + 9y^2) \div (2x + 3y)$

19. $(4x^2 + 17xyz + 15y^2z^2) \div (4x + 5yz)$

20. $(3r^3 - 19r^2 + 27r + 4) \div (r - 4)$

21. $(a^3 + 3a^2b + 3ab^2 + b^3) \div (a + b)$

22. $(2ab + b^2 + a^2) \div (a + b)$

23. $(35y^2 - 62xy + 24x^2) \div (6x - 5y)$

24. $(a^3 - 3a^2b - b^3 + 3ab^2) \div (a - b)$

25. $(8b^3 + 25bc^2 - 12c^3 - 18b^2c) \div (4b - 3c)$

26. $\left(x^2 + x + \dfrac{1}{4}\right) \div (2x + 1)$

27. $\left(x^2 + 2x + \dfrac{8}{9}\right) \div (3x + 2)$

28. $\left(3x^2 + x - \dfrac{8}{25}\right) \div (5x - 1)$

29. $\left(3x^2 + 2x - \dfrac{3}{16}\right) \div (4x + 3)$

30. $\left(3x^3 - 8x^2 + 2x - \dfrac{15}{8}\right) \div (2x - 5)$

In exercises 31–45 divide the first polynomial by the second. In each of these problems there will be a nonzero remainder. Express each answer as a quotient $+ \frac{remainder}{divisor}$ and check by verifying that the dividend is equal to the divisor times the quotient plus the remainder.

Example: $(x^3 - 8) \div (x + 2)$

$$
\begin{array}{r}
x^2 - 2x + 4 \\
x + 2 \overline{)x^3 + 0x^2 + 0x - 8} \\
\underline{x^3 + 2x^2 } \\
-2x^2 + 0x - 8 \\
\underline{-2x^2 - 4x } \\
4x - 8 \\
\underline{4x + 8} \\
-16
\end{array}
$$

Thus $(x^3 - 8) \div (x + 2) = x^2 - 2x + 4 + \dfrac{-16}{x + 2}$.

Check: $(x + 2)(x^2 - 2x + 4) + (-16) = (x^3 + 8) + (-16) = x^3 - 8$

31. $(2x^2 + 3x + 1) \div (x + 6)$
32. $(x^2 + 3x + 7) \div (x + 2)$
33. $(a^2 + ab + b^2) \div (a + b)$
34. $(5x^2 + 6x - 4) \div (x - 5)$
35. $(12x^2 - 4xy - 3y^2) \div (2x - 5y)$
36. $(4m^2 - 4mn + 6n^2) \div (2m - 3n)$
37. $(2x^3 - 3x^2 + 17x + 6) \div (2x + 3)$
38. $(5y^3 - 2y^2 + 5y - 9) \div (5y - 2)$
39. $(5 - 12x^3 - 35x - 43x^2) \div (4x + 5)$
40. $(2x^3 - 13x^2 - 20 + 27x) \div (2x - 3)$
41. $(x^3 - 1) \div (x + 1)$
42. $(x^4 + 16) \div (x + 2)$
43. $(x^3 - y^3) \div (x + y)$
44. $(6a^4 + 4a^3 - 4a^2 + 4a + 7) \div (3a^2 - 4a + 3)$
45. $(4x^4 + 16x^3 + 19x^2 - 15x - 18) \div (x^2 + x - 2)$

In exercises 46–65 divide the first polynomial by the second polynomial. Check by verifying that the dividend equals the divisor times the quotient plus the remainder.

46. $(2x^2 - 4x - 2) \div (2x + 1)$
47. $(3x^2 + 16x - 12) \div (x + 6)$
48. $(x^3 - 3x^2 - 13x + 15) \div (x - 5)$
49. $(2y^3 - 5y^2 + 11y - 12) \div (2y - 3)$
50. $(12a^2 + a - 4) \div (3a - 2)$
51. $(3a^2 - 11a - 18) \div (a - 5)$
52. $(2x^2 - x + 1) \div (2x + 1)$
53. $(8y^3 - 18y^2 - 6 + 11y) \div (4y^2 - 3y + 2)$
54. $(4t^3 - 29t + 9t^2 + 15) \div (t^2 + 3t - 5)$
55. $(6x^2 + 2x - 20) \div (3x + 7)$
56. $(x^4 - y^4) \div (x - y)$
57. $(x^3 - 8) \div (x - 2)$
58. $(6x^3 - 27x - x^2 - 20) \div (2x^2 + 3x - 5)$
59. $(4y^3 - 17y - 12) \div (2y + 3)$
60. $(16m^3 + 3m + 1) \div (4m + 1)$
61. $(9x^3 - 54x - 21) \div (2x^2 + 4x - 4)$
62. $(x^3 - 15x - 18 + 4x^2) \div (2x^2 - 6 - 4x)$
63. $(2a^2 - 4 + 5a^3 - 19a) \div (-8a + 2a^2 - 3)$

64. $\left(8a^3 + \dfrac{64}{125}b^3\right) \div (5a + 2b)$

65. $(7x^3 - 1) \div (2x + 1)$

7.5 Equations Involving Rational Algebraic Expressions

Now we will consider equations involving rational algebraic expressions that are equivalent to linear equations in one variable. The basic device used in solving such equations is to multiply both sides of the equation by the LCM of the denominators. We illustrate this procedure by the following examples.

Example 1: $\dfrac{x}{5} - \dfrac{1}{4} = \dfrac{2x}{15} + \dfrac{1}{12}$

The LCM of the denominators is 60 and we have

$$60\left(\frac{x}{5} - \frac{1}{4}\right) = 60\left(\frac{2x}{15} + \frac{1}{12}\right)$$

$$\frac{60 \cdot x}{5} - \frac{60 \cdot 1}{4} = \frac{60 \cdot 2x}{15} + \frac{60 \cdot 1}{12}$$

$$12x - 15 = 8x + 5$$

$$4x = 20$$

$$x = 5$$

This checks, since if $x = 5$,

$$\frac{x}{5} - \frac{1}{4} = \frac{5}{5} - \frac{1}{4} = \frac{20}{20} - \frac{5}{20} = \frac{15}{20} = \frac{3}{4}$$

and, also,

$$\frac{2x}{15} + \frac{1}{12} = \frac{10}{15} + \frac{1}{12} = \frac{2}{3} + \frac{1}{12} = \frac{8}{12} + \frac{1}{12} = \frac{9}{12} = \frac{3}{4}$$

Example 2: $\dfrac{1}{x} + \dfrac{2}{x + 4} = \dfrac{5}{3x}$

The LCD is $3x(x + 4)$ and we have

$$3x(x + 4) \left(\frac{1}{x} + \frac{2}{x + 4} \right) = 3x(x + 4) \cdot \frac{5}{3x}$$

$$3x(x + 4) \cdot \frac{1}{x} + 3x(x + 4) \cdot \frac{2}{x + 4} = 3x(x + 4) \cdot \frac{5}{3x}$$

$$3(x + 4) + 6x = 5(x + 4)$$

$$3x + 12 + 6x = 5x + 20$$

$$4x = 8$$

$$x = 2$$

Check: $\dfrac{1}{2} + \dfrac{2}{2 + 4} = \dfrac{1}{2} + \dfrac{2}{6} = \dfrac{5}{6}$ and $\dfrac{5}{3 \cdot 2} = \dfrac{5}{6}$

Example 3: $\dfrac{4x - 7}{x - 2} = 3 + \dfrac{1}{x - 2}$

The LCD is $x - 2$ and we have

$$(x - 2) \frac{4x - 7}{x - 2} = (x - 2) \left(3 + \frac{1}{x - 2} \right)$$

$$4x - 7 = 3(x - 2) + 1$$

$$4x - 7 = 3x - 6 + 1$$

$$x = 2$$

But, when we go to check, we find that $x = 2$ makes the denominator, $x - 2$, equal to 0, and we know (Section 2.8) that we cannot divide by 0. Hence, we must reject 2 as a solution of our equation and conclude that its solution set is ∅.

It is always a good idea to check the numbers you obtain as solutions to any equation, as we did in our first two examples. It is *essential* to check to see that no number obtained (as $x = 2$ in our third example) makes any denominator 0. What really happened in Example 3 was that in the very writing of the equation we were assuming $x - 2 \neq 0$ and so the conclusion we reached that $x = 2$ couldn't possibly be a valid one.

As a further safeguard against listing as solutions any members that make a denominator 0, it is a good idea to make a list of prohibited values of the variable to begin with. Thus, in Example 2, the prohibited values are $x = 0$ and $x = -4$. In Example 3 the prohibited value is $x = 2$. Then, whenever any such prohibited values appear as "candidates" for a solution, they can be automatically rejected. We follow this procedure in our next example.

Example 4: $\dfrac{1}{x - 2} + \dfrac{4}{x + 2} = \dfrac{5}{x - 1}$

The prohibited values are $2, -2,$ and 1; the LCD is $(x - 2)(x + 2)(x - 1)$ and we have

$$(x - 2)(x + 2)(x - 1)\left(\frac{1}{x - 2} + \frac{4}{x + 2}\right) = (x - 2)(x + 2)(x - 1) \cdot \frac{5}{x - 1}$$

$$(x + 2)(x - 1) + 4(x - 2)(x - 1) = 5(x - 2)(x + 2)$$

$$x^2 + x - 2 + 4(x^2 - 3x + 2) = 5(x^2 - 4)$$

$$x^2 + x - 2 + 4x^2 - 12x + 8 = 5x^2 - 20$$

$$-11x = -26$$

$$x = \frac{26}{11} \text{ (not on the prohibited list)}$$

Check: If $x = \dfrac{26}{11}$, then

$$\frac{1}{x - 2} + \frac{4}{x + 2} = \frac{1}{\frac{26}{11} - 2} + \frac{4}{\frac{26}{11} + 2} = \frac{1}{\frac{26}{11} - \frac{22}{11}} + \frac{4}{\frac{26}{11} + \frac{22}{11}}$$

$$= \frac{1}{\frac{4}{11}} + \frac{4}{\frac{48}{11}} = \left(1 \div \frac{4}{11}\right) + \left(4 \div \frac{48}{11}\right) = \frac{11}{4} + \frac{44}{48}$$

$$= \frac{11}{4} + \frac{11}{12} = \frac{33}{12} + \frac{11}{12} = \frac{44}{12} = \frac{11}{3}$$

and, also,

$$\frac{5}{x-1} = \frac{5}{\frac{26}{11} - 1} = \frac{5}{\frac{26}{11} - \frac{11}{11}} = \frac{5}{\frac{15}{11}} = 5 \div \frac{15}{11} = \frac{55}{15} = \frac{11}{3}$$

(In a problem like this, a numerical check is hardly worthwhile; it would probably be better to spend the time needed for a numerical check in going over the algebra!)

EXERCISES 7.5

   In exercises 1–45 solve the given equation or show that there are no solutions. Begin with a list of prohibited values of the variable and check to see that none of your candidates for a solution is on this list.

1. $\dfrac{1}{4} + \dfrac{1}{8} = \dfrac{1}{x}$

2. $\dfrac{2}{a} + \dfrac{1}{2} = \dfrac{5}{2a}$

3. $\dfrac{1}{2x} + \dfrac{5}{8} = \dfrac{3}{x}$

4. $\dfrac{1}{y} + \dfrac{2}{y} = 3 - \dfrac{3}{y}$

5. $\dfrac{11}{2x} - \dfrac{2}{3x} = \dfrac{1}{6}$

6. $\dfrac{4}{m} + 2 = \dfrac{14}{m} - 3$

7. $\dfrac{2}{3y} + \dfrac{1}{4} = \dfrac{11}{6y} - \dfrac{1}{3}$

8. $\dfrac{3}{5a} + \dfrac{4}{5} = \dfrac{5}{2a} + \dfrac{1}{6}$

9. $\dfrac{2}{3x} + \dfrac{1}{2x} + 1 = \dfrac{13}{6x}$

10. $\dfrac{x+4}{3x} = \dfrac{3x-8}{4x}$

11. $\dfrac{15}{a} + \dfrac{9a-7}{a+2} = 9$

12. $\dfrac{y}{y-1} - \dfrac{3}{y+1} = 1$

13. $\dfrac{6}{x} - \dfrac{3}{5x} = \dfrac{2}{x}$

14. $\dfrac{4}{x+2} = \dfrac{7}{x+2}$

15. $\dfrac{6b-12}{b+3} + \dfrac{5}{b-2} = 6$

16. $\dfrac{3t-2}{t+1} = 4 - \dfrac{t+2}{t-1}$

17. $\dfrac{3(x-1)}{x+7} = -\dfrac{3}{5}$

18. $\dfrac{x}{x + 1} + 2 = \dfrac{3x}{x + 2}$

19. $\dfrac{2a - 4}{a - 4} - 2 = \dfrac{20}{a + 4}$

20. $\dfrac{x}{x - 1} + 1 = \dfrac{1}{x - 1} - 3$

21. $\dfrac{3x - 4}{x - 2} - 3 = \dfrac{6}{x + 2}$

22. $\dfrac{4}{2a + 6} = \dfrac{3}{4a - 8}$

23. $\dfrac{1}{2(1 - b)} = \dfrac{2}{b(b - 1)} + \dfrac{1}{4b}$

24. $\dfrac{34m - 1}{5m} = \dfrac{2(1 + 8m)}{8m - 1} - \dfrac{3 - 24m}{5m}$

25. $\dfrac{2}{y - 4} + \dfrac{4}{y + 2} = \dfrac{30}{y^2 - 2y - 8}$

26. $\dfrac{8}{x^2 - 9} = \dfrac{2}{x - 3} - \dfrac{4}{x + 3}$

27. $\dfrac{4}{a - 2} = \dfrac{3a - 2}{a^2 - 4} + \dfrac{4}{a + 2}$

28. $\dfrac{2}{x + 2} + \dfrac{x}{x - 2} = \dfrac{x^2 + 4}{x^2 - 4}$

29. $\dfrac{4(9 - x)}{x^2 - 9} - \dfrac{3x + 2}{x + 3} = \dfrac{3x + 2}{3 - x}$

30. $\dfrac{3x + 10}{x^2 + 5x + 6} = \dfrac{x}{x + 3} - \dfrac{x}{x + 2}$

31. $\dfrac{6 - x}{3 - x} - \dfrac{1}{3} = \dfrac{x - 7}{2x - 6}$

32. $\dfrac{2y}{y^2 - 1} + \dfrac{y + 3}{y + 1} = \dfrac{y + 1}{y - 1}$

33. $\dfrac{4b + 1}{b^2 + 3b + 2} + \dfrac{b}{b + 2} = \dfrac{b + 2}{b + 1}$

34. $\dfrac{2x^2 - 15}{x^2 + x - 6} = \dfrac{x - 3}{x - 2} + \dfrac{x + 1}{x + 3}$

35. $\dfrac{2}{a + 4} + \dfrac{1}{a + 2} + \dfrac{a^2 + 3}{a^2 + 6a + 8} = 1$

36. $\dfrac{2x}{x - 1} - \dfrac{3}{x^2 - 1} = 4 - \dfrac{2x - 1}{x + 1}$

37. $\dfrac{a}{a + 4} + \dfrac{4}{4 - a} = \dfrac{a^2 + 16}{a^2 - 16}$

38. $\dfrac{4}{x - 2} - \dfrac{5}{x + 2} = \dfrac{33 - x}{x^2 + 2x}$

39. $\dfrac{5m - 7}{2m - 3} + \dfrac{m + 2}{2m + 3} - \dfrac{6}{4m^2 - 9} = 3$

40. $\dfrac{6x^2 + 14}{4x^2 - 9} - \dfrac{2x + 1}{2x - 3} = \dfrac{x + 1}{2x + 3}$

41. $\dfrac{5a - 2}{5a - 3} = 1 + \dfrac{3}{5a + 3} + \dfrac{2a}{25a^2 - 9}$

42. $\dfrac{3x^2 + 20}{9x^2 - 1} + \dfrac{2x}{3x + 1} = 2 - \dfrac{3x - 2}{3x - 1}$

43. $\dfrac{7x + 5}{x + 2} + \dfrac{x + 1}{2 - x} = 6$

44. $\dfrac{4a + 1}{a^2 - a - 6} = \dfrac{5}{a + 2} - \dfrac{2}{3 - a}$

45. $\dfrac{2m^2 - 25}{m^2 - 3m + 2} + \dfrac{m + 3}{2 - m} = \dfrac{m + 2}{m - 1}$

7.6 Word Problems Involving Rational Algebraic Expressions

   Word problems frequently lead to equations involving rational algebraic expressions.

Example 1: Suppose a swimming pool has two intake pipes. One will fill the pool in 10 hours and the other in 15 hours. If both pipes are open, how long will it take the pool to fill?

Solution: If we let x be the number of hours it takes for both pipes to fill the pool, then in *1 hour* the first pipe will fill $\frac{1}{10}$ of the pool, the second pipe will fill $\frac{1}{15}$ of the pool, and the two pipes together will fill $\frac{1}{x}$ of the pool. So we must have

$$\frac{1}{10} + \frac{1}{15} = \frac{1}{x}$$

Thus

$$30x \left(\frac{1}{10} + \frac{1}{15} \right) = 30x \cdot \frac{1}{x}$$

$$3x + 2x = 30$$

$$5x = 30$$

$$x = 6$$

Hence the two pipes together will fill the pool in 6 hours.

Check: In 6 hours the first pipe will fill $\frac{6}{10}$ of the pool and the second pipe will fill $\frac{6}{15}$ of the pool. Thus, in 6 hours, the two pipes together will fill

$$\frac{6}{10} + \frac{6}{15} = \frac{6}{10} + \frac{2}{5} = \frac{6}{10} + \frac{4}{10} = \frac{10}{10}$$

of the pool; i.e., the entire pool.

Example 2: The denominator of a fraction is 1 more than the numerator. If the numerator is increased by 9 and the denominator is increased by 12, the value of the fraction is unchanged. What is the fraction?

Solution: Let x = the numerator of the first fraction; then $x + 1$ = the denominator of the first fraction and thus the first fraction is $\frac{x}{x+1}$.

For the second fraction we have $x + 9$ = the numerator of the second fraction and $(x + 1) + 12 = x + 13$ = the denominator of the second fraction. Thus the second fraction is $\frac{x+9}{x+13}$. Therefore

$$\frac{x}{x+1} = \frac{x+9}{x+13} \qquad (x \neq -1 \text{ or } -13)$$

Multiplying both sides of this equation by the LCD, $(x + 1)(x + 13)$, we have

$$(x + 1)(x + 13) \cdot \frac{x}{x+1} = (x + 1)(x + 13) \cdot \frac{x+9}{x+13}$$

$$(x + 13) \cdot x = (x + 1)(x + 9)$$

$$x^2 + 13x = x^2 + 10x + 9$$

$$13x = 10x + 9$$

$$3x = 9$$

$$x = 3$$

The first fraction is $\frac{x}{x+1} = \frac{3}{3+1} = \frac{3}{4}$.

Check: $\frac{3+9}{3+13} = \frac{12}{16} = \frac{3}{4}$. Thus $\frac{3}{4} = \frac{12}{16}$ and the solution checks.

Example 3: Divide 77 into two parts so that one part is $\frac{5}{6}$ of the other.

Solution: Let x = the larger part; then $77 - x$ is the smaller part. So

$$\frac{5}{6}x = 77 - x$$

$$6 \cdot \frac{5x}{6} = 6(77 - x)$$

$$5x = 462 - 6x$$

$$11x = 462$$

$$x = 42 \text{ (the larger part)}$$

$$77 - x = 77 - 42 = 35 \text{ (the smaller part)}$$

Check:

$$\frac{5}{6} \cdot 42 \overset{?}{=} 77 - 42$$

$$5 \cdot 7 \overset{?}{=} 35$$

$$35 = 35 \quad \text{(checks)}$$

Example 4: The ones' digit of a two-digit numeral exceeds the tens' digit by 2. If the number is increased by 6 and divided by the sum of the digits, the quotient is 5. What is the number?

Solution: Let x = the tens' digit. Then $x + 2$ = the ones' digit, and the number is $10x + (x + 2) = 11x + 2$.

Thus

$$\frac{(11x + 2) + 6}{x + (x + 2)} = 5$$

$$\frac{11x + 8}{2x + 2} = 5 \quad (x \neq -1)$$

$$(2x + 2) \cdot \frac{11x + 8}{2x + 2} = 5(2x + 2)$$

$$11x + 8 = 10x + 10$$

$$x = 2$$

The tens' digit is 2 and the ones' digit is $2 + 2 = 4$; the number is 24.

Check: $\dfrac{24 + 6}{2 + 4} \overset{?}{=} 5$

$\dfrac{30}{6} \overset{?}{=} 5$

$5 = 5$ (checks)

Example 5: At what time between two and three are the hands of a clock together?

Solution: At two o'clock the hour hand has a 10-minute "lead" on the minute hand. The minute hand travels, of course, at a speed of 1 minute interval per minute, whereas the hour hand travels at a speed of $\frac{5}{60}$ minute intervals per minute. If t is the number of minutes it takes the minute hand to catch up with the hour hand, we have

$$1 \cdot t = \frac{5}{60} t + 10$$

$$t = \frac{1}{12} t + 10$$

$$12t = t + 120$$

$$11t = 120$$

$$t = \frac{120}{11} \text{ minutes} = 10 \frac{10}{11} \text{ minutes}$$

Thus the two hands will be together at $10\frac{10}{11}$ minutes past two o'clock.

EXERCISES 7.6

In exercises 1–30 write an equation for the given situation and then use that equation to solve the problem. Check your answer by seeing if it satisfies the conditions for the given problem.

1. The denominator of a fraction exceeds its numerator by 5. If the numerator is decreased by 3 and the denominator is increased by 1, the resulting fraction is equal to $\frac{2}{5}$. What was the original fraction?

2. Find the number that can be subtracted from both the numerator and the denominator of the fraction $\frac{65}{92}$ so that the result is equal to $\frac{4}{7}$.

3. The denominator of a fraction is 7 less than the numerator, and if 5 is added to the numerator, the fraction is equal to $\frac{9}{5}$. What is the fraction?

4. Divide 480 into two parts so that one part is $1\frac{2}{7}$ times the other.

5. Divide 296 into two parts such that $\frac{3}{11}$ of the greater is $\frac{2}{5}$ of the smaller.

6. If 6 is added to a certain number and the sum thus obtained is divided by 5, the result is 2 less than $\frac{1}{3}$ of the original number. What is the number?

7. The smaller of two numbers is $\frac{2}{3}$ the larger and the sum of their reciprocals is $\frac{1}{6}$. What are the numbers?

8. Alice can type a manuscript on her typewriter in 5 hours. Jack can type the same manuscript on his typewriter in 4 hours. How long would it take them to do the work together?

9. In a factory there is one machine that can turn out 100 of a certain article in 9 hours, a second machine can do 100 articles in 12 hours, and a third machine takes 18 hours for 100 articles. How long will it take to fill a rush order of 100 articles if all three machines work together?

10. In an apartment building one boiler burns 800 gallons of fuel oil in 5 days, and another boiler uses 800 gallons in 8 days. How long will 800 gallons of oil last if the two boilers are run at the same time?

11. The tens' digit of a two-digit numeral exceeds the ones' digit by 4. If the number increased by 39 is divided by the sum of its digits, the quotient is 15. What is the number?

12. The ones' digit of a two-digit numeral is 2 more than the tens' digit. When the number is divided by the sum of the digits, the quotient is 4 and the remainder is 9. Find the number.

13. The sum of the digits of a two-digit numeral is 15. If the number is divided by the ones' digit the quotient is 12 and the remainder is 3. What is the number?

14. Nancy's age is now twice Bill's age. If Nancy's age four years from now is divided by Bill's age at that time, the quotient is $\frac{5}{3}$. What are their ages now?

15. A father is 35 years old and his son is 15 years old. In how many years will the son be $\frac{6}{11}$ as old as his father?

16. Betty's age is $\frac{2}{3}$ the age of Jack. In 5 years Betty's age will be $\frac{3}{4}$ Jack's age. Find their present ages.

17. At what time between 4:00 and 5:00 are the hands of a watch together?

18. At what time between 1:00 and 2:00 are the hands of a watch opposite each other?

19. At what time between 5:00 and 5:30 are the hands of a watch at right angles to each other?

20. *A* and *B* start a business with *B* putting in $\frac{3}{4}$ as much capital as *A*. The first year *A* loses $5000 and *B* gains $\frac{1}{5}$ of his money; the second year *A* gains $\frac{1}{4}$ of the money he had at the beginning of the second year and *B* loses $2050; they now have equal amounts. How much did each have at first?

21. The denominator of certain fraction exceeds its numerator by 6. If the numerator is decreased by 2 and the denominator is increased by 7, the resulting fraction is equal to $\frac{1}{4}$. Find the fraction.

22. One pipe can fill a tank in 6 hours and two other pipes can fill the tank in 3 and 2 hours, respectively. If all three pipes are in use at the same time, how long will it take to fill the tank?

23. Divide 134 into two parts such that one divided by the other will give 3 as a quotient and 26 as a remainder.

24. Jill is $\frac{1}{6}$ as old as her father. In 12 years she will be $\frac{1}{3}$ as old. What are their ages now?

25. The sum of the digits of a two-digit number is 6, and this number divided by the number formed by interchanging the digits is equal to $\frac{5}{17}$. Find the original number.

26. At what time between 9:00 and 10:00 are the hands of a watch together?

27. A tank can be filled by one pipe in 8 hours and emptied by another in 12 hours. If both pipes are turned on when the tank is half full, how many hours will it take to fill the tank?

28. Many years ago a man owned a horse, a carriage worth $100 more than the horse, and a harness. The horse and harness were together worth $\frac{3}{4}$ the value of the carriage, and the carriage and harness were together worth $50 less than twice the value of the horse. Find the value of each.

29. A hundred years ago a grocer bought eggs at the rate of 4 for 7 cents. He sold $\frac{1}{4}$ of them at the rate of 5 for 12 cents, and the remainder at the rate of 6 for 11 cents, making 27 cents on the transaction. How many eggs did he buy?

30. A woman sells half an egg more than half her eggs. She then sells half an egg more than half her remaining eggs. A third time she does the same and now she has sold all her eggs. How many had she at first?

REVIEW EXERCISES

Section 7.1

In exercises 1–10 show how the right-hand expression can be obtained from the left-hand expression.

Example: $\dfrac{5}{x-3} = \dfrac{-5}{3-x}$

Solution: $\dfrac{5}{x-3} = \dfrac{-1(5)}{-1(x-3)} = \dfrac{-5}{-x+3} = \dfrac{-5}{3-x}$

1. $\dfrac{5}{a + b} = -\dfrac{-5}{a + 6}$

2. $\dfrac{3}{a - b} = \dfrac{-3}{b - a}$

3. $\dfrac{x - y}{x + y} = -\dfrac{y - x}{x + y}$

4. $\dfrac{x - y}{3} = \dfrac{y - x}{-3}$

5. $\dfrac{2a + 1}{a} = -\dfrac{2a + 1}{-a}$

6. $\dfrac{3x - 6}{5 - x} = \dfrac{6 - 3x}{x - 5}$

7. $\dfrac{(x + y)\,(x - y)}{x - 2} = \dfrac{(y + x)\,(y - x)}{2 - x}$

8. $\dfrac{x^2 + 3x - 2}{1 - x} = -\dfrac{x^2 + 3x - 2}{x - 1}$

9. $\dfrac{a - b - c}{a - b} = \dfrac{c + b - a}{b - a}$

10. $\dfrac{(x + 2)\,(x - 2)}{(3 - x)\,(5 - x)} = \dfrac{(x + 2)\,(x - 2)}{(x - 3)\,(x - 5)}$

In exercises 11–20 reduce to lowest terms each fraction not already in lowest terms.

11. $\dfrac{2x^2}{x}$

12. $\dfrac{15rs}{3r}$

13. $\dfrac{4x + 4y}{x^2 - y^2}$

14. $\dfrac{2x + 6}{x^2 + 5x + 6}$

15. $\dfrac{2x + 3y}{2x + y}$

16. $\dfrac{1 - x}{x - 1}$

17. $\dfrac{a^2 - 4}{a^2 - 5a + 6}$

18. $\dfrac{3x - 3y}{x^2 - 2xy + y^2}$

19. $\dfrac{9 - 4x^2}{4x^2 - 12x + 9}$

20. $\dfrac{a^2 - (b - c)^2}{(a - b)^2 - c^2}$

Section 7.2

In exercises 21–35 find the product and express it in lowest terms.

21. $\dfrac{x}{y} \cdot \dfrac{3x}{2y}$

22. $\dfrac{2x}{3y} \cdot \dfrac{9x}{14y}$

23. $\dfrac{m - n}{m + n} \cdot \dfrac{m - n}{m + n}$

24. $\dfrac{x + 3}{y} \cdot \dfrac{y}{x - 3}$

25. $\dfrac{2t - 2s}{5} \cdot \dfrac{15r}{t - s}$

26. $\dfrac{y + 4}{y - 2} \cdot \dfrac{y - 2}{3}$

27. $\dfrac{4(x + 3)}{15} \cdot \dfrac{5(x + 2)}{x + 3}$

28. $\dfrac{y + 5}{y - 5} \cdot \dfrac{y^2 - 25}{y}$

29. $\dfrac{7}{x} \cdot \dfrac{x^2 - 4}{x + 7} \cdot \dfrac{x^2 + 7x}{7x - 14}$

30. $\dfrac{x + y}{4x^2} \cdot \dfrac{1}{x^2 - y^2}$

31. $\dfrac{x^2 - y^2}{x + y} \cdot \dfrac{6x + 6y}{3x - 3y}$

32. $\dfrac{9a - 9b}{15a^2 b^2} \cdot \dfrac{21a^3 b^3}{7(a - b)^2}$

33. $\dfrac{x^2 + 3x + 2}{x + 3} \cdot \dfrac{x^2 + 7x + 12}{x^2 + 6x + 5}$

34. $\dfrac{y^2 + 8y + 16}{x + y} \cdot \dfrac{x^2 - y^2}{y^2 - 16}$

35. $\dfrac{x + 3}{x - 7} \cdot \dfrac{2x^2 - x - 10}{x^2 + x - 6} \cdot \dfrac{x - 7}{2x - 5}$

In exercises 36–50 find the quotient and express it in lowest terms.

36. $\dfrac{7}{8} \div \dfrac{3}{11}$

37. $\dfrac{5x}{9y} \div \dfrac{5x}{3y}$

38. $\dfrac{x^2}{y^2} \div \dfrac{y}{x}$

39. $\dfrac{x + 1}{y + 1} \div \dfrac{y}{x}$

40. $4xy \div \dfrac{3x}{2y}$

41. $\dfrac{x + 1}{x - 1} \div \dfrac{x^2 - 1}{x + 1}$

42. $\dfrac{3x - 3}{2x + 2} \div \dfrac{(x - 1)^2}{(x + 1)^2}$

43. $\dfrac{(a - b)^2}{2a} \div (a - b)$

44. $\dfrac{r^2 - 9}{r^2 + 3r} \div \dfrac{r - 3}{5}$

45. $\dfrac{x - 2}{y^2 + y} \div \dfrac{x^2 - 4}{y + 1}$

46. $\dfrac{2m + m^2}{4m - 3} \div \dfrac{4m^2 + 2m^3}{16m - 12}$

47. $\dfrac{x^2 - 6x + 5}{x - 1} \div \dfrac{x - 5}{x - 1}$

48. $\dfrac{r^2 - 4}{r^2 + 4r + 4} \div \dfrac{r^2 + 2r - 8}{2r + 4}$

49. $\left(\dfrac{x + 2y}{x - y} \cdot \dfrac{x - 2y}{y - x} \right) \div \dfrac{x^2 - 4y^2}{x^2 - y^2}$

50. $\left(1 \div \dfrac{y^3 + 2y^2}{x^2 - 3x} \right) \cdot \dfrac{y^2 - y - 6}{x^2 + x - 12}$

Section 7.3

In exercises 51–70 express the given sum or difference as a rational algebraic expression in lowest terms.

51. $\dfrac{3}{3a} - \dfrac{2}{3a} + \dfrac{4}{3a}$

52. $\dfrac{12a}{xy} + \dfrac{a + 1}{xy}$

53. $\dfrac{4x}{w} - \dfrac{x-1}{w}$

54. $\dfrac{x^2 + y^2}{x - y} - \dfrac{2xy}{x - y}$

55. $\dfrac{5 - y}{y^2 - 16} + \dfrac{3 - y^2}{y^2 - 16}$

56. $\dfrac{4}{3y} + \dfrac{4 + x}{6y}$

57. $\dfrac{4}{3ab} + \dfrac{a - 2}{5ab}$

58. $\dfrac{1}{x + y} + \dfrac{x}{x^2 - y^2}$

59. $\dfrac{3u}{2u + 6} + \dfrac{u - 1}{u + 3}$

60. $\dfrac{4y - 7}{y^2 - 3y + 2} + \dfrac{2}{y - 2}$

61. $\dfrac{3}{y^2 + 2y} - \dfrac{5}{3y + 6}$

62. $\dfrac{7}{a - b} + \dfrac{6}{a^2 - 2ab + b^2}$

63. $\dfrac{3 - m^2}{m^2 - 1} + \dfrac{m + 2}{m + 1}$

64. $\dfrac{3t}{t^2 - 2t - 8} + \dfrac{t}{t^2 - 4}$

65. $\dfrac{6}{x^2 + 3x} - \dfrac{x - 2}{x^2 - 9}$

66. $\dfrac{2y}{y - 5} + \dfrac{2y}{y + 2}$

67. $\dfrac{2}{a + b} + \dfrac{3}{a - b} + \dfrac{4}{a^2 - b^2}$

68. $\dfrac{5n + 1}{2n^2 - 2} + \dfrac{7}{6n + 6}$

69. $\dfrac{x^2 + 2}{3x^2} + \dfrac{x - 2}{3x} + \dfrac{4}{3}$

70. $\dfrac{a + 1}{a - 1} - \dfrac{3a^2 - 1}{1 - a^2}$

Section 7.4

In exercises 71–88 divide the first polynomial by the second. If there is a nonzero remainder express the answer as the quotient $+ \dfrac{\text{remainder}}{\text{divisor}}$.

71. $(x^2 + 5x + 6) \div (x + 3)$
72. $(4x^2 - 4x + 1) \div (2x - 1)$
73. $(6x^2 + 5x - 3) \div (2x - 3)$
74. $(2x^3 - 2x^2 + 5) \div (x - 6)$
75. $(16 - 13x + x^2) \div (x - 7)$
76. $(8a^3 - 1) \div (2a - 1)$
77. $(2x^2 - 6x + 3) \div (x - 2)$
78. $(x^2 + 7x + 12) \div (x + 3)$
79. $(t^2 - 6t - 16) \div (t + 2)$
80. $(3x^3 + 11x^2 + 11x + 15) \div (x + 3)$
81. $(2n^3 - 4n^2 + 11n - 12) \div (2n - 3)$
82. $(6c^3 - 5c^2 - 3c + 2) \div (2c^2 - 3c + 1)$
83. $(x^3 - x^2 + x - 1) \div (2x - 1)$
84. $(4x^2 - x + 1) \div (3x - 1)$
85. $(x^4 - x^3 + x - 4) \div (2x - 1)$

86. $(5x^2 + 3x + 1) \div (3x + 2)$

87. $(3x^2 + 2x + 1) \div (4x + 1)$

88. $(3x^3 + 4x^2 - x + 2) \div (2x^2 + 3x + 1)$

Section 7.5

In exercises 89–110 solve the given equation or show that no solution exists. Begin with a list of prohibited values of the variable and check to see that none of your candidates for a solution are on this list.

89. $\dfrac{5}{3} + \dfrac{3}{x} = \dfrac{19}{3x}$

90. $\dfrac{15}{2a} + \dfrac{4}{a} = \dfrac{23}{4}$

91. $\dfrac{13}{2y} - \dfrac{3}{20} = \dfrac{7}{4y} + \dfrac{4}{5}$

92. $\dfrac{4}{x} - \dfrac{5}{12} = \dfrac{1}{2} - \dfrac{3}{2x}$

93. $\dfrac{13}{6x} - \dfrac{2}{3x} = 1 + \dfrac{1}{2x}$

94. $3 - \dfrac{2x - 5}{x} = \dfrac{5}{x}$

95. $3 - \dfrac{3b + 1}{b + 3} = \dfrac{2}{b - 3}$

96. $\dfrac{2}{x - 3} - \dfrac{4}{x + 3} = \dfrac{8}{x^2 - 9}$

97. $\dfrac{y}{y + 5} + 4 = 5$

98. $\dfrac{3x - 2}{x^2 - 4} = \dfrac{4}{x - 2} - \dfrac{4}{x + 2}$

99. $\dfrac{9x + 4}{x^2 + 3x - 4} - \dfrac{3}{x - 1} = \dfrac{5}{x + 4}$

100. $\dfrac{4y + 1}{y^2 + 3y + 2} + \dfrac{y}{y + 2} = \dfrac{y + 2}{y + 1}$

101. $\dfrac{12a + 19}{a^2 + 7a + 12} - \dfrac{3}{a + 3} = \dfrac{5}{a + 4}$

102. $\dfrac{1}{2x + 4} + \dfrac{3}{2x - 4} = \dfrac{x + 5}{x^2 - 4}$

103. $\dfrac{3}{2 - x} - \dfrac{2}{2 + x} = \dfrac{7}{4 - x^2}$

104. $\dfrac{7x + 13}{15 - 8x + x^2} - \dfrac{1}{3 - x} = \dfrac{5}{5 - x}$

105. $\dfrac{3}{t + 4} + \dfrac{2}{4 - t} = \dfrac{17}{16 - t^2}$

106. $2 - \dfrac{2a}{2a - 1} - \dfrac{1}{4a^2 - 1} = \dfrac{2a - 1}{2a + 1}$

107. $\dfrac{2x^2 + 6}{x^2 - 4} = \dfrac{x}{x + 2} + \dfrac{x + 1}{x - 1}$

108. $\dfrac{2 + 3x}{6 - 5x + x^2} - \dfrac{1}{2 - x} = \dfrac{3}{3 - x}$

109. $3 + \dfrac{a}{1 - a} = \dfrac{2a}{a + 1}$

110. $\dfrac{9y + 1}{y^2 - 9} + \dfrac{5}{3 - y} = \dfrac{3}{y + 3}$

Section 7.6

In exercises 111–125 write an equation for the given situation; then use that equation to solve the problem. Check your answer by seeing if it satisfies the conditions for the given problem.

111.   The larger of two numbers is 12 more than the smaller, and the ratio of the larger number to the smaller number is $\frac{5}{3}$. What are the two numbers?

112.   What number exceeds the sum of its third, sixth, and fourteenth parts by 18?

113.   On their vacation, Mr. and Mrs. Matson traveled 1536 miles. They went $\frac{4}{5}$ as many miles by boat as they did by train, and $\frac{5}{12}$ as many miles by bus as by boat. How many miles did they travel in each manner?

114.   A man in a car drove 200 kilometers at a certain rate of speed. The return trip was made over the same route at twice his original speed and took 2 hours less time. Find the speed on the first trip.

115.   The denominator of a certain fraction is 14 more than its numerator. If the numerator and denominator are both increased by 5, the new fraction is equal to $\frac{11}{18}$. What was the original fraction?

116.   Joyce paid $283 for 10 equally priced pictures and 10 equally priced frames. If each picture had cost $4 more and each frame 30¢ less, the price of the frame would have been $\frac{1}{3}$ that of the picture. What did each picture cost?

117.   The tens' digit of a two-digit numeral exceeds the ones' digit by 2. If the number is increased by 6 and divided by the sum of the digits, the quotient is 7. What was the original number?

118.   Are there two consecutive even integers whose quotient is equal to $\frac{1}{36}$? If so, what are they?

119.   A tank can be filled by one pipe in 9 hours and emptied by another in 21 hours. In what time will the tank be filled if both pipes are opened?

120.   A plane travels 5600 kilometers in the same time that a train travels 960 kilometers. If the speed of the plane is 80 kilometers an hour more than 5 times the speed of the train, what is the speed of each?

121.   At what time between 4:00 and 4:30 are the hands of a watch at right angles to each other?

122.   Marian can do a certain job in 15 hours. It takes Felicia 25 hours to do the same job. After Marian has worked for a certain time Felicia completes the job by working 9 hours longer than Marian. How many hours did Marian work?

123.   Pat can shovel the snow from a certain sidewalk in 18 minutes, while Mike can do it in 12 minutes. How long should it take them to shovel the snow from this sidewalk if they work together?

124.   How many minutes must you run at 10 kilometers per hour and how many minutes must you walk at 6 kilometers an hour to cover 5 kilometers in 38 minutes?

125.   A journeyman and 15 helpers working together can complete a certain electrical wiring job in 10 days. The same job could be done in 4 days by 25 journeymen. How long would it take 1 helper working alone to complete the job?

CUMULATIVE REVIEW E

1. $-3 - |-6 - 2(-3 - (-4)| = ?$
2. True or false? $-x$ is always a negative number if $x \neq 0$.

In exercises 3–11 simplify the given expression.

3. $\dfrac{-42x^6}{-3x^2}$

4. $(p - q + r) - (2p - 2q) + (3p - 4r)$

5. $-4(2a + 3) - 5(2a + 3b)$

6. $\sqrt{338}$

7. $3\sqrt{6} \cdot 5\sqrt{10}$

8. $\sqrt{45} \div \sqrt{15}$

9. $\sqrt{\dfrac{4}{3}}$

10. $2\sqrt{24} + \sqrt{54} - \sqrt{96}$

11. $\dfrac{x^4 y^2}{x^{-2} y}$

In exercises 12–26 find the solution set of the given equation or inequality.

12. $x + 2(x - 3) = x + 2$

13. $(3x + 4) + (2x + 7) < 3x + 5$

14. $x - 2(3x + 8) = x - 2(x - 5)$

15. $3x(x - 1) > 3x^2 - 2(x - 4)$

16. $3x - (2x + 6) = 4x - 3$

17. $-5(3 - x) < 5 - (x + 10)$

18. $3 - x(x - 2) = 5 - x^2$

19. $2x + \dfrac{1}{2} < \dfrac{2}{3} + \dfrac{3}{4}x$

20. $\dfrac{4}{x} - \dfrac{5}{3} = \dfrac{1}{5}$

21. $\dfrac{x + 4}{2} - \dfrac{x - 2}{5} = 1$

22. $\dfrac{2x + 1}{x + 2} = \dfrac{2x + 3}{x - 2}$

23. $\dfrac{6x + 7}{15} - \dfrac{2x - 2}{7x - 6} = \dfrac{2x + 1}{5}$

24. $\dfrac{1}{x + 2} + \dfrac{2}{3x + 6} = \dfrac{2}{3}$

25. $\dfrac{3}{x - 4} + \dfrac{5}{3x - 12} = 2$

26. $9 - x - 2(x - 1)(x + 2) = (x - 3)(5 - 2x)$

27. Graph the equation $2x + 5y = 15$.

28. A man invests $8000, part at 8% and the rest at 10%. If he receives an annual interest return of $700, how much was invested at each rate?

29. A woman who can row a boat 6 kilometers per hour in still water rows upstream and back to her starting point. The entire trip takes 3 hours and 45 minutes, and she spends 1 hour and 15 minutes more time rowing upstream than on the return trip. Find the speed of the current.

30. A radiator contains 16 quarts of a solution that is 20% alcohol and 80% water. How much solution should be drained off and replaced by pure alcohol in order to have a solution that is 25% alcohol?

31. A man has a horse, a cow, and a calf; the cow is worth $\frac{2}{3}$ as much as the horse, and the calf is worth $\frac{5}{48}$ as much as the horse. If the cow is worth $81 more than the calf, what is the value of the horse?

32. Divide $2.70 between two people so that $\frac{1}{5}$ of what one receives will be equal to $\frac{1}{4}$ of what the other receives.

33. A baseball team wins 20 out of the first 40 games. It wins $\frac{2}{5}$ of the subsequent games and ends up with a record of 0.450. How many games were played altogether?

34. From the sum of $3x^2 + 6xy - 2y^2$ and $x^2 + 3y^2$ subtract the difference between $2x^2 - 5x - y$ and $x^2 - 6x - 7y$.

35. Multiply: $(x + 2y)(x + 3y)$.

36. Multiply: $(5x^2 - 3x)(5x^2 + 3x)$.

37. Multiply: $(x^2 - xy + 2y^2)(x^2 - 2y^2 + xy)$.

38. Divide: $(x^3 + 4x^2 + 7x + 2) \div (x + 2)$.

In exercises 39–44 factor completely the given polynomial.

39. $4x^2 + 12x - 40$

40. $4x^2 - 16x + 15$

41. $3y^2 + 3y - 126$

42. $9x^4 y - y^3$

43. $x^2 - 11x + 30$

44. $4x^2 - 8adx + 4a^2 d^2$

45. Reduce to lowest terms: $\dfrac{6x^2 + 12xy}{8x^2 + 16xy^2}$.

46. Multiply: $\dfrac{4x^2 - 16}{7x + 7} \cdot \dfrac{14x + 14}{x^2 + 5x + 6}$.

47. Add: $\dfrac{2x + 3}{5} + \dfrac{x - 1}{3}$.

48. Subtract: $\dfrac{6x + 17}{x^2 + 5x + 6} - \dfrac{5}{x + 2}$.

In exercises 49 and 50 solve the given system of equations.

49. $\begin{cases} 2x - 5y = -16 \\ 3x + 7y = 5 \end{cases}$

50. $\begin{cases} \dfrac{x + y}{3} + \dfrac{x + y}{4} = 7 \\ \dfrac{x}{5} - \dfrac{x + y}{15} = \dfrac{8}{5} \end{cases}$

8

FUNCTIONS

8.1 Definitions and Examples

There are many rules in mathematics that tell us how to go from one number to another. The rule for doubling a natural number, for example, takes us from 2 to 4, from 3 to 6, from 5 to 10, and so on.

Applications of mathematics frequently involve such rules. For example, the time, t, in seconds determines the distance that a freely falling object travels, starting from rest. If we neglect air resistance and if we are not far from the surface of the earth, then the distance, s, in meters that an object falls in t seconds is given by $s = 4.9\,t^2$. An object falls 4.9 meters in 1 second, and 19.6 meters in 2 seconds. Thus the rule given by $s = 4.9\,t^2$ takes us from 1 to 4.9, from 2 to 19.6, and so on.

In each of the two preceding examples we have a rule that takes us from one set of numbers D to another (perhaps the same) set of numbers R. The rule together with the sets make up what is called a *function*. Figure 8.1 pictures this situation.

In our first example, D is the set of natural numbers and R is the set of even natural numbers. In our second example, both D and R are the set of nonnegative real numbers. We illustrate these two functions in Figures 8.2 and 8.3, respectively.

The function associates   with   exactly one y in R
each x in D

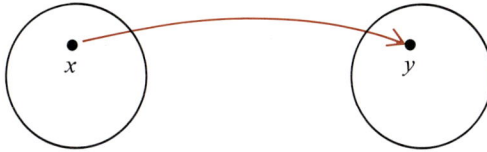

D, the given set of numbers, is called the *domain* of the function.

R, the set of all numbers obtained from the elements of D by the rule, is called the *range* of the function.

Figure 8.1

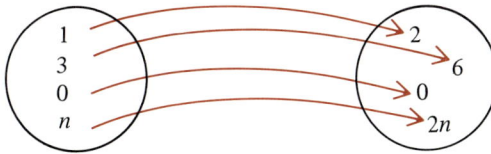

D = set of natural numbers

R = set of even natural numbers

Figure 8.2

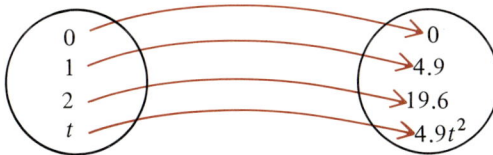

D = set of nonnegative real numbers R = set of nonnegative real numbers

Figure 8.3

So far our examples of functions have been linked with algebraic formulas: $y = 2x$ and $s = 4.9t^2$, respectively. The following table, which gives the height in centimeters of a certain boy at different ages, illustrates a function which does not involve a formula.

Age in years	13	14	15	16	17	18	19	20	21	22
Height in centimeters	155	160	167	172	175	178	178	180	180	180

Here $D = \{13, 14, 15, 16, 17, 18, 19, 20, 21, 22\}$ and $R = \{155, 160, 167, 172, 175, 178, 180\}$. Each element in the set D is associated in a natural way with exactly one element of the set R, as indicated by the arrows below:

$13 \rightarrow 155$	$15 \rightarrow 167$	$17 \rightarrow 175$	$19 \rightarrow 178$	$21 \rightarrow 180$
$14 \rightarrow 160$	$16 \rightarrow 172$	$18 \rightarrow 178$	$20 \rightarrow 180$	$22 \rightarrow 180$

Since the table provides us with a rule that takes us from the set D to the set R, we see that the table and the sets D and R define a function. Note that some elements of the set R are associated with more than one element of the set D: $18 \rightarrow 178$ and also $19 \rightarrow 178$; $20 \rightarrow 180$ and also $21 \rightarrow 180$ and $22 \rightarrow 180$. This is quite permissible. There is nothing in the definition of a function that requires that each element of R correspond to only one element of D — it is only necessary that each element of D correspond to only one element of R.

Post office regulations concerning mailing costs provide us with other examples of functions. For example, the book rate in the United States in 1976 was:

Weight	Cost
Up to 1 pound	21¢
Over 1 pound but not over 2 pounds	30¢
Over 2 pounds but not over 3 pounds	39¢
Over 3 pounds but not over 4 pounds	48¢
Over 4 pounds but not over 5 pounds	57¢

The rule for determining the cost of mailing books defines a function with domain $D = \{x : x$ is a rational number and $0 < x \leqslant 5\}$ and $R = \{21, 30, 39, 48, 57\}$. Some of the correspondences determined by this postage function are shown in the table below.

Weight in pounds	0.5	1	1.75	2	2.8	3	3.1	3.3	4	4.01	5
Postage in cents	21	21	30	30	39	39	48	48	48	57	57

It is often helpful to think of a function as a machine, as suggested by Figure 8.4, that processes elements of the domain to produce elements of the range. The inputs are elements of the domain and the outputs are elements of the range.

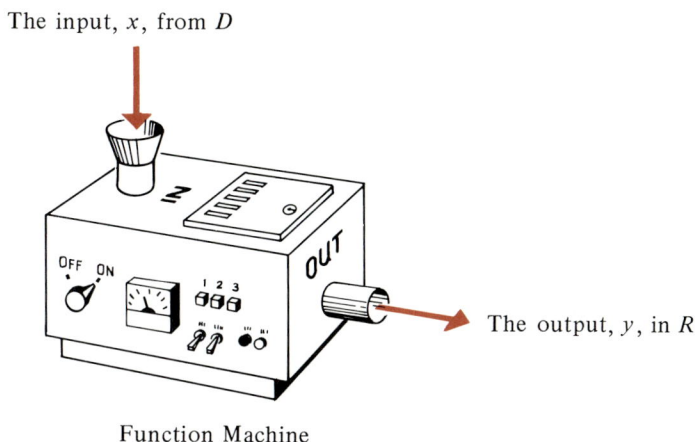

The input, x, from D

Function Machine

Figure 8.4

Any function produces a set of ordered pairs, (x,y), where x is an input and y the corresponding output. The function that associates with each natural number its double produces the set of ordered pairs $\{(x,y) : x$ is a natural number and $y = 2x\}$ which can also be written as $\{(x, 2x) : x$ a natural number$\}$. The function that associated a boy's age with his height produced the set of ordered pairs $\{(13, 155),$ $(14, 160), \ldots, (22, 180)\}$.

It is important to note, however, that not every set of ordered pairs can be considered as arising from a function. The set $\{(2, 1),$ $(2, 2)\}$, for example, could not have been obtained from a function since we would have $2 \to 1$ and also $2 \to 2$ if it had. But our definition of a function says that there is *exactly one* element of the range corresponding to a given element of the domain. This means that we cannot have two ordered pairs (x_1, y_1) and (x_1, y_2) with $y_1 \neq y_2$. This point is illustrated in Figure 8.5.

In the following examples, the domain D and the range R are both taken to be finite sets with only a few elements because it is easy to think about the concept of a function when sets of this kind are used. In most situations of practical importance, however, we have for the domain and range such infinite sets as the set of real numbers, the set of positive real numbers, and the set of rational numbers.

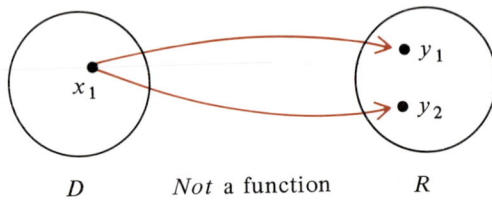

D   *Not* a function   R

Figure 8.5

Example 1. Let $D = \{-2, -1, 0, 1, 2\}$. Then, if our function (rule) is to associate every element x of D with its absolute value, $|x|$, we have $R = \{0, 1, 2\}$ since $|0| = 0$, $|-1| = |1| = 1$, and $|-2| = |2| = 2$. The set of ordered pairs produced is $\{(-2, 2), (-1, 1), (0, 0), (1, 1), (2, 2)\}$.

Example 2. The set of ordered pairs $\{(1, 2), (2, 2), (3, 4), (4, 4), (5, 6), (6, 6)\}$ represents a function that associates the first element in each ordered pair with the second. The association is a function because each element in the domain $\{1, 2, 3, 4, 5, 6\}$ is associated with just one element of the range $\{2, 4, 6\}$.

Example 3. The set of ordered pairs $\{(1, 2), (2, 4) (1, 3)\}$ represents an association of the first element in each ordered pair with the second. The association is *not* a function, however, because 1, which is an element of the domain $\{1, 2\}$, is associated with both 2 and 3.

Example 4. Let $D = \{1, 2, 3, 4\}$ and let each element of D be associated with its square. The association defines a function with range $\{1, 4, 9, 16\}$, and the set of ordered pairs obtained is $\{(1, 1), (2, 4), (3, 9), (4, 16)\}$.

Example 5. Let $D = \{1, 4, 9, 16\}$ and let each element of D be associated with its square roots. Since each element of D is thus associated with two numbers, 1 with 1 and -1, 4 with 2 and -2, 9 with 3 and -3, and 16 with 4 and -4, the association is not a function. However, the association of the elements of D with *either* the positive square roots $(x \to \sqrt{x})$ *or* with the negative square roots $(x \to -\sqrt{x})$ defines a function.

In our definition of a function, we considered D and R to be sets of numbers. It is, however, possible to consider functions whose domain or range are not sets of numbers. Our last example illustrates this fact.

Example 6. Suppose that the first names of five students in a class are Juanita, Bob, Charles, George, and Elsie. Their last names are Ramirez, Jones, Brown, Smith, and Jones, respectively.

We can display this association by means of arrows, as shown in Figure 8.6, or as the set of ordered pairs {(Juanita, Ramirez), (Bob, Jones), (Charles, Brown), (George, Smith), (Elsie, Jones)}. Notice that this association defines a function because all of the students have different first names, and, as required for a function, exactly one element of the range, {Ramirez, Jones, Brown, Smith}, is associated with each element of the domain, {Juanita, Bob, Charles, George, Elsie}.

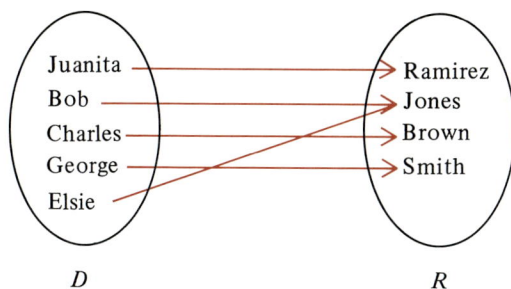

Figure 8.6

EXERCISES 8.1

In exercises 1-5 you are given the input numbers from a domain D and the rule for the function. Find the output numbers in the range R.

Example (a): **Rule**: Add 1.

INPUTS (DOMAIN)	OUTPUTS (RANGE)	
8	9	← 9 = 8 + 1
17	18	← 18 = 17 + 1
24	25	← 25 = 24 + 1
37	38	← 38 = 37 + 1
106	107	← 107 = 106 + 1

Example (b): **Rule:** Multiply by 2 and then subtract 1.

INPUTS (DOMAIN)	−5	−2	0	5	13	50	75
OUTPUTS (RANGE)	↑	↑	↑	↑	↑	↑	↑

$2(-5) - 1 = -11$

$2(-2) - 1 = -5$

$2(0) - 1 = -1$

$2(5) - 1 = 9$

$2(13) - 1 = 25$

$2(50) - 1 = 99$

$2(75) - 1 = 149$

1. Rule: Subtract 5.

INPUTS (DOMAIN)	OUTPUTS (RANGE)
83	78
52	?
21	?
0	?
−3	?
−17	?

2. Rule: Multiply by $2\frac{1}{2}$.

INPUTS (DOMAIN)	OUTPUTS (RANGE)
−7	$-17\frac{1}{2}$
−2	?
0	?
13	?
26	?
102	?

3. Rule: Multiply by 3 and then subtract 7.

INPUTS (DOMAIN)	−11	−8	0	$2\frac{1}{2}$	$7\frac{1}{3}$
OUTPUTS (RANGE)	?	?	?	?	?

4. Rule: Divide by 3 and then add 5.

INPUTS (DOMAIN)	-9	0	$\frac{2}{3}$	$\frac{6}{7}$	15
OUTPUTS (RANGE)	?	?	?	?	?

5. Rule: Multiply by 4 and then subtract $1\frac{3}{4}$.

INPUTS (DOMAIN)	-1	0	2	4	12
OUTPUTS (RANGE)	?	?	?	?	?

In exercises 6-10 you are given the output numbers (range) and the rule. Find the input numbers (domain) as determined by the rule.

6. Rule: Add 1.

INPUTS (DOMAIN)	?	?	?	?	?
OUTPUTS (RANGE)	-5	-2	0	$\frac{7}{6}$	13

7. Rule: Cube the number.

INPUTS (DOMAIN)	?	?	?	?	?
OUTPUTS (RANGE)	1	8	27	0	125

8. Rule: Subtract 1.

DOMAIN	?	?	?	?	?
RANGE	$-\frac{8}{5}$	0	$1\frac{5}{6}$	$\frac{29}{4}$	101

9.   Rule: Multiply by 2 and then add 1.

DOMAIN	?	?	?	?	?
RANGE	-9	-5	$\dfrac{5}{2}$	$\dfrac{23}{5}$	21

10. Rule: Divide by 3 and then subtract 1.

DOMAIN	?	?	?	?	?
RANGE	-5	-1	4	15	32

In exercises 11-20, find the set R of numbers obtained according to the stated rule if the domain is given as $D = \{1, -3, 5, 0, -7\}$.

11.   Add -3 to each element of D.
12.   Multiply each element of D by 5.
13.   Square each element of D.
14.   Associate each number in D with the number 13.
15.   Associate with each negative number in D the number 0, and with each nonnegative number in D the number 1.
16.   Each x in D is associated with $x^2 - 3$ in R.
17.   Each x in D is associated with $- |x|$ in R.
18.   Each x in D is associated with $3x - 5$ in R.
19.   Each x in D is associated with $x^2 + 1$ in R.
20.   Each x in D is associated with $|x - 1|$ in R.

In exercises 21-30, find the set D obtained according to the stated rule if the range is given as $R = \{-8, -1, 0, 1, 8\}$.

21.   Subtract $\frac{3}{2}$ from each element of D.
22.   Divide each element of D by 4.
23.   Cube each element of D.
24.   Multiply each element of D by 3 and then add 1.
25.   Divide each element of D by 3 and then subtract 1.
26.   Each x in D is associated with $\frac{1}{2}x - 3$.
27.   Each x in D is associated with $\frac{x}{3} + 1$.
28.   Each x in D is associated with $3x - 1$.
29.   Each x in D is associated with $-x^3$.
30.   Each x in D is associated with $-x^3$.

In exercises 31-40 give a rule or formula that describes the association. The first row in each exercise gives the domain.

31.

x	1	2	3	4	5	6
y	0	1	2	3	4	5

32.

s	1	2	3	4	5	6
t	3	5	7	9	11	13

33.

a	-1	1	2	-2	3
b	1	1	4	4	9

34.

t	4	5	6	7	12	22
j	8	9	10	11	16	26

35.

s	3	4	5	6	9	13
t	1	2	3	4	7	11

36.

x	-3	-1	1	3	5	7
y	-7	-3	1	5	9	13

37.

A	1	2	3	4	5
B	$1\frac{1}{2}$	2	$2\frac{1}{2}$	3	$3\frac{1}{2}$

38.

x	1	2	3	4	5
y	2	5	10	17	26

39.

u	1	2	3	4	5	6
v	1	2	1	2	1	2

40.

x	2	3	4	5	6	7
y	1	3	6	10	15	21

In exercises 41–50 the given statement implies an association between two sets that defines a function. In each exercise give the domain and range and, if possible, a table, formula, or algebraic equation that defines the association.

Example: The circumference of a circle is π times the length of the diameter.

Solution: The domain and range are both the set of positive real numbers. The formula: $c = \pi d$

41. The area of a square is the square of the length of one side.
42. The area of a rectangle with a width of 5 inches and a certain length is the product of the width and the length.
43. The area of a circle is π times the square of the length of the radius.
44. The perimeter of a square is 4 times the length of a side.
45. Every real number has an absolute value.
46. Every nonnegative real number has a positive square root.
47. In a conversion from the English system of measurement to the metric system of measurement, we use the fact that there are 2.54 centimeters in 1 inch.
48. In a conversion from the metric system of measurement to the English system of measurement, we use the fact that a kilometer is 0.62 miles.
49. The highest temperatures, measured in degrees Celsius, on each of the first ten days of June were, in chronological order, 17, 22, 26, 20, 14, 23, 17, 27, 24, and 21.
50. The lowest temperatures, measured in degrees Celsius, on each of the first ten days of January were, in chronological order, −4, −1, −3, −6, −10, −15, −17, −16, −12, and −4.

8.2 Functional Notation

It is common to denote a function by a letter such as $f, g, h, F, G,$ etc. We can picture a function f as shown in Figure 8.7.

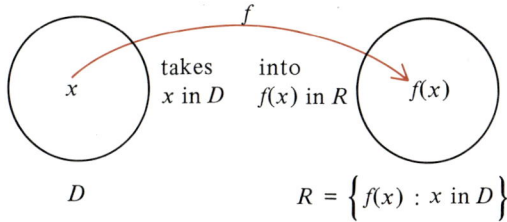

Figure 8.7

We read $f(x)$ as "*f* of *x*" and call $f(x)$ a *functional value.* (Note that $f(x)$ is a symbol for an element of R and does *not* mean the product of f and x!)

We have been using x to designate an element of D. However, the particular symbol used to designate an element of D is unimportant. We also have, for example.

$$R = \{f(y): y \text{ in } D\} = \{f(t): t \text{ in } D\}$$

We can now say that a function f associates with each x in D exactly one element, $f(x)$ in R, and defines the set of ordered pairs $\{(x, f(x)): x \text{ in } D\}$.

Figure 8.8 shows the situation in terms of our machine analogy.

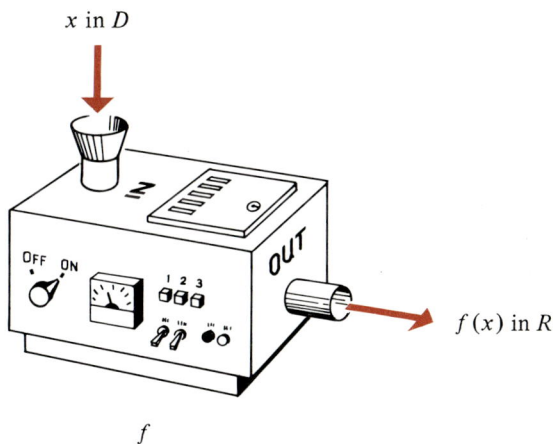

Figure 8.8

The notations f and $f(x)$, g and $g(x)$, F and $F(x)$, etc. help us to describe functions. Thus, for example, we can describe our doubling function as the function f such that

$$f(x) = 2x$$

for each x in N (the set of natural numbers). We have

$$f(1) = 2 \cdot 1 = 2, \; f(2) = 2 \cdot 2 = 4, \; f(3) = 2 \cdot 3 = 6, \text{ etc.}$$

As an abbreviation we will sometimes write, for example:

$$f : f(x) = 2x$$

for "the function f defined by $f(x) = 2x$."

Note carefully the distinction between f (the machine) and $f(x)$ — that which is produced by the machine when x is fed into it. (Surely no one can mistake a sausage *machine* from the sausages it produces!)

Here are some other examples of the use of this *functional notation*.

Example 1: If $D = \{0, 1, 2, 3\}$ and $f(x) = 3x - 5$ for x in D, then

$$f(0) = 3 \cdot 0 - 5 = -5 \qquad f(2) = 3 \cdot 2 - 5 = 1$$
$$f(1) = 3 \cdot 1 - 5 = -2 \qquad f(3) = 3 \cdot 3 - 5 = 4$$

and the range of f is $\{-5, -2, 1, 4\}$.

Example 2: If $f : f(x) = |x|$ with D the set of integers, we have $f(0) = 0$, $f(1) = 1, f(-1) = 1, f(2) = 2, f(-2) = 2$, etc., and the range of f is the set of nonnegative integers.

Example 3: Let g be the function with domain the set of real numbers and such that $g(x) = 0$ if x is any rational number and $g(x) = 1$ if x is any irrational number. Then

$$g(0) = 0, \quad g\left(\frac{1}{2}\right) = 0, \quad g(2.3) = 0, \quad g(-5) = 0$$

whereas

$$g(\sqrt{2}) = 1, \quad g(\pi) = 1, \quad g(1 - \sqrt{3}) = 1, \quad g\left(\frac{\sqrt{5}}{2}\right) = 1$$

and the range of g is $\{0, 1\}$.

It is not always easy to determine the range of a function. This is the case if, for example,

$$F : F(x) = x^3 - 5x^2 + 2x - 1$$

where D is the set of real numbers. We do know, however, that, for each x in D, $F(x)$ is a real number. Thus the range of F is either the set of real numbers or is contained in the set of real numbers.

EXERCISES 8.2

In exercises 1-5 answer the following questions for each table defining a function, f.

a.   What is $f(1)$, $f(4)$, $f(10)$, $f(15)$, $-f(2)$, $f(3) + 2$?

b.   What is the domain of the function?

c.   What is the range of the function?

d.   What is an algebraic expression for $f(x)$?

Example: A function f is defined by the following table:

x	1	2	3	4	5	10	15
y	3	6	9	12	15	30	45

Solution: a. $f(1) = 3$ $f(10) = 30$ $-f(2)$ $= -6$

  $f(4) = 12$ $f(15) = 45$ $f(3) + 2 = 11$

b.   The domain of the function is $\{1, 2, 3, 4, 5, 10, 15\}$.

c.   The range of the function is $\{3, 6, 9, 12, 15, 30, 45\}$.

d. $f(x) = 3x$

1.

x	1	2	3	4	5	6	10	15	20
y	4	5	6	7	8	9	13	18	23

2.

x	-3	-1	0	1	2	3	4	5	10	15
y	-8	-6	-5	-4	-3	-2	-1	0	5	10

3.

x	-10	-5	0	1	2	3	4	5	10	15
y	20	10	0	-2	-4	-6	-8	-10	-20	-30

4.

x	-3	-1	0	1	2	3	4	5	10	15
y	-14	-4	1	6	11	16	21	26	51	76

5.

x	-2	-1	0	1	2	3	4	5	10	15	20
y	2	$2\frac{1}{2}$	3	$3\frac{1}{2}$	4	$4\frac{1}{2}$	5	$5\frac{1}{2}$	8	$10\frac{1}{2}$	13

In Exercises 6–15 find the functional values as indicated for the given function. The domain for all of the functions is to be taken as the set of real numbers.

Example: $g(x) = 2x$; find: $g(3), g(0), -g(-2), g(a), g(\sqrt{2})$

Solution: $g(3) = 6, g(0) = 0, -g(-2) = 4, g(a) = 2a, g(\sqrt{2}) = 2\sqrt{2}$

6. $f(x) = 3x + 1$; find: $f(-2), f(0), f\left(\frac{1}{2}\right), f(5), f(100)$

7. $h(x) = 2 - \frac{x}{2}$; find: $h(0), h\left(\frac{3}{4}\right), -h(5), h(\sqrt{3}), h(a)$

8. $g(u) = 1 - 2u$; find: $g(\sqrt{5}), g\left(\frac{3}{8}\right), g(1) - 2, -g(1), 2g(x)$

9. $f(s) = 4s + 2$; find: $f(-1), f(1) + 2, f(x), \frac{1}{2}f(-2), f(2) - f(-3)$

10. $j(s) = 2s - 1$; find: $j(-3), j(\sqrt{7}), j(x), j(-3) - j(2), j(2t)$

11. $h(t) = 5t^2 - 1$; find: $h(\sqrt{2}), h(-2), -h(5) - 25, -h(1) \cdot 3h(-1)$

12. $t(x) = x + 2$; find: $t(-6), t(-10) - 2, 3t(5) - 20, 3t(-2) + 5t(0), \frac{t(-1)}{t(1)}$

13. $g(x) = 2x - 1$ and $f(x) = \frac{x}{2}$; find: $g(2) \cdot f(2), \frac{g(3)}{f(3)}, \frac{g(1)}{f(1)},$
     $-[g(-2) + f(0)], g(s) + f(2s)$

14. $g(t) = \frac{1}{2}t - 1$ and $h(t) = 2t + 1$; find: $g(1) \cdot h(0), g(2) - 2h(1),$
     $g(2x) - h(x), \frac{g(0)}{h(0)}$

15. $f(g) = \frac{3g}{2} + 1$ and $h(g) = 5 - g$; find: $f(6) + h(6), \frac{f(2)}{f(1)}, f(1) - h(2),$
     $f(0) \cdot h(0), f(-3) \cdot f(-5)$

16. If $g: g(x) = 2x$ has as its domain the set of integers, what is the range of g?

17. If $f: f(a) = \frac{a}{2}$ has as its domain the set of integers, what is the range of f?
18. If $g: g(b) = b + 1$ has its domain the set of positive integers, what is the range of g?
19. If $f: f(x) = x^2$ has as its domain the set of integers, what is the range of f?
20. A function H with domain the set of positive integers is defined by $H(n) = 1$ if n is odd and $H(n) = -1$ if n is even. What is the range of H?
21. A function G with domain $D = \{1, 2, 3, 4, 5\}$ is defined by $G(t) = \frac{1}{2} t$ if $t \in D$ is odd and $G(t) = t$ if $t \in D$ is even. What is the range of G?

In exercises 22–27 let f be a function such that $f(x) = \sqrt{x^2 - 16}$ with the domain of f the set of real numbers x such that $|x| \geqslant 4$. Find:

22. $f(-5)$
23. $f(4)$
24. $f(a)$ if $|a| \geqslant 4$
25. $f(a - 1)$ if $|a - 1| \geqslant 4$
26. $f(\pi)$
27. the range of f

28. Let G be a function defined by $G(y) = 2y$ for all integers y and let H be the function defined by $H(t) = 2t$ for all integers t. Are G and H the same function? Explain.

29. Let f be the function defined by $f(x) = x$ for all rational numbers x and let g be the function defined by $g(t) = \frac{t^2}{t}$ for all rational numbers $t \neq 0$. Are f and g the same function? Explain.

In exercises 30–35 let $g(x) = x - 2$ for every real number x. Express the given $f(x)$ in exercises 30–35 in terms of $g(x)$.

Example: Express $f(x) = -x + 2$ in terms of $g(x)$.

Solution: $f(x) = -g(x)$

30. $f(x) = |x + 2|$
31. $f(x) = x$
32. $f(x) = x + 2$
33. $f(x) = |x| + 2$
34. $f(x) = -x - 2$
35. $f(x) = x^2 - 2$

8.3 Graphs of Functions

When the domain and range of a function are sets of real numbers, the ordered pairs produced by a function f can be plotted in a coordinate plane as discussed in Section 4.2. The set of all points corresponding to these ordered pairs is called the *graph* of the function.

In Section 4.3 we considered graphs of linear equations, $ax + by = c$, and saw that these graphs were straight lines. Since, if

$b \neq 0$, $ax + by = c$ is equivalent to $y = -\frac{a}{b}x + \frac{c}{b}$ we can consider the equation $ax + by = c$ as defining a function

$$f: f(x) = -\frac{a}{b}x + \frac{c}{b}$$

called, naturally, a *linear* function.

Graphs of functions aid us in understanding their nature. We give here several examples of such graphs.

Example 1: $F : F(x) = x + 2, D = \{1, 2, 3, 4\}$

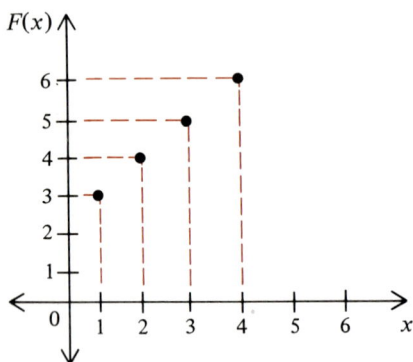

Example 2: $g : g(x) = \frac{1}{2}x + 1, D = \{x : x \text{ is a real number and } 2 \leqslant x < 6\}$

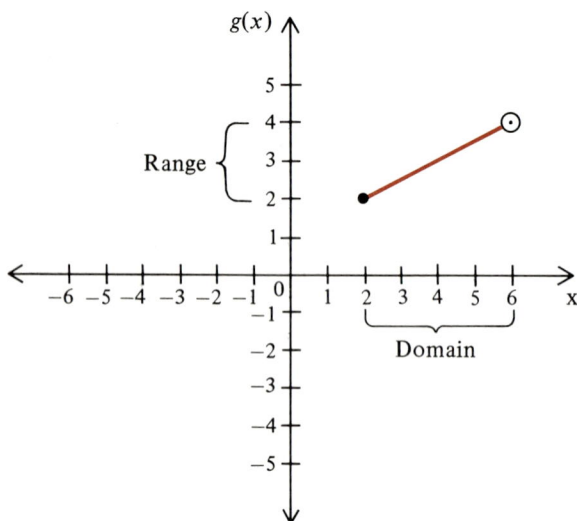

Example 3: $f: f(x) = |x|, D = \{x : x \text{ is a real number and } -6 \leqslant x \leqslant 6\}$

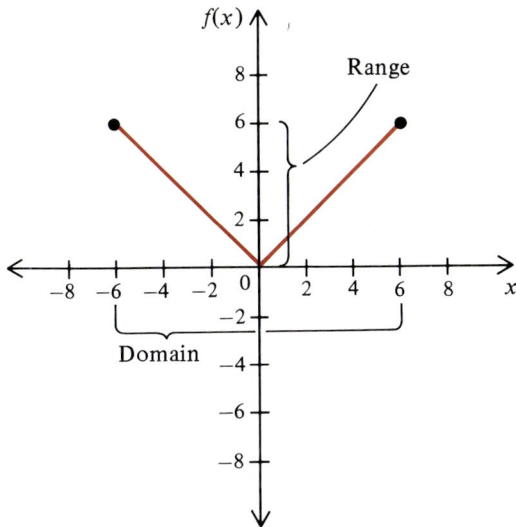

So far our examples have involved domains that are proper subsets of the real numbers; that is, the domains were not the entire set of real numbers. When the entire set of real numbers is taken as the domain we cannot actually draw the entire graph but we can often indicate how it would look if extended. This is shown in the next examples.

Example 4: $f: f(x) = 3 - x, D = \text{set of real numbers}$

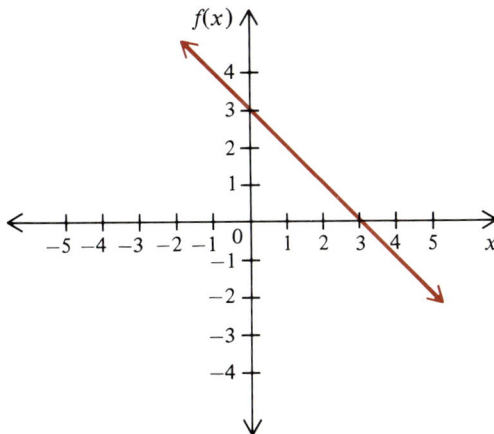

(Note the arrows indicating that we have drawn only a portion of the line.)

Example 5: $h : h(x) = 3, D = $ set of real numbers

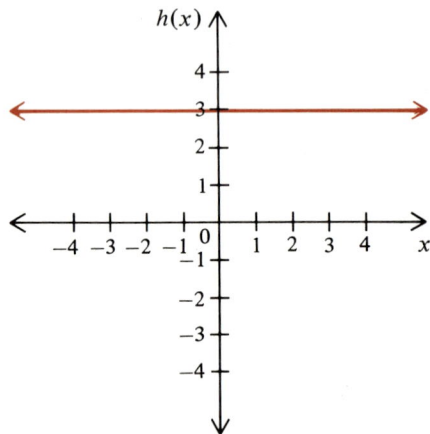

Any function F such the $F(x) = c$ where c is a real number is called a *constant* function.

Example 6: $f : f(x) = x^2, D = $ set of real numbers

We know from Chapter 4 that to graph a linear function we need only plot two points (with a third usually included as a check). When a nonlinear function is graphed, however, it is not always easy to tell how many points must be plotted to reveal clearly the shape of the graph. We begin here with a partial table of functional values:

x	0	1	-1	2	-2	3	-3
$f(x) = x^2$	0	1	1	4	4	9	9

Plotting these points as shown below, we draw a smooth curve through the points to obtain a portion of a curve known as a *parabola.* As a check we take $x = \frac{3}{2}$, get $f(x) = \frac{9}{4}$, and note that the point corresponding to $(\frac{3}{2},\frac{9}{4})$ does appear to lie on the graph.

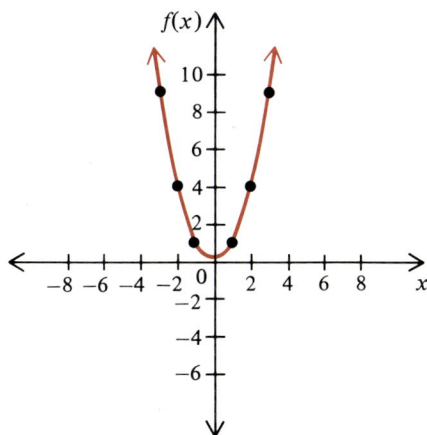

EXERCISES 8.3

In exercises 1-25 use $f(x)$ notation to define the function implied by the given equation.

Example: $2x + 3y = 5$

Solution: Solve for y in terms of x: $3y = 5 - 2x$

$$y = \frac{5}{3} - \frac{2x}{3}$$

Since $f(x) = y$ and $y = \frac{5}{3} - \frac{2x}{3}$, $f(x) = \frac{5}{3} - \frac{2x}{3}$.

1. $x + y = 7$
2. $x - y = 3$
3. $x + 2y = 6$
4. $3x - 4y = 5$
5. $y = \frac{1}{2}$

6. $x - 3 = y$
7. $3x - 5y = 6$
8. $4x - 3y = 2$
9. $x + 7y = 10$
10. $2x + y = -5$

11. $7x - 3y = -2$

12. $x - y = 0$

13. $x + y = 0$

14. $x = y$

15. $xy = 0$

16. $2x^2 - 3y = 3$

17. $3xy - 1 = 2$

18. $y = x^2 + 2x + 3$

19. $y = x^3 + 3x^2 + 3x - 1$

20. $x^2 + 2y = -3$

21. $xy - y = -1$

22. $xy + 2 = y$

23. $2xy - 3y = 5$

24. $5x^2 + 2xy = 2$

25. $5x^2 - 2xy + 1 = 0$

In exercises 26–50 draw a graph of the given function. Be sure to label each axis of your graph and indicate the scale. (Remember that a variable such as t represents the first coordinate and $f(t)$ the second coordinate.)

26. $f : f(x) = x + 2$

27. $f : f(t) = t$

28. $g : g(x) = x - 2$

29. $g : g(t) = 2t - 1$

30. $h : h(x) = 3x + 5$

31. $f : f(x) = -x$

32. $g : g(x) = -2x$

33. $h : h(a) = a^2$

34. $f : f(x) = x^2 + 1$

35. $g : g(t) = t^2 - 1$

36. $h : h(b) = 2b + 3$

37. $F : F(x) = -2x + 1$

38. $G : G(x) = -2x - 1$

39. $H : H(s) = -s - 5$

40. $f : f(x) = |x|$

41. $g : g(s) = |s| + 1$

42. $h : h(t) = |t| - 1$

43. $g : g(x) = |x + 1|$

44. $f : f(x) = |x - 1|$

45. $g : g(y) = 2\,|y|$

46. $h : h(x) = -\,|x|$

47. $j : j(x) = -2\,|x|$

48. $f : f(x) = 2x^2$

49. $f : f(x) = \frac{1}{2}x^2$

50. $f : f(b) = b^2 + 2b - 3$

In exercises 51–57 draw a graph of the given function. Find the range of each function, given that the domain is the set of real numbers.

51. $f : f(x) = |x| + 2$

52. $g : g(x) = |x| - 2$

53. $h : h(x) = -\,|x| + 2$

54. $F : F(x) = \begin{cases} x & \text{for } x < 0 \\ 3 & \text{for } x \geqslant 0 \end{cases}$

55. $g : g(a) = \begin{cases} a - 1 & \text{for } a > 1 \\ 3 & \text{for } a \leqslant 1 \end{cases}$

56. $H : H(t) = \begin{cases} -1 & \text{for } t < -3 \\ 0 & \text{for } -3 \leqslant t \leqslant 3 \\ 1 & \text{for } t > 3 \end{cases}$

57. $f : f(x) = \begin{cases} 3x & \text{for } 0 \leqslant x < 4 \\ x & \text{for } 4 \leqslant x \leqslant 10 \end{cases}$

58. Graph the function defined by this rule: Postage for letters is 13 cents for each ounce or fraction thereof.

59. Graph the function defined by this situation: A salesman is paid a 5% commission on the first $10,000 of monthly sales and 7% on all sales in excess of $10,000.

60. The Metro Water District published the following rates:
    First 500 cubic feet of water or less . . . . $5.00 minimum charge
    Next 5000 cubic feet . . . .$.55 per 100 cubic feet
    Over 5500 cubic feet . . . .$.50 per 100 cubic feet
    Graph the function described by the rates above.

REVIEW EXERCISES

Section 8.1

In exercises 1–9 you are given a set of numbers to be used as inputs and a rule for determining the output. Find the corresponding output number for each input number.

Example: Inputs: $\{-2, -1, 0, 1, 2\}$; Rule: Add 5.

Solution: Outputs: $\{3, 4, 5, 6, 7\}$

1. Inputs: $\{-2, -1, 0, 1, 2\}$; Rule: Subtract -9.
2. Inputs: $\{-10, -5, 0, 5, 10\}$; Rule: Multiply by 2 and then divide by 5.
3. Inputs: $\{-7, -2, 5, 8, 13\}$; Rule: Multiply by 3 and then subtract 2.
4. Inputs: $\{-9, -7, -3, 1, 5\}$; Rule: Add 3 and then divide by 2.
5. Inputs: $\left\{\frac{1}{2}, \frac{2}{3}, \frac{3}{4}, \frac{4}{5}, \frac{5}{6}\right\}$; Rule: Divide by 2.
6. Inputs: $\left\{-\frac{1}{2}, -\frac{2}{3}, -\frac{3}{4}, -\frac{4}{5}, -\frac{5}{6}\right\}$; Rule: Multiply by $-\frac{3}{2}$.
7. Inputs: $\left\{-13, -7\frac{1}{2}, -5\frac{3}{4}, 2\frac{1}{2}, 3\frac{5}{8}\right\}$; Rule: Subtract $\frac{1}{2}$.
8. Inputs: $\{-1.5, -1.2, 0.5, 1.8, 2.4\}$; Rule: Add 1 and then divide by 2.
9. Inputs: $\{100, 1000, 10,000, 100,000, 1,000,000\}$; Rule: Multiply by 100 and then divide by 2.

In exercises 10–20 you are given a set of numbers to be used as outputs, and a rule for determining the output. Find the corresponding input number for each output number.

10. Outputs: {50, 26, 10, 2, 1}; Rule: Square the number and add 1.
11. Outputs: {20, 29.5, 38, 51, 62}; Rule: Add 13.
12. Outputs: $\left\{-2\frac{1}{2}, 1\frac{1}{4}, 5\frac{1}{8}, 7\frac{3}{8}, 15\right\}$; Rule: Subtract $2\frac{3}{8}$.

13. Outputs: {14, 35, 84, 119, 749}; Rule: Multiply by 7.
14. Outputs: $\left\{-12, 4\frac{1}{9}, 5, 7\frac{1}{3}, 12\right\}$; Rule: Divide by 9.

15. Outputs: {7, 17, 21, 47, 199}; Rule: Multiply by 2 and then subtract 3.
16. Outputs: $\left\{-3, 2, 4\frac{1}{2}, 5\frac{1}{2}, 11\frac{1}{2}\right\}$; Rule: Divide by 2 and then add 2.

17. Outputs: {0, 1, 2, 3, 4}; Rule: Multiply by $\frac{1}{2}$ and then add 2.

18. Outputs: {-4, -3, -1, 3, 4}; Rule: Multiply by $\frac{1}{3}$ and then subtract 1.

19. Outputs: $\left\{-4, -\frac{8}{3}, -\frac{4}{3}, 0, \frac{4}{3}, \frac{8}{3}\right\}$; Rule: Multiply by $\frac{4}{3}$.

20. Outputs: $\left\{-\frac{4}{3}, 0, \frac{4}{3}, \frac{8}{3}, 4\right\}$; Rule: Multiply by $\frac{2}{3}$.

In exercises 21-30, given a table of input and output values, find a rule that determines these numbers.

21.

Inputs	1	2	3	4	5	Rule ?
Outputs	3	6	9	12	15	

22.

Inputs	1	2	3	4	5	Rule ?
Outputs	15	30	45	60	75	

23.

Inputs	-4	-5	6	7	8	Rule ?
Outputs	-1	-2	9	10	11	

24.

Inputs	1	2	3	4	5	Rule ?
Outputs	3	5	7	9	11	

25.

Inputs	1	2	4	5	7	Rule ?
Outputs	5	8	14	17	23	

26.

Inputs	3	5	7	10	12	Rule ?
Outputs	11	19	27	39	47	

27.

Inputs	2	4	6	9	12
Outputs	5	11	17	26	35

Rule ?

28.

Inputs	2	4	7	10	14
Outputs	1	5	11	17	25

Rule ?

29.

Inputs	2	6	8	14	20
Outputs	4	6	7	10	13

Rule ?

30.

Inputs	12	15	21	27	30
Outputs	0	1	3	5	6

Rule ?

Section 8.2

In exercises 31–40 let $f(x) = 2x + 1$ and $g(x) = -x$.

31. Find $f(3)$.
32. Find $g(-6)$.
33. Find $f(p)$.
34. Find $g(y + 3)$.
35. Find $f(3) + g(3)$.

36. Find $f(3) \cdot g(3)$.
37. Find $f(a) + g(a)$.
38. Find $f(t) \cdot g(t)$.
39. Find $\dfrac{f(-11)}{g(7)}$.
40. Find $f(n) - g(n)$.

In exercises 41–50 let $f(x) = x + 2$ and $g(x) = x^2 - 3$.

41. Find $f(-2)$.
42. Find $f(3)$.
43. Find $f(a - 1)$.
44. Find $g(-3)$.
45. Find $g(-5)$.

46. Find $g(2b)$.
47. Find $g(a + 2)$.
48. Find $-5f(-4)$.
49. Find $g(5) - f(5)$.
50. Find $\dfrac{f(3)}{g(1)}$.

Section 8.3

In exercises 51–60 use $f(x)$ notation to define the function implied by the equation.

51. $2x + y = -3$

52. $x - 3y = -6$

53. $2y = 3x$

54. $10x = 90 + 3y$

55. $5x - 7 = 2y + 15$

56. $x = 2y + 3$

57. $\dfrac{x + 9}{6} + \dfrac{y}{9} = \dfrac{1}{3}$

58. $3x - \dfrac{y + 4}{8} = -8$

59. $\dfrac{y}{x} + 5 = \dfrac{3}{2x}$

60. $\dfrac{12}{x + y} = 2$

In exercises 61–80 graph the given function.

61. $f : f(x) = 2x + 5$
62. $f : f(x) = -3x + 5$
63. $f : f(x) = 5x + 2$

64. $f : f(x) = -x - 1$

65. $g : g(x) = -4x$

66. $g : g(x) = \dfrac{2}{3}x + 2$

67. $g : g(x) = \dfrac{1}{2}x + 5$

68. $g : g(x) = -\dfrac{1}{2}x + 5$

69. $h : h(x) = x^2$
70. $h : h(x) = -x^2$

71. $f : f(a) = a^2 + 2$
72. $g : g(c) = c^2 - 2$
73. $h : h(t) = 2t^2$

74. $f : f(b) = \dfrac{1}{2}b^2$

75. $g : g(x) = x^2 + 4x - 4$

76. $h : h(x) = |x|$

77. $f : f(a) = 2|a|$

78. $g : g(w) = |w + 1|$

79. $t : t(x) = |x| + 1$
80. $f : f(x) = -2|x|$

CUMULATIVE REVIEW F

In exercises 1–8 simplify the given expression.

1. $27ab + 6b + 9ab + 2b$
2. $(2x + y - z) - (-2x + y - z)$
3. $(-2a^4)(-3a^2 b)$
4. $(x^3 + 2x^2 + 4x + 8)(x - 2)$
5. $(x - 1)(x - 2) + (x - 2)(x - 3) - (x - 7)(x - 1)$
6. $-\dfrac{1}{2}\left(-\dfrac{1}{2}a + \dfrac{3}{4}b - \dfrac{1}{6}\right) - \left(-\dfrac{5}{16}a - \dfrac{2}{3}b - \dfrac{1}{4}\right)$

7. $\left(\frac{4}{5}x^2y^3\right)\left(-\frac{5}{9}xyz\right)$

8. $\left(x-\frac{3}{4}\right)\left(x+\frac{3}{4}\right)$

In exercises 9–14 factor completely the given polynomial.

9. $x^2 + x - 56$
10. $5x^2 - 75 + 10x$
11. $1 - x^8$
12. $36x^2 - 6x - 2$
13. $6x^2 - 19x + 10$
14. $2x^2 + 13x - 24$

15. Simplify: $\dfrac{8a^3 - 16a^2b}{16a^2b - 32ab^2}$

16. Simplify: $\dfrac{12xyz^2 - 36x^2y^2z^2 + 24xyz}{-6xyz}$

17. Simplify: $\dfrac{5x^2 - 3xy - 2y^2}{x - y}$

In exercises 18–23 perform the indicated operations.

18. $(x^3 + 4x^2 - 17x - 60) \div (x + 5)$

19. $\dfrac{x^2 + 6x + 8}{6x + 30} \cdot \dfrac{4x + 20}{x^2 - x - 6}$

20. $\dfrac{x^2 + xy}{x^2 + y^2} \cdot \dfrac{x^2 - y^2}{x^3 + x^2y} \cdot \dfrac{x^2 - xy}{x^2 - 2xy + y^2}$

21. $\dfrac{3x - 2}{xy} - \dfrac{2x + 3}{xy}$

22. $\dfrac{5}{2x + 1} - \dfrac{6}{x - 3}$

23. $\dfrac{3a - 21}{5a + 15} \div \dfrac{7a - 49}{5a - 10}$

24. Write with positive exponents: $\dfrac{3x^2}{x^{-1}y^{-2}}$

In exercises 25–29 simplify the given expression.

25. $\sqrt{192}$

26. $\sqrt[3]{\dfrac{27}{8}}$

27. $5\sqrt{12} \cdot \dfrac{1}{2}\sqrt{3}$

28. $3\sqrt{50} - 2\sqrt{98} + 5\sqrt{12}$

29. $\dfrac{\sqrt{8x^6}}{\sqrt{12x^4}}$

In exercises 30–36 find the solution set of the given equation or inequality.

30. $5(x + 2) = 4(x - 3)$

31. $x(x - 2) = x^2 - (x + 6)$

32. $x(x + 3) - 6 > x^2 + 6$

33. $\dfrac{4}{x} - \dfrac{5}{3} = \dfrac{1}{5}$

34. $\dfrac{3x^2 - 2x - 8}{5} = \dfrac{(7x - 2)(3x - 6)}{35}$

35. $\dfrac{3}{4}x + \dfrac{-5}{3} < \dfrac{1}{2} + \dfrac{9x}{12}$

36. $\dfrac{4}{2x + 8} - \dfrac{3}{x + 4} = \dfrac{1}{4}$

37.   In the formula $a = p + prt$, solve for p.

38.   If $f(x) = x^2 - 3x + 2$, find $f(-7)$.

39.   If $g(x) = x + 5$ and $h(x) = x^2 + 6x + 5$, find $\dfrac{h(5)}{g(5)}$.

40.   Graph $f(x) = \dfrac{3}{4}x - 5$.

41.   Graph $x + 2y = 5$.

42.   The table

x	-2	-1	0	1	2	3
$f(x)$	-6	$-5\dfrac{1}{2}$	-5	$-4\dfrac{1}{2}$	-4	$-3\dfrac{1}{2}$

checks if $f(x) = $?

In exercises 43–45 solve the given system of equations.

43. $\begin{cases} 5x + 4y = 22 \\ 3x + y = 9 \end{cases}$

44. $\begin{cases} 7x - 2y = 31 \\ 4x + 3y = -3 \end{cases}$

45. $\begin{cases} 4x - \dfrac{y}{2} = 7 \\ 6x - 5y = 0 \end{cases}$

46. Two-thirds of a number is 11 less than 10 more than the number. What is the number?

47. An automobile left a service station traveling at a rate of 72 kilometers per hour. Two hours later a second automobile left the same service station going in the same direction at 96 kilometers per hour. How long will it take the second car to overtake the first car? (Assume that the rates remain constant.)

48. A collection of 50 coins consists of nickels and dimes. If there is a total of $3.40 in the collection, how many coins of each denomination are there?

49. A chemistry student wishes to reduce 60 cubic centimeters of a 40% carbonate solution to a 15% carbonate solution by adding water. How much water should be added?

50. What is the average rate for a round trip in which one travels to his destination at 50 kilometers per hour for 50 kilometers and returns at 70 kilometers per hour?

9

QUADRATIC EQUATIONS AND INEQUALITIES

9.1 Quadratic Equations

The word problems that we have considered up to now have all led to linear equations. Other word problems, however, can lead to non-linear equations. Consider, for example, the following problem:

Find the dimensions of a rectangle whose width is 7 centimeters less than the length and whose area is 60 square centimeters.

Solution: Let x = the length in centimeters. Then $x - 7$ = the width in centimeters and the area is $x(x - 7)$ square centimeters. Thus we have $x(x - 7) = 60$ or $x^2 - 7x - 60 = 0$.

The equation $x^2 - 7x - 60 = 0$ is an example of a *quadratic* equation — an equation of the form

$$ax^2 + bx + c = 0$$

where a, b, and c are real numbers with $a \neq 0$.

Some quadratic equations can be solved very simply. Thus the equation $x^2 - 25 = 0$ is equivalent to the equation $x^2 = 25$ and so $x = \sqrt{25} = 5$ or $x = -\sqrt{25} = -5$. Hence its solution set is $\{5, -5\}$. Similarly, the equation $x^2 - 7 = 0$ has the solution set $\{\sqrt{7}, -\sqrt{7}\}$ and, in general, the equation $x^2 - k = 0$ with $k > 0$ has the solution set $\{\sqrt{k}, -\sqrt{k}\}$.

Now consider the equation

$$(x - 3)^2 = 25$$

Clearly, we have either

$$x - 3 = \sqrt{25} \quad \text{or} \quad x - 3 = -\sqrt{25}$$
$$x - 3 = 5 \quad \text{or} \quad x - 3 = -5$$
$$x = 8 \quad \text{or} \quad x = -2$$

Thus the solution set of $(x - 3)^2 = 25$ is $\{8, -2\}$. (Note that we wrote "$x = 8$ *or* $x = -2$" and not, incorrectly, "$x = 8$ *and* $x = -2$." Certainly we cannot have $x = 8$ *and*, also, $x = -2$. Also note that we sometimes abbreviate such statements as "$x - 3 = 5$ or $x - 3 = -5$" to "$x - 3 = \pm 5$.")

Generalizing this example by replacing 3 by n and 25 by k, we see that the equation

$$(x - n)^2 = k$$

with $k > 0$ has the solution set $\{n + \sqrt{k}, n - \sqrt{k}\}$.

Our problem now, given a quadratic equation,

$$ax^2 + bx + c = 0$$

is to find an equation of the form $(x - n)^2 = k$ equivalent to the equation $ax^2 + bx + c = 0$; that is, to find an equation of the form $(x - n)^2 = k$ with the same solution set as the equation $ax^2 + bx + c = 0$. The following examples show how to perform this task by writing a series of equations equivalent to the given equation.

Example 1: $x^2 - 6x - 7 = 0$
$$x^2 - 6x = 7$$
$$x^2 - 6x + 9 = 7 + 9$$
$$(x - 3)^2 = 16$$

From the last equation we get

$$x - 3 = 4 \quad \text{or} \quad x - 3 = -4$$
$$x = 7 \quad \text{or} \quad x = -1$$

so that the solution set is $\{7, -1\}$.

The key step in our example was the addition of 9 to both sides of the equation $x^2 - 6x = 7$. We did this, of course, to form the trinomial perfect square $x^2 - 6x + 9$, and the process of adding the proper number to obtain a trinomial perfect square to obtain a solution to the quadratic equation is called *completing the square.* Note that we have $9 = (\frac{-6}{2})^2$ where -6 is the coefficient of x in the given quadratic equation.

Example 2: $x^2 + 10x - 6 = 0$

$$x^2 + 10x = 6$$

$$x^2 + 10x + 25 = 6 + 25 \qquad \text{(Note that } \left(\frac{10}{2}\right)^2 = 25.\text{)}$$

$$(x + 5)^2 = 31$$

$$x + 5 = \sqrt{31} \quad \text{or} \quad x + 5 = -\sqrt{31}$$

Solution set is $\{\sqrt{31} - 5, -\sqrt{31} - 5\}$.

Example 3: $x^2 + 5x - 2 = 0$

$$x^2 + 5x = 2$$

$$x^2 + 5x + \frac{25}{4} = 2 + \frac{25}{4} \qquad \text{(Note that } \left(\frac{5}{2}\right)^2 = \frac{25}{4}.\text{)}$$

$$\left(x + \frac{5}{2}\right)^2 = \frac{33}{4}$$

$$x + \frac{5}{2} = \sqrt{\frac{33}{4}} = \frac{\sqrt{33}}{2} \quad \text{or} \quad x + \frac{5}{2} = -\frac{\sqrt{33}}{2}$$

Solution set is $\left\{\frac{1}{2}(\sqrt{33} - 5), -\frac{1}{2}(\sqrt{33} + 5)\right\}$.

Example 4: $3x^2 + 7x - 8 = 0$

When the coefficient of x^2 is other than 1, our first step involves dividing each term of the equation by the coefficient of x^2. In this case we have

$$x^2 + \frac{7}{3}x = \frac{8}{3}$$

$$x^2 + \frac{7}{3}x + \left(\frac{7}{6}\right)^2 = \frac{8}{3} + \left(\frac{7}{6}\right)^2 \qquad \left(\text{Note that } \frac{7}{3} \div 2 = \frac{7}{6}.\right)$$

$$\left(x + \frac{7}{6}\right)^2 = \frac{96}{36} + \frac{49}{36} = \frac{145}{36}$$

$$x + \frac{7}{6} = \sqrt{\frac{145}{36}} \quad \text{or} \quad x + \frac{7}{6} = -\sqrt{\frac{145}{36}}$$

Solution set is $\left\{ \frac{1}{6} (\sqrt{145} - 7), -\frac{1}{6}(\sqrt{145} + 7) \right\}$.

There is another way of solving quadratic equations that is very similar to the completion of the square method and which avoids the use of fractions. In this alternate procedure applied to the equation $3x^2 + 7x - 8 = 0$, we begin by multiplying both sides of the equation by 4 times the coefficient of x^2, i.e., $4 \cdot 3 = 12$ in this case. We have

(1) $12(3x^2 + 7x - 8) = 12 \cdot 0$

(2) $36x^2 + 84x - 96 = 0$

(3) $36x^2 + 84x = 96$

Now we add the square of 7, the coefficient of x in the original equation, to both sides of Equation (3) to get

(4) $36x^2 + 84x + 49 = 96 + 49$

(5) $(6x + 7)^2 = 145$

(6) $6x + 7 = \sqrt{145} \quad \text{or} \quad 6x + 7 = -\sqrt{145}$

(7) $6x = -7 + \sqrt{145} \quad \text{or} \quad 6x = -7 - \sqrt{145}$

(8) $x = \dfrac{-7 + \sqrt{145}}{6} \quad \text{or} \quad x = \dfrac{-7 - \sqrt{145}}{6}$

We will find this alternate procedure and the slightly different form in which we have written the answer useful in deriving a general formula for solving quadratic equations in the next section.

Example 5: Solve $\dfrac{4x^2 - 5}{3} = \dfrac{2x^2 - 3}{2}$.

Solution: $2(4x^2 - 5) = 3(2x^2 - 3)$

$$8x^2 - 10 = 6x^2 - 9$$

$$8x^2 - 6x^2 = 10 - 9$$

$$2x^2 = 1$$

$$x^2 = \frac{1}{2}$$

$$x = \sqrt{\frac{1}{2}} \quad \text{or} \quad -\sqrt{\frac{1}{2}}$$

$$x = \frac{1}{2}\sqrt{2} \quad \text{or} \quad -\frac{1}{2}\sqrt{2}$$

Solution set $= \left\{ \frac{1}{2}\sqrt{2}, -\frac{1}{2}\sqrt{2} \right\}$.

Check:

$$\frac{4\left(\frac{1}{2}\sqrt{2}\right)^2 - 5}{3} \overset{?}{=} \frac{2\left(\frac{1}{2}\sqrt{2}\right)^2 - 3}{2}$$

$$\frac{\left(4 \cdot \frac{1}{2}\right) - 5}{3} \overset{?}{=} \frac{\left(2 \cdot \frac{1}{2}\right) - 3}{2}$$

$$\frac{-3}{3} \overset{?}{=} \frac{-2}{2}$$

$$-1 = -1 \qquad \text{(checks)}$$

$$\frac{4\left(-\frac{1}{2}\sqrt{2}\right)^2 - 5}{3} \overset{?}{=} \frac{2\left(-\frac{1}{2}\sqrt{2}\right)^2 - 3}{2}$$

$$\frac{\left(4 \cdot \frac{1}{2}\right) - 5}{3} \overset{?}{=} \frac{\left(2 \cdot \frac{1}{2}\right) - 3}{2}$$

$$\frac{-3}{3} \overset{?}{=} \frac{-2}{2}$$

$$-1 = -1 \qquad \text{(checks)}$$

Not all quadratic equations have real number solutions. Thus, for example, the equation $x^2 + 4 = 0$ (which is equivalent to $x^2 = -4$) has no real number solutions since the square of a real number is never negative. Similarly, $x^2 + 6x + 13 = 0$ is equivalent to $(x + 3)^2 = -4$ and has no real number solution. Both these equations, as well as all other quadratic equations, do, however, have solutions in the set of complex numbers — a set considered in more advanced mathematics courses.

EXERCISES 9.1

In exercises 1–35 find the solution set of the given equation. If the answer involves a radical, write it in the simplest form. If no real number solution exists, write $\emptyset$ for the solution set.

1. $x^2 = 36$
2. $x^2 = 64$
3. $y^2 = 18$
4. $g^2 = 27$
5. $4x^2 = 49$
6. $25x^2 = 100$
7. $x^2 - 81 = 0$
8. $y^2 + 9 = 0$
9. $t^2 - 8 = 0$
10. $a^2 + 7 = 8$
11. $x^2 - 3 = 6$
12. $n^2 - 34 = 15$
13. $3x^2 - 17 = 58$
14. $5x^2 + 13 = x^2 - 33$
15. $(x - 3)^2 = 16$
16. $(x - 2)^2 = 25$
17. $(y + 4)^2 = 9$
18. $(m + 3)^2 = 8$
19. $(x - 5)^2 - 9 = 0$
20. $(s - 4)^2 + 13 = 9$
21. $x^2 + 6x + 9 = 16$
22. $y^2 - 4y + 4 = 9$
23. $a^2 - 2a + 1 = 64$
24. $x^2 + 2x + 1 = 7$
25. $x^2 + 25x + 25 = 8$
26. $s^2 + 12s + 36 = 49$
27. $2a^2 - 12a + 18 = 54$
28. $5y^2 + 10y + 5 = 20$
29. $\dfrac{2x^2 - 3}{2} = \dfrac{x^2 - 3}{3}$
30. $\dfrac{y^2 - 3y}{3} = \dfrac{4y^2 - 5y - 8}{5}$

31. $\dfrac{x^2 - 1}{4} - \dfrac{2x^2 - 3}{15} - \dfrac{x^2 - 4}{5} = 0$

32. $x + 7 - \dfrac{5x}{4} = \dfrac{x^2 - x}{4}$

33. $\dfrac{(x - 1)^2}{6} + \dfrac{(x + 1)^2}{6} = 1$

34. $x^2(x + 1) = (x + 2)(x^2 - 2x + 4)$

35. $(x - 1)(x^2 - 2x - 2) = (x - 2)(x^2 + x + 2)$

In exercises 36–45 find the constant term that is needed in order to make the given trinomial a trinomial perfect square.

36. $x^2 - 6x + $ ____?____

37. $x^2 + 4x + $ ____?____

38. $x^2 + 20x + $ ____?____

39. $a^2 - 14a + $ ____?____

40. $y^2 - 8y + $ ____?____

41. $x^2 + 5x + $ ____?____

42. $x^2 - \dfrac{1}{2}x + $ ____?____

43. $a^2 - \dfrac{2}{5}a + $ ____?____

44. $t^2 + \dfrac{8}{3}t + $ ____?____

45. $y^2 - \dfrac{4}{7}y + $ ____?____

In exercises 46–80 find the solution set of the given quadratic equation by completing the square. If the answer involves a radical, write it in the simplest form. If no real number solution exists, write $\varnothing$ for the solution set.

46. $x^2 - 2x = 15$

47. $y^2 - 6y = -5$

48. $x^2 + 2x = 5$

49. $a^2 + 8a + 12 = 0$

50. $x^2 + 4x - 60 = 0$

51. $t^2 + 8t + 15 = 0$

52. $x^2 - 6x + 5 = 0$

53. $x^2 - 3x = 4$

54. $x^2 - \dfrac{1}{2}x = \dfrac{8}{16}$

55. $a^2 = 5a - 4$

56. $y^2 = y + 30$

57. $2b^2 - 5b = 3$

58. $2x^2 + x - 6 = 0$

59. $3y^2 + 12y = 15$

60. $3x^2 - 4x - 4 = 0$

61. $3a^2 - 8a + 5 = 0$

62. $12a^2 - 7a - 10 = 0$

63. $9x^2 + 12x + 4 = 0$

64. $x^2 + 8x = 4$
65. $x^2 + 4x = 8$
66. $y^2 - 6y - 6 = 0$
67. $x^2 - x - 1 = 0$
68. $2x^2 - 4x + 1 = 0$
69. $9x^2 - 12x - 8 = 0$
70. $4x^2 + 4x - 1 = 0$
71. $12x^2 - 14x + 3 = 0$
72. $2r^2 - 7r - 6 = 0$

73. $8x^2 - 30x - 27 = 0$
74. $5x^2 - 15x = 7$
75. $6m^2 + 7m + 2 = 0$
76. $(x + 5)(x + 4) = 10$
77. $y(y - 6) = 11$
78. $x - 2 = x^2 + 7$
79. $x(x + 6) = x$
80. $(2x + 3)(x - 5) = 1$

9.2   The Quadratic Formula

In deriving a formula for solving any quadratic equation $ax^2 + bx + c = 0$ $(a \neq 0)$ we will use the procedure given in Section 9.1 for the alternate solution to the equation $3x^2 + 7x - 8 = 0$. In the work shown below we repeat this solution along with the corresponding steps for the solution of $ax^2 + bx + c = 0$.

(1) $12(3x^2 + 7x - 8) = 12 \cdot 0$ $4a(ax^2 + bx + c) = 4a \cdot 0$

(Multiplying both sides by 4 times the coefficient of x.)

(2) $36x^2 + 84x - 96 = 0$ $4a^2x^2 + 4abx + 4ac = 0$

(Using the distributive property and the multipli-cation property of 0.)

(3) $36x^2 + 84x = 96$ $4a^2x^2 + 4abx = -4ac$

(Adding the negative of the constant term to both sides of the equation.)

(4) $36x^2 + 84x + 49 = 96 + 49$ $4a^2x^2 + 4abx + b^2 = -4ac + b^2$

(Adding the square of the coefficient of x in the original equation.)

(5) $(6x + 7)^2 = 145$ $(2ax + b)^2 = b^2 - 4ac$

(Recognizing a trinomial perfect square.)

(6) $6x + 7 = \sqrt{145}$ $2ax + b = \sqrt{b^2 - 4ac}$

or   or

$6x + 7 = -\sqrt{145}$ $2ax + b = -\sqrt{b^2 - 4ac}$

(Taking square roots.)

(7) $6x = -7 + \sqrt{145}$ $2ax = -b + \sqrt{b^2 - 4ac}$

or   or

$6x = -7 - \sqrt{145}$ $2ax = -b - \sqrt{b^2 - 4ac}$

(Adding the opposite of 7 or b
to both sides of the equation.)

(8) $x = \dfrac{-7 + \sqrt{145}}{6}$ $x = \dfrac{-b + \sqrt{b^2 - 4ac}}{2a}$

or   or

$x = \dfrac{-7 - \sqrt{145}}{6}$ $x = \dfrac{-b - \sqrt{b^2 - 4ac}}{2a}$

(Dividing by the coefficient of x.)

(Of course, we must have $b^2 - 4ac \geqslant 0$ in order to have $\sqrt{b^2 - 4ac}$ a real number.)

Thus the solution set of $ax^2 + bx + c = 0$ is

$$\left\{ \frac{-b + \sqrt{b^2 - 4ac}}{2a}, \frac{-b - \sqrt{b^2 - 4ac}}{2a} \right\}$$

if $b^2 - 4ac \geqslant 0$ and $a \neq 0$.

For convenience we often write the solution set in the condensed form

$$\left\{ \frac{-b \pm \sqrt{b^2 - 4ac}}{2a} \right\}$$

and write the *quadratic formula* as

$$x = \frac{-b \pm \sqrt{b^2 - 4ac}}{2a}$$

Here are some examples of the use of the quadratic formula.

Example 1: $x^2 - 10x + 21 = 0$. Here $a = 1$, $b = -10$, and $c = 21$.

By substitution in the quadratic formula we have

$$x = \frac{-(-10) \pm \sqrt{(-10)^2 - 4 \cdot 1 \cdot 21}}{2 \cdot 1} = \frac{10 \pm \sqrt{100 - 84}}{2}$$

$$= \frac{10 \pm \sqrt{16}}{2} = \frac{10 \pm 4}{2}$$

Thus

$$x = \frac{10 + 4}{2} = \frac{14}{2} = 7 \quad \text{or} \quad x = \frac{10 - 4}{2} = \frac{6}{2} = 3$$

and so the solution set is $\{7, 3\}$.

Example 2: $3x^2 - 6x = 5$. Here we first write $3x^2 - 6x - 5 = 0$ and then observe that $a = 3$, $b = -6$, and $c = -5$. So the quadratic formula gives us

$$\frac{-(-6) \pm \sqrt{(-6)^2 - 4 \cdot 3 (-5)}}{2 \cdot 3} = \frac{6 \pm \sqrt{36 + 60}}{6}$$

$$= \frac{6 \pm \sqrt{96}}{6} = \frac{6 \pm 4\sqrt{6}}{6} = 1 \pm \frac{2}{3}\sqrt{6}$$

and so the solution set is $\left\{1 + \frac{2}{3}\sqrt{6}, \ 1 - \frac{2}{3}\sqrt{6}\right\}$.

Example 3: $x^2 + 6x + 9 = 0$. Here $a = 1$, $b = 6$, and $c = 9$. So the quadratic formula gives us

$$\frac{-6 \pm \sqrt{6^2 - 4 \cdot 1 \cdot 9}}{2} = \frac{-6 \pm \sqrt{36 - 36}}{2} = \frac{-6 \pm 0}{2} = \frac{-6}{2} = -3$$

and so the solution set is $\{-3\}$.

Example 4: Solve: $x^2 - 2x + 2 = 0$. Here $a = 1$, $b = -2$, and $c = 2$. So the quadratic formula gives us

$$x = \frac{-(-2) \pm \sqrt{(-2)^2 - 4 \cdot 1 \cdot 2}}{2 \cdot 1} = \frac{2 \pm \sqrt{4 - 8}}{2} = \frac{2 \pm \sqrt{-4}}{2}$$

Here $b^2 - 4ac = -4 < 0$, contrary to our stated requirement that $b^2 - 4ac \geqslant 0$. The number -4 has no real number square root and so the quadratic equation $x^2 - 2x + 1 = 0$ has no real number solution; its solution set is $\emptyset$.

Note that the expression $b^2 - 4ac$, called the ⎯⎯⎯⎯⎯⎯ of the quadratic equation $ax^2 + bx + c = 0$, plays a particularly important role in the solution of $ax^2 + bx + c = 0$. The fact that $b^2 - 4ac > 0$ in the first two examples tells us that the equations have exactly two real number solutions, whereas, in our third example, the fact that $b^2 - 4ac = 0$ tells us that there is only one real number solution to the equation. Finally, the fact that $b^2 - 4ac < 0$ in our fourth example tells us that the quadratic equation has no real number solution.

Summarizing the above remarks, we have
The equation $ax^2 + bx + c = 0$ $(a \neq 0)$ has
two real number solutions if $b^2 - 4ac > 0$;
one real number solution if $b^2 - 4ac = 0$; and
no real number solutions if $b^2 - 4ac < 0$.

EXERCISES 9.2

Use the quadratic formula to find the solution sets of the following quadratic equations. If no real number solutions exist for an equation, write ø.

1. $x^2 + 6x + 5 = 0$

2. $x^2 + 10x - 24 = 0$

3. $x^2 + 7x + 10 = 0$

4. $x^2 + 4x - 21 = 0$

5. $x^2 + 3x - 40 = 0$

6. $x^2 + 9x + 20 = 0$

7. $x^2 + 12x + 35 = 0$

8. $x^2 + x - 2 = 0$

9. $x^2 + 16x + 55 = 0$

10. $x^2 + 11x + 24 = 0$

11. $y^2 - 6y + 9 = 0$

12. $m^2 - 8m + 16 = 0$

13. $t^2 - t = 6$

14. $x^2 - 7x = 18$

15. $2x^2 + x - 15 = 0$

16. $3x^2 - 7x - 6 = 0$

17. $2y^2 + 3y - 20 = 0$

18. $2a^2 + 2a - 12 = 0$

19. $9x^2 - 3x = 2$

20. $3m^2 - 10m + 3 = 0$

21. $6x^2 - 17x + 12 = 0$

22. $6x^2 - 13x = 5$

23. $12y^2 - y - 1 = 0$

24. $20x^2 + 7x = 6$

25. $4x^2 - 1 = 0$

26. $9x^2 - 25 = 0$

27. $16y^2 - 9 = 0$

28. $7x^2 - 14x = -7$

29. $64x^2 - 112x + 49 = 0$

30. $y^2 - 6 = 0$

31. $x^2 - 13 = 0$

32. $x^2 - 6x + 10 = 0$

33. $y^2 + 7y + 1 = 0$

34. $t^2 + 5t = 2$

35. $b^2 + 11b + 1 = 0$

36. $x^2 + x - 31 = 0$

37. $y^2 + 3y = 9$

38. $2x^2 + 8x + 3 = 0$

39. $4b^2 + 2b - 3 = 0$

40. $3t^2 - 6t + 1 = 0$

41. $3a^2 - 4a + 3 = 0$

42. $3x^2 + 2x + 1 = 0$

43. $\dfrac{x^2}{4} + \dfrac{x}{2} = 2$

44. $\dfrac{x}{6} + \dfrac{x(x+3)}{2} = 5\dfrac{1}{3}$

45. $\dfrac{1}{2}x(x+1) - \dfrac{1}{3}x(x-1) = \dfrac{1}{2}$

46. $\dfrac{1}{2}x^2 - 5x = \dfrac{1}{3}$

47. $x^2 + 0.2x = 3.6$

48. $x^2 + x = -0.21$

49. $x^2 + x - 2.64 = 0$

50. $2x^2 - 9x - 5 = 0$

In exericses 51-60, to be done *only* if your calculator has a square root key, each quadratic equation has been taken from a scientific, engineering, trade, or business application in which the equation has been determined and is now ready to be solved. Solve each equation by using the quadratic formula.

*51. $\dfrac{(128.8)\,(4)\,(w^2)}{4(32.2)} = 32.2w + 16.8$

*52. $1.418 = 1.454 - 0.0015T + 0.00001T^2$

*53. $1280 = 200 - 40t + 0.2t^2$

*54. $100 = 22.5w + \dfrac{21w^2}{2}$

*55. $65(15 + s) - \dfrac{20s^2}{2} = 0$

*56. $70.7(10 + s) - 7.07(10 + s) - \dfrac{1000s^2}{2} = 0$

*57. $2000s - 10s^2 = 96{,}600$

*58. $42.3 = (4 - y^2)\left(\dfrac{9}{y}\right)$

*59. $\dfrac{(1000)\,(12 + x)^2}{2.24} + \dfrac{2(500x^2)}{24} = \dfrac{64{,}400}{64.4}(10^2)$

*60. $0 = 5.245 + \dfrac{200w^2}{64.4} + \dfrac{800w^2}{64.4} + 200w + 120w$

9.3 Solving Quadratic Equations by Factoring

For some quadratic equations there exists a convenient shortcut to the general procedures we have given. The justification for this shortcut is based on the following two properties of real numbers:

1. If a is any real number, then $a \cdot 0 = 0$.

2. If a and b are any two real numbers such that $a \cdot b = 0$, then at least one of the numbers a or b must be zero.

To see how these properties can be applied to the solution of certain quadratic equations, consider the equation $x^2 - x - 6 = 0$. In factored form this can be written as

$$(x + 2)\,(x - 3) = 0$$

Using the property that $a \cdot 0 = 0$ we know that $(x + 2)\,(x - 3) = 0$ if either

$$x + 2 = 0 \quad \text{or} \quad x - 3 = 0$$

that is, if

$$x = -2 \quad \text{or} \quad x = 3$$

Conversely, the second property given above says that

$$(x + 2)(x - 3) = 0 \quad \textit{only if} \quad x + 2 = 0 \quad \text{or} \quad x - 3 = 0$$

that is, *only* if

$$x = -2 \quad \text{or} \quad x = 3$$

Hence the solution set of the equation $x^2 - x - 6 = 0$ is $\{-2, 3\}$.

Here are three other examples of the use of factoring to solve quadratic equations.

Example 1: $x^2 - 19x + 34 = 0$

$$(x - 17)(x - 2) = 0$$

Thus $x - 17 = 0$ or $x - 2 = 0$ and hence the solution set is $\{17, 2\}$.

Example 2: $10x^2 + 11x = 6$

$$10x^2 + 11x - 6 = 0$$

$$(5x - 2)(2x + 3) = 0$$

Thus $5x - 2 = 0$ or $2x + 3 = 0$. Hence

$$x = \frac{2}{5} \quad \text{or} \quad x = -\frac{3}{2}$$

and so the solution set is $\left\{ \frac{2}{5}, -\frac{3}{2} \right\}$.

Example 3: $x^2 - 6x + 9 = 0$

$$(x - 3)^2 = 0$$

and so the solution set is $\{3\}$.

The use of factoring provides the easiest method of solving the quadratic equation $x^2 - 7x - 60 = 0$ associated with the word problem given in the beginning of the chapter. We have $x^2 - 7x - 60 = (x - 12)(x + 5) = 0$ and so $x = 12$ or $x = -5$. Clearly -5 is impossible as a length and so the length is 12 centimeters and the width is $(12 - 7)$ centimeters $= 5$ centimeters. As a check, we note that $5 \cdot 12 = 60$ gives the area in square centimeters.

EXERCISES 9.3

In exercises 1-10 find the solution set of the given equation.

1. $(x - 2)(x - 3) = 0$
2. $(x + 1)(x - 5) = 0$
3. $x(x - 4) = 0$
4. $(3x - 2)(3x + 2) = 0$
5. $(4x + 1)(5x - 2) = 0$
6. $(2x - 3)(4x + 5) = 0$
7. $(x - 3)(x - 5)(x + 4) = 0$
8. $x(x + 2)(x + 2) = 0$
9. $x(3x - 1)(2x + 5)(3x - 5) = 0$
10. $(x - 3)^2 = 0$

In exercises 11-45 solve the given quadratic equation by factoring.

11. $x^2 - 8x + 7 = 0$
12. $x^2 - 12x + 27 = 0$
13. $x^2 - 5x + 4 = 0$
14. $x^2 - 7x + 12 = 0$
15. $x^2 + 9x + 20 = 0$
16. $y^2 + 10y + 21 = 0$
17. $t^2 - 15t + 50 = 0$
18. $a^2 - 8a + 15 = 0$
19. $x^2 - 18x + 77 = 0$
20. $x^2 - 5x - 24 = 0$
21. $2y^2 + 5y - 3 = 0$
22. $2x^2 + 5x + 2 = 0$
23. $3y^2 - y - 4 = 0$
24. $2a^2 - 5a - 12 = 0$
25. $5x^2 + 12x + 4 = 0$
26. $8x^2 - 26x + 15 = 0$
27. $3t^2 - 2t = 5$
28. $3b^2 - b = 4$

29. $6x^2 = 11x + 10$
30. $6m^2 - 7m = 20$
31. $6x^2 - x = 0$
32. $3x^2 = 2x$
33. $x^2 - 64 = 0$
34. $y^2 - 49 = 0$
35. $4x^2 - 25 = 0$
36. $9x^2 - 16 = 0$
37. $4x^2 - 12x + 9 = 0$
38. $4x^2 - 4x + 1 = 0$
39. $x^2 + 26x + 169 = 0$
40. $y^2 + 2y + 1 = 0$
41. $\frac{1}{3}x^2 + \frac{1}{2}x + \frac{1}{6} = 0$
42. $x^2 - \frac{8}{3}x = 1$
43. $\frac{x^2}{2} + \frac{3x}{4} = 11$
44. $\frac{y^2}{6} - \frac{y}{3} = 20$
45. $0.2x^2 - 1.75x + 1.2 = 0$

In exercises 46–55 solve the given quadratic equation using any convenient method, or show that there are no real number solutions.

46. $(x + 2)(x - 2) = 2x^2 - 13$ 51. $(x + 1)(2x - 3) = 4x^2 - 9$

47. $x^2 + 2x = 8$ 52. $x(x + 2) + 2(x - 2) = 1$

48. $0.16x^2 + 1.6x - 12 = 0$ 53. $49x^2 + 1 = 14x$

49. $x^2 - 2x + 2 = 0$ 54. $x^2 = 0.1x + 0.12$

50. $16x^2 + 12x = -9$ 55. $12x^2 + 42x + 36 = 0$

9.4 Word Problems Leading to Quadratic Equations

Word problems, as we have seen, sometimes lead to quadratic equations. Here are two additional examples.

Example 1: If the square of a number is increased by 3 times the number, the result is 28. Find the number.

Solution: Let x = the number. Then

$$x^2 + 3x = 28$$

$$x^2 + 3x - 28 = 0$$

$$(x + 7)(x - 4) = 0$$

Solution set = $\{-7, 4\}$.

There are two numbers that satisfy the problem: -7 and 4.

Check:

Squaring -7: $(-7)^2 = 49$ Squaring 4: $4^2 = 16$

Adding 3 times -7: $49 + (-21) = 28$ Adding 3 times 4: $16 + 12 = 28$

Result is 28 (this checks).   Result is 28 (this checks).

Example 2: The length of a rectangle is 40 feet more than the width. The area is 4500 square feet. Find the length and width of the rectangle.

Solution: Let W = the width of the rectangle in feet; then $W + 40 =$ the length of the rectangle in feet. Thus

$$W(W + 40) = 4500$$

$$W^2 + 40W = 4500$$

Solving by completing the square:

$$W^2 + 40W + 400 = 4500 + 400$$
$$(W + 20)^2 = 4900$$
$$W + 20 = \pm 70$$

Hence $W = -20 + 70 = 50$ or $W = -20 - 70 = -90$. Note, however, that -90 as the number of feet in the width is meaningless and thus the width of the rectangle must be 50 feet. If the width is 50 feet, then the length is $(50 + 40)$ feet = 90 feet.

Check: The length is 40 feet more than the width.

$$90 \text{ feet} - 50 \text{ feet} = 40 \text{ feet}$$

The area is 4500 square feet.

$$(50 \text{ feet}) \times (90 \text{ feet}) = 4500 \text{ square feet}$$

EXERCISES 9.4

1. When 5 times a certain number is added to the square of the number, the result is 36. Find the number.

2. When a certain number is subtracted from its square, the difference is 30. Find the number.

3. One number is 8 more than another. If their product is 65, find the two numbers.

4. One number is 6 more than twice another. If their product is 20, find the two numbers.

5. The sum of two whole numbers is 18. Their product is 45. Find the numbers.

6. The product of two numbers is 104. If one number is 5 more than the other, find the numbers.

7. The product of two consecutive integers is 156. Find the integers. (Note: If x is the smaller of two consecutive integers, then $x + 1$ is the other.)

8. The product of two consecutive integers is 210. Find the integers.

9. The sum of the reciprocals of two consecutive integers is $\frac{7}{12}$. What are the integers?

10. One-sixth of a certain integer equals the reciprocal of the next consecutive integer. Find that integer.

11. The product of two consecutive odd integers is 143. Find the integers. (Note: If x is the smaller of the two consecutive odd integers, then $x + 2$ is the other.)

12. The product of two consecutive odd integers is 195. Find the integers.

13. The length of a rectangle is 12 yards greater than the width. If the area of the rectangle is 85 square yards, what are the dimensions?

14. The width of a rectangle is 3 centimeters less than the length. The area is 70 square centimeters. Find the length and width of the rectangle.

15. The perimeter of a rectangle is 22 inches and its area is 24 square inches. Find the length and width of the rectangle.

16. The number of square meters in the area of a square exceeds the number of meters in its perimeter by 32. What is its area?

17. The altitude of a triangle is 4 feet shorter than its base. If the area is 126 square feet, find the base and altitude of the triangle.

18. The base of a triangle exceeds its altitude by 2 centimeters and its area is 112 square centimeters. Find its altitude.

19. If the lengths of two parallel sides of a square are each increased by 6 inches and the lengths of the other two parallel sides are decreased by 1 inch, a rectangle is formed having twice the area of the square. How long is each side of the square?

20. The perimeter of a rectangle is 26 meters. If the length is increased by 4 meters and the width is increased by 3 meters, the area of the new rectangle is 96 square meters. Find the length and width of the original rectangle.

21. A man bought a number of shares of a certain stock for $900. The price advanced $15 a share, and he then sold all but two shares for $900. How many shares did he buy? (Disregard any brokerage fees!)

22. Susan lives 3 miles from work. When she rides her bicycle she travels 6 miles per hour faster than when she walks, and thus saves 27 minutes. What is her average hourly rate of walking?

23. Jane's Uncle Harry is 3 years more than twice as old as Jane. The product of their ages, in years, is 560. How old is Jane?

24. The ones' digit of a two-digit number is 2 less than the tens' digit. The product of the two digits is 28 less than the number. What is the number? (Note: If u is the units' digit and t the tens' digit, then the number is expressed as $10t + u$.)

25. The sum of the digits of a two-digit number is 9 and the number itself is equal to twice the product of the digits. Find the number.

9.5 Some Equations Equivalent to Quadratic Equations

We begin by considering a somewhat different kind of word problem from those we have considered before:

Mary's father can paint their garage in 5 hours less time than Mary can. If they work together, they can paint the garage in $3\frac{1}{3}$ hours. How long would it take Mary to do the job alone?

Solution: Time to paint garage together is $3\frac{1}{3}$ or $\frac{10}{3}$ hours.

Part of garage painted in 1 hour is $\frac{3}{10}$.

Let x = the number of hours Mary would take alone.

Then $\frac{1}{x}$ = the part of garage Mary paints in 1 hour.

Time for father to paint garage = $(x - 5)$ hours.

Part of garage father paints in 1 hour is $\frac{1}{x - 5}$.

Equation: $\dfrac{1}{x} + \dfrac{1}{x - 5} = \dfrac{3}{10}$

Using the principles discussed in Section 7.5, we conclude that, if $x \neq 0$ and $x - 5 \neq 0$, this equation is equivalent to the equation

$$10x\,(x - 5)\left(\frac{1}{x} + \frac{1}{x - 5}\right) = 10x\,(x - 5) \cdot \frac{3}{10}$$

which gives

$$10(x - 5) + 10x = 3x(x - 5)$$
$$10x - 50 + 10x = 3x^2 - 15x$$
$$3x^2 - 35x + 50 = 0$$
$$(3x - 5)\,(x - 10) = 0$$

so that $x = 10$ or $x = \dfrac{5}{3}$.

As always, we must check to see that our answers not only satisfy the equation we obtained but also the actual conditions of the given word problem. It turns out, as you should verify, that both 10 and $\frac{5}{3}$ are solutions of the equation we began with, but only 10 satisfies the conditions of our problem. For if Mary were able to paint the garage alone in $\frac{5}{3}$ hours, her father would have to be able to paint it in $x - 5 = \frac{5}{3} - 5 = -3\frac{1}{3}$ hours!

Here is another example of a problem that leads to an equation involving rational algebraic expressions that is equivalent to a quadratic equation.

Example: A saleswoman travels by car a distance of 640 kilometers from home. On her return trip she averages 16 kilometers per hour faster and takes 2 hours less time for the trip. What was her average speed on the return trip?

Solution: Let x = speed in kilometers per hour on the way out; then $x + 16$ = speed in kilometers per hour on the return trip.

$$\text{time out} = \frac{640}{x}, \qquad \text{time back} = \frac{640}{x + 16}$$

and so

$$\frac{640}{x} - 2 = \frac{640}{x + 16}$$

$$640(x + 16) - 2x(x + 16) = 640x$$

$$640x + 10{,}240 - 2x^2 - 32x = 640x$$

$$2x^2 + 32x - 10{,}240 = 0$$

$$x^2 + 16x - 5120 = 0$$

$$(x - 64)(x + 80) = 0$$

$$x = 64 \text{ or } -80$$

We obviously reject -80 as an answer to our problem and hence get $x + 16 = 64 + 16 = 80$ for the average speed in kilometers per hour on the return trip.

(a) Check in the original equation:

Original equation is $\dfrac{640}{x} - 2 = \dfrac{640}{x + 16}$.

Substituting 64 for x we obtain $\dfrac{640}{x} - 2 = \dfrac{640}{64} - 2 = 10 - 2 = 8$

and also

$$\frac{640}{x + 16} = \frac{640}{64 - 16} = \frac{640}{80} = 8$$

(b) Check in the problem:

$$\text{time out} = \frac{640}{x} = \frac{640}{64} = 10 \text{ hours}$$

$$\text{time back} = \frac{640}{x + 16} = \frac{640}{64 + 16} = \frac{640}{80} = 8 \text{ hours}$$

This checks with the problem situation, as the time back must be 2 hours less than the time going and 8 is 2 less than 10. Thus the solution to the problem is 64 kilometers per hour for her average speed on the return trip.

EXERCISES 9.5

In exercises 1–42 solve the given equation or show that there are no real number solutions. Remember to check to see that no candidate for a solution makes any denominator in the equation equal to zero.

1. $\dfrac{x}{3} = \dfrac{3}{x}$

2. $\dfrac{x+3}{2} = \dfrac{8}{x-3}$

3. $\dfrac{6-y}{4} = \dfrac{5}{6+y}$

4. $\dfrac{3x+4}{3} = \dfrac{3}{3x-4}$

5. $\dfrac{2-3a}{3a} - \dfrac{2a}{2+3a} = 0$

6. $\dfrac{4}{t+5} = \dfrac{t}{t+3}$

7. $\dfrac{1}{x+3} = \dfrac{3-x}{7+x}$

8. $x = \dfrac{10}{x-3}$

9. $\dfrac{t}{2t+1} = \dfrac{3t+2}{4t+3}$

10. $\dfrac{y}{y+2} = \dfrac{y+4}{3y}$

11. $\dfrac{6}{x+3} = \dfrac{x+2}{5}$

12. $\dfrac{2}{m} + 4 = m - \dfrac{3}{m}$

13. $\dfrac{1}{x} - \dfrac{1}{x+1} = \dfrac{1}{6}$

14. $\dfrac{4}{a} + \dfrac{2}{3-a} = 4$

15. $\dfrac{1}{x-3} + \dfrac{1}{x+4} = \dfrac{1}{12}$

16. $\dfrac{8}{x} + \dfrac{4}{5+x} = 2$

17. $\dfrac{12}{x+5} + \dfrac{21}{x+10} = 1$

18. $\dfrac{1}{y} + \dfrac{4}{y+3} = 1$

19. $\dfrac{2}{b} + \dfrac{2}{b-2} = \dfrac{16}{15}$

20. $\dfrac{9}{x-1} + \dfrac{4}{x+3} = 1$

21. $\dfrac{1}{x} + \dfrac{1}{x-1} = \dfrac{7}{12}$

22. $\dfrac{4}{c} + \dfrac{4}{c-5} = \dfrac{2}{3}$

23. $\dfrac{1}{3x-4} - \dfrac{3}{3x+4} - \dfrac{2}{5x} = 0$

24. $\dfrac{x}{2x-1} = \dfrac{2x+3}{15}$

25. $\dfrac{1}{4x^2+3} + \dfrac{1}{4x^2-3} = \dfrac{1}{16x^4-9}$

26. $\dfrac{x-5}{x} + \dfrac{x}{x+5} = \dfrac{1}{2}$

27. $\dfrac{3}{x^2} + \dfrac{5}{x} = 8$

28. $\dfrac{4}{y} - \dfrac{1}{y-2} = \dfrac{7}{15}$

29. $\dfrac{3}{x-2} + \dfrac{2}{x} = \dfrac{31}{35}$

30. $\dfrac{x-4}{2x+1} - \dfrac{3x-1}{x-3} = \dfrac{x^2-1}{2x^2-5x-3}$

31. $\dfrac{x}{x-1} + \dfrac{5}{x-5} = \dfrac{3x+5}{x^2-6x+5}$

37. $\dfrac{2x}{3x+1} = \dfrac{x-2}{2x-1}$

32. $\dfrac{1}{x} + \dfrac{2}{x+1} + \dfrac{3}{x+2} = 0$

38. $\dfrac{t+6}{2t+1} = \dfrac{2t-3}{t+2}$

33. $\dfrac{1}{x^2} + 5 = \dfrac{6x-7}{x}$

39. $\dfrac{5}{t} + \dfrac{10}{t-5} = \dfrac{3}{2}$

34. $\dfrac{a}{a+1} + \dfrac{a+1}{a+2} = 1$

40. $\dfrac{1}{2y-1} + \dfrac{3}{y+2} + \dfrac{2}{2y+4} = 1$

35. $1 + \dfrac{12}{y^2-4} = \dfrac{3}{y-2}$

41. $\dfrac{6}{x-2} + \dfrac{6x}{3} = \dfrac{7}{2}$

36. $\dfrac{a^2+1}{a-1} + 1 + \dfrac{1+3a}{2-2a} = 0$

42. $\dfrac{x+2}{x-1} - \dfrac{x+3}{x+1} = \dfrac{2x^2-13}{1-x^2}$

In exercises 43–62 (a) write an equation for the given situation; (b) solve the equation; and (c) check the candidates for solution both in the equation and in the problem situation.

43. A train travels 240 kilometers at a uniform rate. If the speed had been 8 kilometers per hour more, the journey would have taken 1 hour less. Find the speed of the train.

44. A cyclist traveling a distance of 24 miles would have arrived 1 hour earlier if he had ridden 4 miles per hour faster. What was his speed?

45. An airplane travels 1500 miles west. On the return flight the plane travels 100 miles per hour faster and takes a half hour less flight time. What was the average speed of the return trip?

46. Joe can mow a lawn in 2 hours less time than his brother. If they work together using two mowers, they can mow all but $\frac{7}{15}$ of the lawn in an hour. How long would it take Joe to mow the lawn by himself?

47. Betty working alone can do a piece of work in 5 days less than the time required by Jim alone to do the job. Working together, they complete the work in 6 days. How long would it take each to do the job alone?

48. A cistern can be filled by two pipes together in 20 minutes; the larger pipe alone will fill the cistern in 9 minutes less than the smaller one. Find the time required by the larger pipe alone to fill the cistern.

49. A tank can be filled by two pipes running together in $3\frac{3}{4}$ hours. The larger pipe, by itself, will fill it 4 hours faster than the small pipe. What time will each pipe separately take to fill it?

50. The sum of the reciprocals of two consecutive even integers is $\frac{7}{24}$. What are the integers?

51. Find a number whose square diminished by 169 is equal to 10 times the excess of the number over 5.

52. If the product of three consecutive numbers is divided by each of them in turn, the sum of the three quotients is 74. What are the numbers?

53. One number is 2 less than a second number. If the reciprocal of the smaller number is added to the larger number, the sum is $\frac{36}{5}$. Find the two numbers.

54. A man bought some pigs for $300. If each pig had cost $3 more he would have obtained 5 fewer pigs for the same amount of money. How many pigs did he buy?

55. A farmer bought a number of sheep for $378. Having lost 6, he sold the remainder for $10 a head more than they cost him and gained $42. How many did he buy?

56. A woman paid $990 for some horses. By selling all but 4 of them at a profit of $10 each, she received the amount she paid for all the horses. How many horses did she buy?

57. The telephone poles along a certain road are at equal intervals. If there were 2 more in each mile the interval would be decreased by 20 feet. How many poles are there per mile?

58. The perimeter of a rectangle is 38 meters and its area is 84 square meters. Find its dimensions.

59. Twice the reciprocal of a natural number is subtracted from 3 times the reciprocal of the next successive natural number and the result is $\frac{1}{10}$. What are the numbers?

60. Sally can row 8 kilometers downstream and back again in 3 hours. If the rate of the current is 2 kilometers per hour, how fast could she row in still water?

61. When 24 is divided by a certain number, the quotient is 3 more than when 36 is divided by 2 more than the number. Find the number.

62. At what price per dozen were eggs selling for if, when the price was raised 5¢ per dozen, you would receive 12 fewer for a dollar?

9.6 Other Equations Equivalent to Linear or Quadratic Equations

There are other types of equations which are neither linear equations nor quadratic equations but which are equivalent to either linear or quadratic equations. Here are some examples.

Example 1: Find the solution set of the equation $\sqrt{x - 5} = 2$.

Since if $a = b$, then $a^2 = b^2$, we know that

$$\sqrt{x - 5} = 2 \text{ implies that}$$
$$(\sqrt{x - 5})^2 = 2^2$$

so that

$$x - 5 = 4 \quad \text{and} \quad x = 9$$

Check: $\sqrt{9 - 5} = \sqrt{4} = 2$. Solution set is $\{9\}$.

Example 2: Find the solution set of the equation $\sqrt{x + 12} = x$.

Now we have

$$(\sqrt{x + 12})^2 = x^2$$
$$x + 12 = x^2$$
$$x^2 - x - 12 = (x - 4)(x + 3) = 0$$

and so $x = 4$ or $x = -3$.

Check: If $x = 4$, $\sqrt{x + 12} = \sqrt{16} = 4 = x$. However, if $x = -3$, we have $\sqrt{x + 12} = \sqrt{9} = 3 \neq -3$. Hence the solution set is $\{4\}$ and not $\{4, -3\}$. What has gone wrong? Recall that we said that if $a = b$, then $a^2 = b^2$. We did *not* say that if $a^2 = b^2$, then $a = b$ for the very good reason that this statement is not true:

$$(-3)^2 = 3^2, \text{ for example, but } -3 \neq 3$$

Thus when we square both sides of an equation (1) to obtain an equation (2), all of the solutions (if any) of equation (1) must be solutions of equation (2) but there *may* be solutions of equation (2) that are not solutions of equation (1). Hence, when this squaring process is used, a check of the possible solutions is absolutely essential.

Example 3: Find the solution set of the equation

$$\sqrt{4x - 3} + x = 0$$

We first "isolate" the radical by writing

$$\sqrt{4x - 3} = -x$$

Then, squaring both sides, we have

$$(\sqrt{4x - 3})^2 = (-x)^2$$
$$4x - 3 = x^2$$
$$x^2 - 4x + 3 = (x - 3)(x - 1) = 0$$
$$x = 3 \quad \text{or} \quad x = 1$$

Check: If $x = 3, \sqrt{4x - 3} + x = \sqrt{12 - 3} + 3 = \sqrt{9} + 3 = 3 + 3 = 6 \neq 0$; if $x = 1, \sqrt{4x - 3} + x = \sqrt{4 - 3} + 1 = \sqrt{1} + 1 = 2 \neq 0$. Hence the solution set of the equation is $\emptyset$.

Example 4: Find the solution set of the equation

$$\sqrt{2x + 3} - \sqrt{x + 2} = 0$$

Here we can "isolate" each radical to obtain

$$\sqrt{2x + 3} = \sqrt{x + 2}$$

Then, squaring both sides, we have

$$(\sqrt{2x + 3})^2 = (\sqrt{x + 2})^2$$

$$2x + 3 = x + 2$$

$$x = -1$$

Check:
If $x = -1, \sqrt{2x + 3} - \sqrt{x + 2} = \sqrt{2(-1) + 3} - \sqrt{-1 + 2} = \sqrt{1} - \sqrt{1} = 0$.

Hence the solution set is $\{-1\}$.

Example 5: Find the solution set of the equation

$$\sqrt{3 - 2x} + \sqrt{2x + 2} = 3$$

In this problem we can't "isolate" both radicals, *but*, "isolating" $\sqrt{3 - 2x}$, we can write

$$\sqrt{3 - 2x} = 3 - \sqrt{2x + 2}$$

Then, squaring both sides, we have

$$(\sqrt{3 - 2x})^2 = (3 - \sqrt{2x + 2})^2$$

$$3 - 2x = 9 - 2 \cdot 3\sqrt{2x + 2} + (\sqrt{2x + 2})^2$$

$$3 - 2x = 9 - 6\sqrt{2x + 2} + 2x + 2$$

Now we "isolate" the remaining radical and then square both sides once again. Thus we have

$$(3 - 2x) - 2x - 2 - 9 = -6\sqrt{2x + 2}$$

$$-4x - 8 = -6\sqrt{2x + 2}$$

$$4x + 8 = 6\sqrt{2x + 2}$$

$$2x + 4 = 3\sqrt{2x + 2}$$

$$(2x + 4)^2 = (3\sqrt{2x + 2})^2$$

$$4x^2 + 16x + 16 = 9(2x + 2)$$

$$4x^2 + 16x + 16 = 18x + 18$$

$$4x^2 - 2x - 2 = 0$$

$$2x^2 - x - 1 = (2x + 1)(x - 1) = 0$$

$$2x + 1 = 0 \quad \text{or} \quad x - 1 = 0$$

$$x = -\frac{1}{2} \quad \text{or} \quad x = 1$$

Check:

If $x = -\frac{1}{2}$, $\sqrt{3 - 2x} + \sqrt{2x + 2} = \sqrt{3 - 2(-\frac{1}{2})} + \sqrt{2(-\frac{1}{2}) + 2} = \sqrt{3 + 1}$

$+ \sqrt{-1 + 2} = \sqrt{4} + \sqrt{1} = 2 + 1 = 3$ so that $-\frac{1}{2}$ is a solution. If $x = 1$,

$\sqrt{3 - 2x} + \sqrt{2x + 2} = \sqrt{3 - (2 \cdot 1)} + \sqrt{(2 \cdot 1) + 2} = \sqrt{3 - 2} + \sqrt{2 + 2}$
$= \sqrt{1} + \sqrt{4} = 1 + 2 = 3$ so that 1 is also a solution. Hence the solution set is $\{-\frac{1}{2}, 1\}$.

Example 6: Find the solution set of $|x + 2| = 5$.

By the definition of absolute value it follows that

$$x + 2 = 5 \quad \text{or} \quad x + 2 = -5$$

Hence $x = 3$ or $x = -7$.

Check: If $x = 3$, $|x + 2| = |3 + 2| = |5| = 5$; if $x = -7$, $|x + 2|$
$= |-7 + 2| = |-5| = 5$. Hence the solution set is $\{3, -7\}$.

Example 7: Find the solution set of $|2x - 3| = x - 5$.

By the definition of absolute value it follows that

$$2x - 3 = x - 5 \quad \text{or} \quad 2x - 3 = -(x - 5)$$

Hence

$$x = -2 \quad \text{or} \quad x = \frac{8}{3}$$

We must, however, check these solutions. If $x = 2$ we have

$$|2 \cdot 2 - 3| \overset{?}{=} 2 - 5$$

$$|1| \overset{?}{=} -3$$

$$1 \neq -3$$

and if $x = \frac{8}{3}$,

$$\left| 2 \cdot \frac{8}{3} - 3 \right| \overset{?}{=} \frac{8}{3} - 5$$

$$\left| \frac{7}{3} \right| \overset{?}{=} -\frac{7}{3}$$

$$\frac{7}{3} \neq -\frac{7}{3}$$

Hence the solution set is ∅.

EXERCISES 9.6

In exercises 1–50 find the solution set of the given equation. Be sure to check your candidates for solutions.

1. $\sqrt{x - 3} = 5$
2. $\sqrt{x + 7} = 3$
3. $\sqrt{2x + 1} = 5$
4. $\sqrt{4x - 3} = 0$
5. $7 = \sqrt{2x + 5}$
6. $\sqrt{4y + 5} = y$
7. $x + \sqrt{x + 6} = 0$
8. $\sqrt{2a + 1} = a + 1$
9. $\sqrt{x - 3} = \sqrt{x} - 3$
10. $\sqrt{x - 4} = \sqrt{x} + 1$
11. $t = \sqrt{7t - 12}$

12. $\sqrt{x + 1} = x$
13. $\sqrt{6x - 17} = 2x - 9$
14. $3\sqrt{y + 13} = y + 9$
15. $\sqrt{x - 1} = \sqrt{x^2 + 3}$
16. $\sqrt{7 + x} = 1 + x$
17. $\dfrac{\sqrt{6x + 27}}{x} = 1$
18. $\dfrac{\sqrt{x + 3}}{\sqrt{x + 2}} = 1$
19. $\dfrac{\sqrt{x + 1}}{x + 1} = 2$
20. $3 + 2\sqrt{a + 2} = 6$
21. $\sqrt{19 - s} = s - 7$
22. $1 + \sqrt{4 + x} = 0$

23. $\sqrt{x^2 + 9} = x + 3$
24. $4\sqrt{5b^2 + 5} = 20$
25. $\sqrt{a + 3} = \sqrt{a} + \sqrt{3}$
26. $\sqrt{y + 4} + \sqrt{y - 4} = 4$
27. $\sqrt{x + 7} = 2x - 1$
28. $\sqrt{4y + 1} = y - 5$
29. $\dfrac{\sqrt{12x + 5}}{3} = x$
30. $\sqrt{x - 2} = \sqrt{x - 12}$
31. $\sqrt{x + 4} = 3 - \sqrt{x}$
32. $\sqrt{5x + 10} - \sqrt{5x} = 2$
33. $\sqrt{x - 1} + \sqrt{3x + 3} = 4$
34. $|x - 1| = 5$
35. $|x - 2| = 4$
36. $7 = |3 - x|$

37. $7 = |3 + x|$
38. $|7 + x| = 5$
39. $|x + 5| = 3$
40. $|y + 3| + 1 = 0$
41. $|a - 4| + 3 = 0$
42. $|2x - 5| - 1 = 0$
43. $0 = |3x - 4| - 5$
44. $|2x + 3| = x$
45. $|y| = 2y - 7$
46. $|5x| = 3x - 2$
47. $|2x - 7| = x$
48. $2|x + 1| = 5$
49. $\left|\dfrac{x + 4}{3}\right| = \dfrac{1}{2}$
50. $|2x + 1| + 3 = 0$

In exercises 51–60:

(a) write an equation for each problem;
(b) solve the equation;
(c) check your candidates for solutions to see that they satisfy the condition for the problem; and
(d) answer the question asked in the problem.

Example: The square root of 1 more than 10 times a certain number is 9. Find the number.

Let x = the number.

(a) Write an equation for the problem. $\sqrt{10x + 1} = 9$

(b) Solve equation. $(\sqrt{10x + 1})^2 = 9^2$
$$10x + 1 = 81$$
$$10x = 80$$
$$x = 8$$

(c) Check candidate for solution to equation. $\sqrt{(10 \cdot 8) + 1} = \sqrt{81} = 9$

(d) Answer question.   The number is 8.

51. The square root of 2 more than 17 times a number is 6. Find the number.

52. If the square root of twice a certain number is subtracted from the number and the result is 4, what is the number?

53. If the square root of 5 less than 6 times a certain number is divided by the number, the quotient is 1. Find the number.

54. If 4 is subtracted from a number, the resulting number is equal to the square root of 2 more than the number. Find the number.

55. If the square root of 2 less than 6 times a number is divided by 2 less than the number, the quotient is 2. Find the number.

56. If 2 is the quotient of the square root of 1 more than a number divided by 1 greater than the number, what is the number?

57. Find a number such that the square root of 1 less than the number is equal to the square root of 3 less than the number squared.

58. One side of a right triangle is 12 feet long. The hypotenuse is 3 feet more than twice as long as the other side. What are the dimensions of the triangle? (Note: In a right triangle, if a and b are the sides of the triangle and c is the hypotenuse, then $\sqrt{a^2 + b^2} = c$.)

59. The hypotenuse of a right triangle is 4 centimeters less than twice as long as one of the other sides. If the third side is 16 centimeters long, find the length of the hypotenuse. (Use the formula given in exercise 58.)

60. Show, by writing an equation and solving it, that there are two numbers such that each is equal to its own square root.

9.7 Quadratic Inequalities

The solution of quadratic inequalities is based on the following properties of real numbers:

A. If a and b are real numbers such that $a \cdot b > 0$, then either

$$(1) \quad a > 0 \text{ and } b > 0 \text{ or}$$

$$(2) \quad a < 0 \text{ and } b < 0.$$

(That is, if the product of two numbers is positive, then both numbers are positive or both are negative.)

B. If a and b are real numbers such that $a \cdot b < 0$, then either

$$(1) \quad a > 0 \text{ and } b < 0 \text{ or}$$

$$(2) \quad a < 0 \text{ and } b > 0.$$

(That is, if the product of two numbers is negative, then one of the numbers must be positive and the other negative.)

Here are three examples of the solution of quadratic inequalities.

Example 1: $x^2 - x - 12 > 0$

$(x - 4)(x + 3) > 0$

So, using Property A, we conclude that either

(*i*) $x - 4 > 0$ and $x + 3 > 0$ so that $x > 4$ and $x > -3$, or

(*ii*) $x - 4 < 0$ and $x + 3 < 0$ so that $x < 4$ and $x < -3$.

Now in (*i*), if $x > 4$, we certainly also have $x > -3$ so that (*i*) becomes simply

$$x > 4$$

On the other hand, in (*ii*), if $x < -3$, we certainly also have $x < 4$. Thus we can combine (*i*) and (*ii*) to get

(*iii*) $x > 4$ or $x < -3$.

Pictured on a number line, our solution set, $\{x : x > 4$ or $x < -3\}$, looks like this:

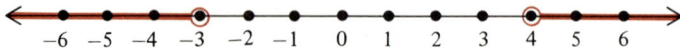

Example 2: $x^2 - 5x < -6$

$x^2 - 5x + 6 < 0$

$(x - 3)(x - 2) < 0$

Here we use property B to conclude that either

(*i*) $x - 3 > 0$ and $x - 2 < 0$ so that $x > 3$ and $x < 2$, or

(*ii*) $x - 3 < 0$ and $x - 2 > 0$ so that $x < 3$ and $x > 2$.

But condition (*i*) is impossible; there is no real number x such that $x > 3$ *and* $x < 2$.

On the other hand, condition (*ii*) can be written as

$$2 < x < 3$$

(read "x is greater than two and less than three"). Thus the solution set is $\{x : 2 < x < 3\}$ which can be pictured on a number line as shown below.

Example 3: $6x^2 + x - 1 \leqslant 0$

This says that

$$6x^2 + x - 1 = 0 \quad \text{or} \quad 6x^2 + x - 1 < 0$$

Since $6x^2 + x - 1 = (3x - 1)(2x + 1)$ we see that $\frac{1}{3}$ and $-\frac{1}{2}$ are in our solution set. Turning to $6x^2 + x - 1 < 0$ we have, by property B, either

(i) $3x - 1 > 0$ and $2x + 1 < 0$ so that $x > \frac{1}{3}$ and $x < -\frac{1}{2}$, or

(ii) $3x - 1 < 0$ and $2x + 1 > 0$ so that $x < \frac{1}{3}$ and $x > -\frac{1}{2}$.

But condition (i) is impossible since we cannot have both $x > \frac{1}{3}$ and $x < -\frac{1}{2}$. This leaves condition (ii), which can be written as

$$-\frac{1}{2} < x < \frac{1}{3}$$

Thus the solution set of the given inequality is $\{x : -\frac{1}{2} \leqslant x \leqslant \frac{1}{3}\}$.

EXERCISES 9.7

In each of these exercises:

(a)   solve the inequality;
(b)   describe the solution set using set builder notation; and
(c)   draw a graph of the solution set.

1. $x^2 - 9 < 0$ 6. $x^2 - 4 > 12$
2. $x^2 - 9 > 0$ 7. $(x + 5)(x - 2) > 0$
3. $x^2 \geqslant 16$ 8. $(x - 3)(x + 4) < 0$
4. $x^2 \geqslant \frac{1}{4}$ 9. $(x + 2)(x + 3) > 0$
5. $x^2 - 4 < 12$ 10. $(x - 2)(x + 3) \leqslant 0$

11. $x^2 - 5x - 14 > 0$
12. $x^2 - 7x - 18 \geqslant 0$
13. $x^2 + 3x - 18 \leqslant 0$
14. $3x^2 - 27 > 0$
15. $2x^2 - 32 < 0$
16. $x^2 - 8 > 0$
17. $x^2 - 12 < 0$
18. $2x^2 + x - 6 < 0$
19. $6x^2 + x - 12 > 0$
20. $15x^2 - 19x + 6 \geqslant 0$
21. $x^2 < 5x - 4$
22. $x^2 \geqslant x + 30$
23. $2x^2 - 5x > 3$
24. $2x^2 + x \leqslant 6$
25. $3x^2 - 4x - 4 < 0$

26. $3x^2 - 8x + 5 > 0$
27. $6x^2 + 2 \geqslant -7x$
28. $8x^2 - 3 < 2x$
29. $12x^2 + 7x - 12 \leqslant 0$
30. $24x^2 + 34x - 45 \geqslant 0$
31. $x^2 + 10x + 30 < 5$
32. $x^2 - 14x + 50 > 1$
33. $x^2 + 9 \geqslant 6x$
34. $8x^2 - 30x - 27 > 0$
35. $12x^2 - 7x < 10$
36. $9x^2 + 12x > -4$
37. $8x^2 \leqslant 30x - 27$
38. $x^2 + 6x + 14 < 0$
39. $x^2 + 7 < 4x$
40. $x^2 - 2x + 2 < 0$

9.8 Inequalities Involving Absolute Value

Sometimes we need to consider inequalities involving absolute values.

Example: Solve and graph $|x + 1| \leqslant 2$.

From the definition of absolute value we know that

$x + 1 \geqslant 0$ and $x + 1 \leqslant 2$ or $x + 1 < 0$ and $-(x + 1) \leqslant 2$

$x \geqslant -1$ and $x \leqslant 1$ or $x < -1$ and $-x - 1 \leqslant 2$

$x < -1$ and $-x \leqslant 3$

$x < -1$ and $x \geqslant -3$

$-1 \leqslant x \leqslant 1$ or $-3 \leqslant x < -1$

Hence the solution set is $\{x: -1 \leqslant x \leqslant 1$ or $-3 \leqslant x < -1\}$, which is equivalent to $\{x: -3 \leqslant x \leqslant 1\}$, and the graph is as shown below.

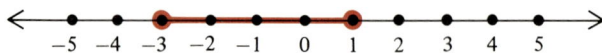

Partial Check:

If $x = -3$, then $|x + 1| = |-3 + 1| = |-2| = 2$, and $2 \leqslant 2$. (True)

If $x = 0$, then $|x + 1| = |0 + 1| = |1| = 1$, and $1 \leqslant 2$. (True)

If $x = 1$, then $|x + 1| = |1 + 1| = |2| = 2$, and $2 \leqslant 2$. (True)

If $x = -4$, then $|x + 1| = |-4 + 1| = |-3| = 3$, and $3 \leqslant 2$. (False)

If $x = 2$, then $|x + 1| = |2 + 1| = |3| = 3$, and $3 \leqslant 2$. (False)

EXERCISES 9.8

Find and graph the solution set of the following inequalities.

1. $|x| > 2$
2. $|x| \leqslant 5$
3. $-|x| < 6$
4. $-|x| > 6$
5. $-|x| \geqslant -3$
6. $|x + 2| \leqslant 3$
7. $|x - 3| \geqslant 5$
8. $|x + 5| < 7$

9. $|x - 5| > 4$
10. $|2x - 3| < 2$
11. $|2x + 1| > 5$
12. $|x + 3| \leqslant 4$
13. $|x + 5| \geqslant 36$
14. $|x| + 2 < 5$
15. $|x| - 3 \geqslant 0$
16. $|x + 1| \geqslant |1 - x|$

REVIEW EXERCISES

Section 9.1

In exercises 1–20 find the solution set of the given equation. If an answer involves a radical, write it in simplest form.

1. $x^2 = 9$
2. $b^2 = 49$
3. $y^2 - 4 = 0$
4. $a^2 - 25 = 0$
5. $x^2 - 20 = 0$
6. $x^2 - 8 = 0$
7. $x^2 + 16 = 0$
8. $4x^2 - 16 = 0$

9. $5y^2 - 20 = 0$
10. $4x^2 - 25 = 0$
11. $2y^2 - 36 = 0$
12. $5x^2 - 40 = 0$
13. $3t^2 + 9 = 57$
14. $9x^2 + 1 = 2$
15. $1 = \dfrac{1}{3}x^2$
16. $(x + 4)(x - 6) = 25 - 2x$

17. $(x - 2)^2 = 36$

18. $(y + 3)^2 = 9$

19. $\dfrac{4}{y^2 - 7} = 1$

20. $\dfrac{4}{x - 3} = \dfrac{1}{3} + \dfrac{4}{x + 3}$

In exercises 21–40 find the solution set of the given quadratic equation by completing the square.

21. $x^2 - 12x = -27$

22. $a^2 - 2a = 35$

23. $x^2 + 2x - 24 = 0$

24. $x^2 - 6x - 40 = 0$

25. $t^2 + 14t + 48 = 0$

26. $a^2 + 4a - 96 = 0$

27. $x^2 + 2x - 48 = 0$

28. $x^2 + 4x = -3$

29. $y^2 + 8y = 0$

30. $2x^2 - 5x - 12 = 0$

31. $3y^2 - 5y + 2 = 0$

32. $6x^2 - x - 2 = 0$

33. $3y^2 - 5y + 1 = 0$

34. $x^2 - 4x - 7 = 0$

35. $x^2 - 14x + 8 = 0$

36. $3y^2 - 16y + 19 = 0$

37. $3a^2 + 5a + 1 = 0$

38. $\dfrac{12}{x} - \dfrac{12}{x + 2} = 3$

39. $\dfrac{4}{x - 7} + \dfrac{18}{x + 3} = 2$

40. $\dfrac{x + 3}{x - 2} - \dfrac{x - 3}{x + 2} = \dfrac{15}{8}$

Section 9.2

In exercises 41–65 find the solution set of the given quadratic equation by using the quadratic formula.

41. $x^2 - 5x + 4 = 0$

42. $x^2 - 6x + 9 = 0$

43. $3x^2 + 8x + 5 = 0$

44. $2x^2 + 2x - 24 = 0$

45. $4y^2 - 8y - 13 = 0$

46. $5a^2 + 26a + 5 = 0$

47. $x^2 + x - 1 = 0$

48. $x^2 - x - 1 = 0$

49. $m^2 + 6m = 55$

50. $8y^2 - 2y = 1$

51. $x^2 - 3x = 2$

52. $3x^2 - 7x + 2 = 0$

53. $2x^2 + 5x - 2 = 0$

54. $x^2 + x - 2 - 0$

55. $2b^2 = 1 - 3b$

56. $3a = 2 - 9a^2$

57. $x^2 + x + 1 = 0$

58. $1 - 4y = 8y^2$

59. $t^2 + 6t + 9 = 0$

60. $x^2 - 49 = 0$

61. $3x^2 + 4x - 6 = 0$

62. $11y^2 - 12y = -3$

63. $x^2 - 4x - 1 = 0$

64. $x^2 - 4x + 4 = 0$

65. $7x^2 - 12x + 6 = 0$

Section 9.3

In exercises 66–90 solve the given quadratic equation by factoring.

66. $x^2 - 10x + 24 = 0$

67. $x^2 + 8x + 7 = 0$

68. $x^2 - 3x + 2 = 0$

69. $3x^2 - 5x - 2 = 0$

70. $3x^2 - 5x + 2 = 0$

71. $a^2 + 5a = 24$

72. $b^2 + 5b = 66$

73. $x^2 - 2x = 15$

74. $y - 6 = -y^2$

75. $15t^2 = t + 6$

76. $x^2 - 9 = 0$

77. $4x^2 - 64 = 0$

78. $2x^2 + 5 + 11x = 0$

79. $a^2 + 12a = -32$

80. $x^2 - 10 = 3x$

81. $z^2 - 4z = 0$

82. $x^2 - x = 90$

83. $x^2 - 2x = 80$

84. $y(y - 7) = 144$

85. $a(a + 11) + 18 = 0$

86. $x^2 - 36 = 0$

87. $4y^2 - 49 = 0$

88. $3x^2 - 6x + 3 = 0$

89. $2b^2 - 20b + 18 = 0$

90. $7x^2 + 7 = 50x$

In exercises 91–120 use any convenient method to solve the given quadratic equation or to show that no real number solution exists.

91. $2x^2 - 50 = 0$

92. $x^2 - x - 90 = 0$

93. $2x^2 + 7x = 0$

94. $y^2 - 8y + 17 = 0$

95. $a^2 - 3a - 28 = 0$

96. $3x^2 + 13x = 10$

97. $9b^2 - 4 = 0$

98. $m^2 + 10m - 7 = 0$

99. $2x^2 - 3x - 20 = 0$

100. $x^2 + 2x - 1 = 0$

101. $y^2 + 6y - 7 = 0$

102. $x^2 = 10x$

103. $9x^2 + 6x + 1 = 0$

104. $5x^2 - x = 4$

105. $x^2 - 2x - 2 = 0$

106. $2a^2 + 2a + 1 = 0$

107. $3y^2 - 6y + 6 = 0$

108. $2x^2 - 8x + 1 = 0$

109. $4x^2 - 100 = 0$

110. $5x^2 - 3x = 0$

111. $10x^2 - 33x - 7 = 0$

112. $b^2 = 2b + 7$

113. $x^2 + 2x + 15 = 0$

114. $m^2 - 9m + 14 = 0$

115. $x^2 + 6x + 3 = 0$

116. $x^2 - 8x = 16$

117. $x^2 - 4x = 1$

118. $4a^2 + 3a = 10$

119. $4b^2 + 8b + 1 = 0$

120. $9x^2 - 6x = 1$

Section 9.4

In exercises 121–140 solve each problem by translating the problem situation into a quadratic equation, finding the solution set for the equation, and then using that solution set to answer the question asked in the problem.

121. Separate 16 into two parts whose product is 63.

122. Separate 90 into two parts such that the first is the square of the second.

123. If 4 times a number is added to 3 times its square, the sum is 95. Find the number.

124. Find two consecutive whole numbers such that the sum of their squares is 145.

125. A chicken yard 20 by 30 feet is to be enlarged by adding the same amount to the length and width. How much will have to be added to each to double the area of the yard?

126. The length of a rectangle is 5 meters longer than its width, and its area is 266 square meters. How long and how wide is the rectangle?

127. The altitude of a triangle is 11 centimeters less than its base. If the area is 105 square centimeters, what are the lengths of the base and altitude?

128. A side of one square is 1 inch longer than a side of another square. The sum of the areas is 113 square inches. Find the area of each square.

129. The perimeter of a square in meters and its area in square meters are expressed by the same number. Find the side of the square.

130. How wide a strip must be mowed around a rectangular grass field 40 feet wide and 60 feet long for two-thirds of the mowing to be done?

131. A rectangular garden plot is 30 meters long and 9 meters wide. A walk of uniform width is built around the plot. If the area of the walk is 74 square meters, how wide is the walk?

132. A rectangular sheet of metal is 3 times as long as it is wide. From each corner a 2-inch square is cut out and the ends are turned up to form a box. If the volume of the box is 512 cubic inches, find the original dimensions of the sheet of metal.

133. A side of one square is 1 centimeter longer than the side of another square, and the sum of the areas is 113 square centimeters. Find the dimensions of each square.

134. If an object is thrown downward with a velocity of v feet per second, the distance in feet that it will fall in t seconds is given by the formula

$d = 16t^2 + vt$. If a baseball is thrown downward with an initial velocity of 16 feet per second from the top of a building 320 feet high, how long will it take to reach the ground?

135. How long would it take the baseball in exercise 134 to reach the ground if it were dropped (initial velocity then is zero)?

136. The maximum depth of the Grand Canyon in Arizona is approximately 5488 feet. At a point where the depth is maximum, how long would it take a rock to fall to the bottom if it were possible to drop one off the edge? (See exercises 134 and 135.)

137. In the metric system the formula for a falling body is $d = 490t^2 + vt$, where d is the distance in centimeters, t is the time in seconds, and v is the initial velocity in centimeters per second. When a skydiver left his airplane, the altimeter read 3562.5 meters. If he fell freely to an altitude of 500 meters before he opened his parachute, how long did he fall? (1 meter = 100 centimeters.)

138. If you drop a stone from a cliff 122.5 meters high, how long would it be before you hear it strike the base of the cliff? (Sound travels approximately 330 meters per second.)

139. The formula $h = vt - 490t^2$ gives the height in centimeters at the end of t seconds if an object is projected straight up with an initial velocity of v centimeters per second. How long will it take an object to reach a height of 127.5 meters if it is projected upward with an initial velocity of 50 meters per second?

140. If an arrow is shot vertically upward into the air with an initial velocity of 29.7 meters per second, how long will it take the arrow to reach a height of 45 meters?

Section 9.5

In exercises 141–160 find the solution set of the given equation. Remember to check all candidates for solutions.

141. $\dfrac{12}{x} - \dfrac{12}{x + 2} = 3$

142. $\dfrac{x}{3} + \dfrac{3}{x} = \dfrac{5}{2}$

143. $\dfrac{x - 1}{2x} = \dfrac{2}{x + 3}$

144. $\dfrac{4}{a - 7} + \dfrac{18}{a + 3} = 2$

145. $\dfrac{2y + 1}{1 - 2y} = \dfrac{5y - 38}{14}$

146. $\dfrac{2t}{2t - 3} - \dfrac{2t - 3}{2t} = -\dfrac{3}{2}$

147. $\dfrac{x - 1}{2} - \dfrac{5}{2} = \dfrac{2}{1 - x}$

148. $\dfrac{a + 3}{a - 2} = \dfrac{15}{8} + \dfrac{a - 3}{a + 2}$

149. $\dfrac{7 - 3x}{x} + \dfrac{7}{2} = \dfrac{4x - 10}{x + 5}$

150. $\dfrac{3y - 5}{9y} - \dfrac{1}{3} = y$

151. $\dfrac{2x}{x-1} - \dfrac{19}{x^2-1} = \dfrac{x+3}{x+1}$

156. $y + \dfrac{5}{2y} + \dfrac{7}{2} = 0$

152. $\dfrac{1}{4} - \dfrac{1}{2(x-1)} = \dfrac{3}{x^2-1}$

157. $\dfrac{x}{x-5} - \dfrac{x-5}{x} = -\dfrac{5}{4}$

153. $\dfrac{1}{x^2-1} - \dfrac{x}{x+1} = \dfrac{2x-1}{x-1}$

158. $\dfrac{5}{b+2} + 4 = \dfrac{3}{b-2}$

154. $\dfrac{3}{x-6} + \dfrac{1}{2} = \dfrac{2}{x-5} + 1$

159. $\dfrac{x-7}{3x-7} - \dfrac{x+3}{2x+2} = \dfrac{7}{4}$

155. $\dfrac{8}{y-1} - \dfrac{4}{3y+1} = 1$

160. $x + 2 = \dfrac{2x-4}{3x-1}$

In exercises 161–175 solve the problem by:

(a) writing an equation for the situation;
(b) finding the solution set for the equation; and
(c) checking the candidates for solution in both the equation and the problem situation.

161. The time going 180 kilometers was 1 hour more than the time returning. Find the rate in each direction if the rate returning was 15 kilometers an hour more than the rate going.

162. The number of hours necessary to travel 1260 kilometers was 12 less than the rate in kilometers per hour. Find the rate.

163. A motorist travels 420 miles. Returning by the same route, she increases her speed 10 miles per hour and gets back in 2 hours and 40 minutes less time. How fast did she travel each way?

164. At what air speed must a plane fly to complete a round trip flight between two cities 600 miles apart in 5 hours if there is a 40 mile per hour head wind going and a 30 mile per hour tail wind coming back?

165. The speeds of two trains differ by 14 miles per hour. The slower train requires 2 hours longer to run 560 miles. Find the speed of each.

166. A man traveled 836 miles by train and 1210 miles by plane. The speed of the plane was 6 times that of the train. When he returned, because of a storm and track conditions, the speed of the train was 50 miles per hour less, and that of the plane was the same. The return journey took $12\frac{1}{2}$ hours more time. How fast did the train and plane travel each time?

167. Susan drives her car a distance of 150 kilometers and then returns, the round trip taking 8 hours and 45 minutes. Her speed returning was 10 kilometers per hour faster than her speed going. What was her speed on the trip out?

168. Two pumps can fill a tank in 2 hours and 24 minutes. One can fill the tank in 2 hours less time than the other. Find the number of minutes each pump requires to fill the tank alone.

169. Carla and Maria together require $4\frac{4}{5}$ days to do a certain job. Carla alone requires 4 days more than Maria. Find the number of days each would take to do it alone.

170. Ron and Linda can do a job together in 12 hours. It takes Ron 10 hours more than Linda to do the same kind of job. How long would it take Linda to do the job alone?

171. An electrician and her helper work on a job for 3 days. The electrician leaves and the helper finishes the job in 5 more days. If it takes the helper 3 more days than the electrician to do the job when each works alone, how long would it take the electrician alone?

172. At one time, if oranges cost 6¢ more a dozen, one would get 6 fewer oranges for 90¢. What was the price per dozen for oranges then?

173. The cost of a field trip was $24. Had there been 4 more in the group the cost would have been $1 less for each one. How many went on the field trip?

174. A resort owner rented a certain number of cabins for $288. When he reduced the price $2 per cabin he rented 4 more cabins, but the total receipts were $48 less. How many cabins did he rent at the higher rate?

175. The circumference of the rear wheel of a wagon is 1 meter more than that of the front wheel. One wheel makes 75 revolutions more than the other in traveling 900 meters. Find the circumference of each wheel.

Section 9.6

In exercises 176–197 find the solution set of the given equation. Be sure to check your candidates for solutions.

176. $\sqrt{x + 7} = 5$

177. $\sqrt{3x - 11} - 5 = 0$

178. $\sqrt{3y - 5} - 2 = 6$

179. $\sqrt{5x + 7} = \sqrt{7x + 5}$

180. $2\sqrt{3a + 1} = \sqrt{7a + 29}$

181. $\sqrt{3x + 7} = x - 1$

182. $x - 4 = \sqrt{x - 2}$

183. $2\sqrt{2b^2 + 112} = 3\sqrt{3b^2 - 3}$

184. $x - \sqrt{x^2 - 3} = 1$

185. $t - \sqrt{4t - 11} = 4$

186. $\sqrt{m + 4} = m - 8$

187. $11 + \sqrt{2x + 6} = 3x$

188. $\sqrt{x} - \sqrt{x - 5} = \sqrt{5}$

189. $\sqrt{x - 9} + \sqrt{x} = 1$

190. $\sqrt{x + 7} = \sqrt{x - 7}$

191. $\sqrt{x - 5} + \sqrt{x + 5} = 5$

192. $|8 + x| = 5$

193. $|y + 2| = 9$

194. $2|x + 5| = 6$

195. $\left|\dfrac{y + 1}{3}\right| = \dfrac{3}{4}$

196. $|3x + 5| + 2 = 7$

197. $2|x - 2| = |x|$

In exercises 198–207 translate the problem into a mathematical equation and then solve the equation to find the answer to the problem.

198. The square root of 10 times a number is equal to the number. Find the number.

199. Two more than a number is equal to the square root of 7 more than 6 times that number. What is the number?

200. One more than the square root of a number is equal to the square root of 1 less than 5 times the number. What is the number?

201. When 2 is subtracted from the square root of 1 more than 8 times a number, the result is the square root of 3 less than 4 times the number. What is the number?

202. The sum of a number and the square root of 5 less than twice the number is 4. What is the number?

203. If the square root of 3 more than 3 times a number is subtracted from the square root of 7 more than 6 times the same number, the result is 1. What is the number?

204. Find a number such that the square root of 5 more than 8 times the number is equal to 2 more than the square root of twice the number.

205. The square root of 6 more than 5 times a number added to the square root of 2 less than 3 times the number is equal to 6. What is the number?

206. What number, when 11 is added to the square root of 1 more than 3 times the number, is equal to 3 times the number?

207. Find a number such that the square root of 3 less than the number is equal to 15 less than the number.

Section 9.7

In exercises 208–221 (a) solve the inequality, (b) describe the solution set using set builder notation, and (c) graph each of the solution sets on a number line.

208. $x^2 - 4x + 3 < 0$

209. $x^2 + 5x + 4 > 0$

210. $y^2 + y - 6 \geq 0$

211. $2a^2 + 7a + 5 \leq 0$

212. $t^2 - 16 < 0$

213. $m^2 - 6m + 9 \leq 0$

214. $x^2 - 6x + 8 < 0$

215. $-x^2 + 2x + 3 \leq 0$

216. $5x < 2 - 3x^2$

217. $2(a + 1) \leq 4a^2$

218. $6(y^2 + 1) > 13y$

219. $-x^2 - 4x - 5 < 0$

220. $4x^2 + 1 \geq 4x$

221. $-x^2 + x \geq 0$

Section 9.8

In exercises 222–227 (a) solve the inequality, (b) describe the solution set using set builder notation, and (c) graph the solution set on a number line.

222. $|x| \geqslant 7$

223. $|x| + 5 \leqslant 7$

224. $|x + 1| < 10$

225. $|x - 3| \geqslant 13$

226. $|x + 1| \leqslant 5$

227. $|x + 5| < |5 - x|$

CUMULATIVE REVIEW G

In exercises 1–6 simplify the given expression.

1. $\left(\frac{7}{8}x^2 + \frac{2}{3}x - \frac{1}{2} \right) - \left(\frac{1}{4}x^2 - \frac{1}{4}x - 3 \right) = ?$

2. $\left(\frac{5}{6}xy \right) \left(-\frac{2}{3}x^2 z \right) = ?$

3. $\left(a + \frac{1}{2} \right) \left(a - \frac{1}{2} \right) = ?$

4. $\dfrac{4a^2 b + 4ab^2}{6a^3 b^2 + 6a^2 b^3} = ?$

5. $\dfrac{-18a^2 b^2 c^2 + 40a^2 bc - 2ab^2 c^2}{-2abc} = ?$

6. $\dfrac{2x^3 + 3x^2 - 8x + 3}{x + 3} = ?$

In exercises 7–12 factor the given polynomial completely.

7. $6a + 9 + 2ab + 3b$

8. $4xy^2 - 6xy + 2x^2 y$

9. $\frac{1}{2}a^2 + \frac{7}{2}a + 6$

10. $3k^2 - 7k - 20$

11. $x^4 - 16$

12. $27a^2 b - 9ab$

In exercises 13–16 perform the indicated operation.

13. $\dfrac{x^2 - 5x - 24}{x^2 + 8x + 15} \cdot \dfrac{x^2 + 4x - 5}{x^2 + x - 2} = ?$

14. $\dfrac{2x + 3}{5} + \dfrac{x - 1}{3} = ?$

15. $\dfrac{6x + 17}{x^2 + 5x + 6} - \dfrac{5}{x + 2} = ?$

16. $\dfrac{5x - 10}{4x + 2} \div \dfrac{5x + 15}{12x + 6} = ?$

In exercises 17–32 find the solution set of the given equation or inequality.

17. $4 - (4x - 2) = 27 - (8x - 3)$

18. $(x + 3)(x + 7) = (x - 2)(x + 3)$

19. $\dfrac{5}{y} - \dfrac{1}{4} = \dfrac{4}{-5}$

20. $(5x + 2) - (3x + 8) < 26 - (2x + 4)$

21. $-12x + 4 + 2x > - 6$

22. $\dfrac{3}{x - 4} + \dfrac{3}{3x - 12} = 2$

23. $\dfrac{3x + 2}{x - 3} = \dfrac{3x + 1}{x + 1}$

24. $\dfrac{3}{8x + 12} + \dfrac{5}{2x + 3} = \dfrac{3}{4}$

25. $x^2 + 5x - 6 = 0$

26. $3x^2 + 5x + 3 = 0$

27. $\dfrac{1}{x - 1} - \dfrac{1}{x + 1} = \dfrac{1}{24}$

28. $\dfrac{3}{5} - x > \dfrac{2}{3} + \dfrac{1}{2}x$

29. $x^2 - 6x + 8 < 0$

30. $|x + 4| \leqslant 4$

31. $\sqrt{x - 5} = \sqrt{x} + 2$

32. $1 + \sqrt{x} = \sqrt{x + 3}$

33. In the formula $R = T\left(\dfrac{S_1 S_2}{S_1 + S_2}\right)$, solve for S_1.

In exercises 34–36 solve the given system of equations.

34. $\begin{cases} 5x + 3y = -9 \\ 3x - 4y = -17 \end{cases}$

35. $\begin{cases} 5x - 8y = 60 \\ 6x + 7y = -11 \end{cases}$

36. $\begin{cases} \dfrac{2x}{3} - \dfrac{3y}{4} = -\dfrac{35}{12} \\[2mm] \dfrac{x}{4} - \dfrac{2y}{5} = -\dfrac{29}{20} \end{cases}$

In exercises 37–41 simplify the given expression, leaving only rational numbers in the denominators. Assume all variables are nonnegative.

37. $\sqrt{27y^3} + \sqrt{12y^3}$

38. $\sqrt[5]{-32} \cdot \sqrt[4]{\dfrac{16}{625}}$

39. $\dfrac{2}{\sqrt{3}}$ 41. $2\sqrt{6} \cdot 3\sqrt{10}$

40. $\dfrac{-\sqrt{50x}}{\sqrt{3}}$

42. Graph the following: $2x - 3y + 6 = 0$.

43. Graph the following: $f(x) = x^2 - 4x + 3$.

44. Three times the larger of two numbers exceeds 4 times the smaller by 3, and the sum of the numbers exceeds 5 times their difference by 4. What are the numbers?

45. A tank can be filled by two pipes, one running 6 hours and the other 12 hours; or by the same two pipes with the first running 9 hours and the second 8 hours. How long would it take each pipe alone to fill the tank?

46. A small plane at its usual speed took 8 hours less time for a trip of 480 miles than an automobile traveling at $\frac{1}{3}$ the speed of the plane. What was the speed of each?

47. The number of seats in each row of an auditorium is 10 less than the number of rows. How many seats are there in each row if there are 600 seats in the auditorium?

48. The length of a rectangle is 3 yards less than twice its width, and the perimeter is 60 yards. What is the length and width of the rectangle?

49. A wholesaler wants to make up a 150-pound mixture of bulk coffee that he can sell for 96¢ a pound. He has some coffee worth $1.28 a pound and some worth 80¢ a pound. How many pounds of each should he use in his mixture?

50. The denominator of certain fraction is 2 more than the numerator. If the numerator is increased by 7 and the denominator is increased by 1, the resulting fraction equals the reciprocal of the original fraction. What is the original fraction?

Table I Metric-English Measurement Equivalents

Length
1 centimeter (cm) = 0.39 inches
1 meter (m) = 1.09 yards
1 kilometer (km) = 0.62 miles
(100 cm = 1 m and 1000 m = 1 km)

Volume
1 liter (1) = 1.06 quarts

Weights
1 gram (g) = 0.007 ounces
1 kilogram (kg) = 0.45 pounds
(1000g = 1 kg)

Table II Table of Powers and Roots

n	n^2	$\sqrt{n}$	n	n^2	$\sqrt{n}$
1	1	1.0000	51	2601	7.1414
2	4	1.4142	52	2704	7.2111
3	9	1.7321	53	2809	7.2801
4	16	2.0000	54	2916	7.3485
5	25	2.2361	55	3025	7.4162
6	36	2.4495	56	3136	7.4833
7	49	2.6458	57	3249	7.5498
8	64	2.8284	58	3364	7.6158
9	81	3.0000	59	3481	7.6811
10	100	3.1623	60	3600	7.7460
11	121	3.3166	61	3721	7.8102
12	144	3.4641	62	3844	7.8740
13	169	3.6056	63	3969	7.9373
14	196	3.7417	64	4096	8.0000
15	225	3.8730	65	4225	8.0623
16	256	4.0000	66	4356	8.1240
17	289	4.1231	67	4489	8.1854
18	324	4.2426	68	4624	8.2462
19	361	4.3589	69	4761	8.3066
20	400	4.4721	70	4900	8.3666
21	441	4.5826	71	5041	8.4261
22	484	4.6904	72	5184	8.4853
23	529	4.7958	73	5329	8.5440
24	576	4.8990	74	5476	8.6023
25	625	5.0000	75	5625	8.6602
26	676	5.0990	76	5776	8.7178
27	729	5.1962	77	5929	8.7750
28	784	5.2915	78	6084	8.8318
29	841	5.3852	79	6241	8.8882
30	900	5.4772	80	6400	8.9443
31	961	5.5678	81	6561	9.0000
32	1024	5.6569	82	6724	9.0554
33	1089	5.7446	83	6889	9.1104
34	1156	5.8310	84	7056	9.1652
35	1225	5.9161	85	7225	9.2195
36	1296	6.0000	86	7396	9.2736
37	1369	6.0828	87	7569	9.3274
38	1444	6.1644	88	7744	9.3808
39	1521	6.2450	89	7921	9.4340
40	1600	6.3246	90	8100	9.4868
41	1681	6.4031	91	8281	9.5394
42	1764	6.4807	92	8464	9.5917
43	1849	6.5574	93	8649	9.6437
44	1936	6.6332	94	8836	9.6954
45	2025	6.7082	95	9025	9.7468
46	2116	6.7823	96	9216	9.7980
47	2209	6.8557	97	9409	9.8489
48	2304	6.9282	98	9604	9.8995
49	2401	7.0000	99	9801	9.9499
50	2500	7.0711	100	10,000	10.0000

ANSWERS

CHAPTER 1

1. $72°F$ 3. Chirps/min.

Chirps/min.	28	44	60	88	100	128	144	160
Temp. °F	47	51	55	62	65	72	76	80

5. $64°F$ 7. Warm, $86°F$

9.

Celsius	0	25	40	50	60	75	100
Fahrenheit	32	77	104	122	140	167	212

11. $40°C$

13. 15   15. 1220   17. There are several ways. One way is as follows: Fill the 3-qt. from the 8-qt.; pour 3-qt. into 5-qt.; fill 3-qt. from 8-qt.; fill 5-qt. from 3-qt.; pour 5-qt. into 8-qt.; pour 3-qt. into 5-qt.; fill 3-qt. from 8-qt.; pour 3-qt. into 5-qt.   19. 48 miles. It will take Sue and Bill 2 hrs. to meet. The fly travels 2 hrs. at 24 mph.
21. South Pole   23. 6 min. Engine must travel 2 miles before last car is out of tunnel. 2 miles at 20 mph gives us $\frac{1}{10}$ hr.   25. One possible solution is $1 + 1 + 3 + 3 + 3 + 3 + 3 + 3 = 20$. Another solution is $1 + 1 + 1 + 1 + 1 + 3 + 5 + 7 = 20$.   27. A 50¢ piece and a nickel. The 50¢ piece is not a nickel.   29. At a penny the first day and double your wage each day afterwards, the total salary for 3 weeks is $327.67. At $10.00 per day, the total salary for 3 weeks is $150.00.

CHAPTER 2

Exercises 2.1

1. 3764   3. $a + b = b + a$ 5. 628   7. 70,556
9. $a \times b = b \times a$ 11. 23,805   13. 2958
15. $a \times (b + c) = (a \times b) + (a \times c)$ 17. When $c = 0$ 19. 1196
21. When $a = b$ and $a, b \neq 0$ 23. $a + 0 = a$ 25. $a \times 1 = a$

Exercises 2.2

1. False   3. True: Associative property of addition   5. True: Associative property of addition   7. True: Commutative property of multiplication   9. True: Commutative property of multiplication
11. True: Distributive property of multiplication over addition

13. False   15. True: Multiplication property of 1   17. True:
Multiplication property of 1   19. True: Multiplication property of 0
21. True: Commutative property of addition   23. 9   25. 5
27. 30, 20   29. 8, 137   31. 3   33. 16   35. 350   37. 23
39. 128   41. 80   43. $4 \cdot (2 + 3) = (4 \cdot 2) + (4 \cdot 3)$
45. $(8 + 4) \cdot 9 = (8 \cdot 9) + (4 \cdot 9)$ 47. $3(a + b) = 3a + 3b$
49. $(13 \cdot 3) + (9 \cdot 3) = (13 + 9) \cdot 3$ 51. $(a \cdot 3) + (b \cdot 3) = (a + b) \cdot 3$

Exercises 2.3

1. $\dfrac{79}{33}$ 3. $\dfrac{5}{4}$ 5. $\dfrac{38}{15}$ 7. $\dfrac{1}{12}$ 9. $\dfrac{31}{60}$ 11. $\dfrac{5}{9}$ 13. $\dfrac{45}{8}$

15. $\dfrac{12}{5}$ 17. $\dfrac{14}{5}$ 19. $\dfrac{40}{9}$ 21.

			$\dfrac{13}{12}$
$\dfrac{5}{12}$	$\dfrac{11}{12}$		
			1
			$\dfrac{1}{12}$

23. $\dfrac{8}{3}, \dfrac{3}{4}, \dfrac{1}{3}$ 25. $6\frac{1}{4}, 4\frac{3}{8}, 10\frac{1}{2}$ 27. $0, 0, 0$

Exercises 2.4

1. $^-3$

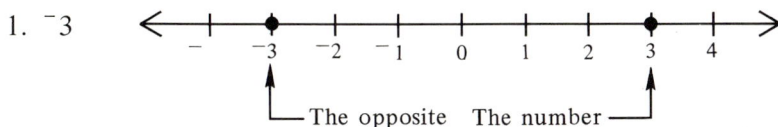
The opposite   The number

3. 5

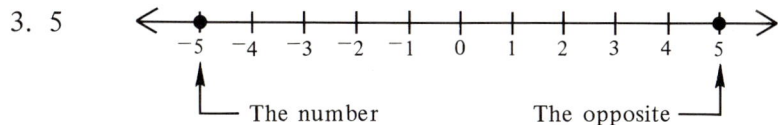
The number   The opposite

5. 0

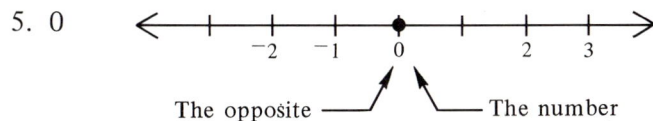
The opposite   The number

7. $\frac{3}{4}$

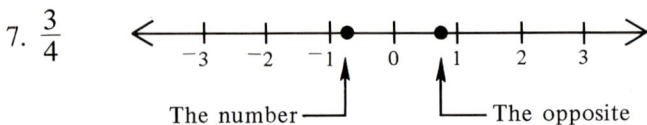

The number ⎯ The opposite

9. $^-\left(\frac{8}{5}\right)$

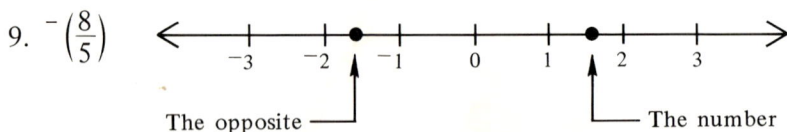

The opposite ⎯ The number

11. $6\frac{1}{5}$

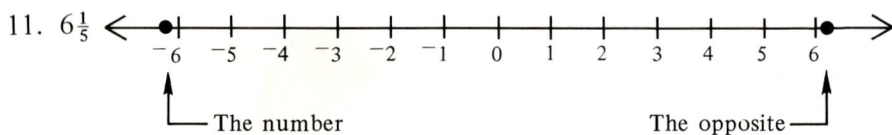

The number ⎯ The opposite

13. $^-1.5$

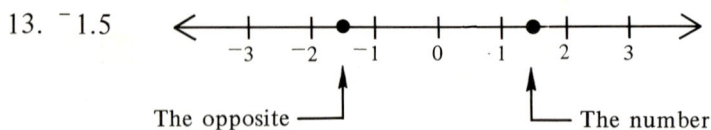

The opposite ⎯ The number

15. $^-8$

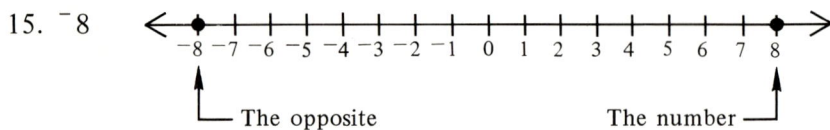

The opposite ⎯ The number

17. 5   19. $^-24$ 21. 0   23. $^-100$ 25. 13   27. $^-3\frac{1}{2}$

29. $^-\left(\frac{37}{5}\right)$ 31. (a) Negative, (b) positive, (c) 0   33. Yes

35. Yes   37. No   39. Yes

Exercises 2.5

1. 5   3. 3   5. 1   7. -34   9. -4   11. $-\frac{1}{12}$ 13. $-\frac{17}{12}$

15. $-\dfrac{5}{2}$ 17. $-\dfrac{22}{5}$ 19. $-\dfrac{4}{3}$ 21. 6   23. -6 25. 1

27. -5 29. -1 31. -2 33. $-\dfrac{1}{2}$ 35. 0   37. 0   39. $-\dfrac{24}{5}$

41. -5 43. -8 45. 0   47. 0   49. $-6, 58$

Exercises 2.6

1. -3 3. 0   5. $\dfrac{1}{2}$ 7. $-\dfrac{4}{5}$ 9. $-\dfrac{3}{4}$ 11. $\dfrac{8}{7}$ 13. $-\dfrac{2}{3}$ 15. $\dfrac{5}{3}$

17. 5   19. $-\dfrac{3}{4}$ 21. (a) Negative integer, (b) positive integer, (c) 0

23. You don't know until you know the replacement for a (see #21).

25. 6   27. 30   29. -30 31. 20   33. $\dfrac{7}{20}$ 35. $-\dfrac{11}{10}$

37. $-7\dfrac{3}{4}$ 39. $\dfrac{1}{4}$ 41. $\dfrac{17}{12}$ 43. $-\dfrac{1}{4}$

Exercises 2.7

1. 20   3. -21 5. 20   7. 0   9. 0   11. -1 13. $-\dfrac{1}{8}$

15. $-\dfrac{5}{7}$ 17. $\dfrac{24}{25}$ 19. $\dfrac{1}{2}$ 21. -30 23. -300 25. 4   27. 0

29. $\dfrac{1}{8}$ 31. -36 33. -20 35. -9 37. -12 39. -2

41. 0   43. -10 45. 72   47. $\dfrac{-13}{2}, -\dfrac{13}{2}$ 49. $-\dfrac{5}{3}, \dfrac{5}{-3}$

51. False, $-\dfrac{15}{5} = \dfrac{15}{-5}$ or $-\dfrac{15}{5} = \dfrac{-15}{5}$ 53. True   55. True   57. $\dfrac{1}{5}$

59. $-\dfrac{1}{27}$ 61. Does not exist.   63. $-\dfrac{5}{8}$

Exercises 2.8

1. $\dfrac{4}{9}$ 3. $-\dfrac{3}{4}$ 5. $-\dfrac{77}{48}$ 7. $-\dfrac{8}{7}$ 9. -20 11. $-\dfrac{22}{5}$ 13. $\dfrac{3}{4}$

15. $-\dfrac{21}{40}$

Exercises 2.9

1. Commutative property of multiplication   3. Commutative property of addition   5. Multiplication property of 1   7. Additive inverse property   9. Commutative property of addition
11. Distributive property   13. Commutative property of addition
15. Multiplicative inverse property   17. True: Commutative property of addition   19. True: Associative property of multiplication
21. True: Commutative property of addition   23. True: Distributive property   25. True: Commutative property of multiplication
27. True: Multiplicative inverse property   29. True: Distributive property   31. True: Commutative property of addition   33. True: Commutative property of multiplication   35. True: Distributive property   37. $14xy$ 39. $225abc$ 41. $\frac{3}{8}pq$ 43. $\frac{1}{8}xy$

45. abc 47. $36 + 9m$ 49. $36 + 9m$ 51. $\frac{1}{3}x + \frac{1}{2}$

53. $6x + 8y + 4z$ 55. $abx + 2aby$ 57. $11y$ 59. $\frac{3}{2}y$

61. $5(3a + 4b)$ 63. $3(2x + 3y + 4z)$ 65. $2x$ 67. $4p$
69. $a(bc + bd + cd)$

Exercises 2.10

1. 0.35   3. 1.08   5. 0.285   7. 0.425   9. 0.09375   11. $0.\overline{5}$; 9 has a factor of 3   13. $0.\overline{142857}$; 7 has a factor of 7   15. $0.8\overline{3}$; 6 has a factor of 3   17. $0.\overline{8}$; 9 has a factor of 3   19. $0.5\overline{71428}$; 7 has a factor of 7   21. $0.3\overline{5}$ 23. $0.\overline{213}$ 25. $0.1\overline{3}$ 27. $0.\overline{5005}$

29. $0.00\overline{5}$ 31. $\frac{1}{3} = 0.\overline{3}, \frac{2}{3} = 0.\overline{6}, \frac{1}{3} + \frac{2}{3} = 1, 0.\overline{3} + 0.\overline{6} = 0.\overline{9}$; therefore $0.\overline{9} = 1$ 33. 36 maximum in repeating cycle, $0.\overline{702}$; 3 digits in the repeating cycle

35. (a) 1. $2 \cdot 2 \cdot 5$
           2. $2 \cdot 2 \cdot 2$
           3. $5 \cdot 5$
           4. $2 \cdot 2$
           5. $2 \cdot 2 \cdot 2 \cdot 5 \cdot 5$
           6. $5 \cdot 5 \cdot 5$
           7. $2 \cdot 2 \cdot 2 \cdot 5$
           8. $2 \cdot 2 \cdot 2 \cdot 2$
           9. $2 \cdot 2 \cdot 2 \cdot 2 \cdot 2$
          10. $2 \cdot 2 \cdot 2 \cdot 2 \cdot 5$

     (b) 11. $3 \cdot 3$
          12. 11
          13. 7
          14. 3
          15. $2 \cdot 3$
          16. $2 \cdot 2 \cdot 3$
          17. $3 \cdot 3$
          18. $2 \cdot 2 \cdot 2 \cdot 2 \cdot 3$
          19. 7
          20. 37

(c) If the denominator of a fraction in lowest terms has 1, 2, or 5 as its only factors, then the decimal will be a terminating decimal.

Exercises 2.11

1. $a < b, c > d$ 3. $<$ 5. $>$ 7. $>$ 9. $>$ 11. $>$ 13. $<$
15. $>$ 17. True 19. True 21. True 23. True 25. False
27. $-4 < -3, -12 < -10$ 29. $0 > -7, -9 < 12$

31. $-1 > \dfrac{-9}{8}, \dfrac{1}{4} < \dfrac{5}{16}$ 33. $2x - 1 > -3x - 1, -2x < 3x$

35. $2x - 7 < 3x, -2x + 10 > -6x - 4$ 37. $9 > c + 2$; Addition property of inequalities 39. $m - 2 < n - 2$; Addition or subtrac-

tion property of inequalities 41. $-3, -\dfrac{20}{7}, -\dfrac{5}{2}, -\dfrac{37}{41}, 0, \dfrac{2}{5}, 0.41, \dfrac{4}{5}$,

$\dfrac{49}{60}, \dfrac{5}{6}, \dfrac{7}{3}, 2.3$ 43. Smallest, -156; largest, 434. 45. Smallest,

-47; no largest. 47. Smallest 1; no largest. 49. No smallest, no largest.

Exercises 2.12

1. 5 3. $-\dfrac{1}{2}$ 5. 5 7. 12 9. 0 11. -15 13. 5 15. 2

17. Positive, negative 19. 4 21. -10 23. $-\dfrac{1}{4}$ 25. 0

27. -17 29. 29 31. False 33. True 35. True 37. False
39. True 41. $=$ 43. $<$ 45. $=$ 47. $=$ 49. $=$ 51. $\leqslant$

Review Exercises: Chapter 2

1. True: Commutative property of addition 3. True: Commutative property of multiplication 5. True: Associative property of addition 7. True: Distributive property 9. True: Distributive property
11. $9(x + y)$ 13. $5a$ 15. $8y + 24$ 17. $4c + 4d$

19. $2x + 3x$ 21. $\dfrac{1}{14}$ 23. $\dfrac{74}{55}$ 25. $\dfrac{3}{20}$ 27. $\dfrac{3}{5}$ 29. $\dfrac{15}{32}$

31. 7 33. 3 35. $-\dfrac{3}{4}$ 37. 2 39. -2 41. $-\dfrac{7}{6}$ 43. $2\frac{11}{12}$

45. $\dfrac{58}{55}$ 47. $\dfrac{41}{63}$ 49. $-\dfrac{47}{40}$ 51. $\dfrac{5}{28}$ 53. $\dfrac{47}{24}$ 55. $-\dfrac{5}{56}$

57. -35 59. 35 61. $\dfrac{2}{5}$ 63. $-\dfrac{3}{4}$ 65. -6 67. $\dfrac{5}{-6}$ or $-\dfrac{5}{6}$

69. $\dfrac{2}{3}$ or $-\dfrac{-2}{3}$ or $-\dfrac{2}{-3}$ 71. $-\dfrac{15}{28}$ 73. -1 75. $\dfrac{16}{9}$ 77. $-\dfrac{121}{49}$

79. $-\dfrac{4}{3}$ 81. Associative property of addition   83. Associative property of multiplication   85. Distributive property
87. Commutative property of multiplication   89. Commutative property of addition   91. $9a + 9b$ 93. $x + 2y$ 95. $5(c + 2d)$

97. $\dfrac{1}{2}(a + 4b)$ 99. $4ab + 6ac$ 101. $\dfrac{4}{5}x + \dfrac{3}{4}y$ 103. $abx + aby$

105. $3a(2c + 3d)$ 107. Repeating, $0.41\overline{6}$ 109. Repeating, $0.\overline{571428}$ 111. Terminating, 0.1875 113. Terminating, 1.6
115. Terminating, 0.625 117. $=$ 119. $>$ 121. $>$ 123. $<$
125. $>$ 127. $3 < 6, -18 > -27$ 129. $-8 < -7, 15 > 10$

131. $\dfrac{10}{12} > \dfrac{9}{12}, \dfrac{1}{16} > \dfrac{1}{18}$ 133. $x + 2 > y + 2, 2x > 2y$

135. $x + 4 < a + 6, 2x + 4 < 2a + 8$ 137. 5 139. $-\dfrac{1}{4}$ 141. 7

143. $\dfrac{1}{5}$ 145. -1

CHAPTER 3

Exercises 3.1

1. False, $7 + 8 < 17$ 3. False, $9 + 8 = 8 + 9$ 5. False, $-3 \times -6 = 15 + 3$ 7. True   9. True   11. False,

$\dfrac{1}{8} \div \dfrac{2}{7} < \left(\dfrac{1}{2} \times 3\right) \div 2$ 13. False, $6.07 + 7.006 < 13.13$ 15. $\{7\}$

17. $\{1, 2, 3, 4, 5, 6, 7, 8, 9\}$ 19. $\{9\}$ 21. $\{1, 2, 3, 4, 6, 7, 8, 9\}$
23. $\{7, 8, 9\}$ 25. $\{1, 2, 3, 4, 5, 6, 7\}$ 27. $\{1, 2\}$
29. $\{1, 2, 3, 4, 5\}$ 31. $\{5\}$ 33. $\{-5, -4, -3, -2\}$
35. $\{-2, -1, 0, 1, 2, 3, 4, 5\}$ 37. $\{-5, -4, -3, -2\}$ 39. $\varnothing$ 41. $\varnothing$
43. $\{2\}$

Exercises 3.2

1. 7   3. 10   5. 6   7. 20   9. 21   11. 2   13. 3   15. 2

17. 4   19. −1   21. 18   23. 6   25. $\frac{11}{5}$ 27. 12   29. −2

31. 20   33. 6   35. 9   37. 2   39. 0   41. 21   43. 3

45. $\frac{1}{13}$ 47. 1   49. $6\frac{3}{5}$ 51. 9   53. 2   55. 398,677.6

57. 4.802744   59. 0.21601   61. 159.73775   63. 9.9618433
65. 104.07993

Exercises 3.3

1. $\frac{4}{9}$ 3. $\frac{7}{9}$ 5. $\frac{8}{11}$ 7. $\frac{1}{11}$ 9. $\frac{15}{37}$ 11. $\frac{32}{9}$ 13. $\frac{1}{110}$

15. $\frac{89}{4950}$ 17. $\frac{5}{18}$ 19. $\frac{76033}{166500}$ 21. $0.\bar{7}, 0.\overline{707}, 0.\overline{70}, 0.707,$
0.70

Exercises 3.5

1. (*b*)   3. (*c*)   5. (*a*)   7. (*i*)   9. (*j*)   11. (*d*)   13. (*f*)
15. (*a*)   17. (*i*)   19. (*b*)   21. $5x + 3$ 23. $x - 2$

25. $5(x + y)$ 27. $xy + 4$ 29. $\frac{x + y}{2}$ 31. $2x + 12$ 33. $5x$

35. $60x$ 37. $\frac{x + 10}{2}$ 39. $2n(x + y)$ 41. $50 - h$ 43. $\frac{540}{n}$

45. $330h$ 47. $2n + 1$ or $2n - 1$ 49. $2n + (2n + 2)$ or $2n + (2n - 2)$

Exercises 3.6

1. (*e*)   3. (*i*)   5. (*b*)   7. (*c*)   9. (*j*)   11. $1.15, knife, 35¢;
ball costs 40¢, knife costs 75¢   13. 210, 18, wide; length is $61\frac{1}{2}$ feet,
width is $43\frac{1}{2}$ feet   15. $6000, 4%, 6%, $320; $2000 at 4%, $4000
at 6%   17. 2, 16, units'; the number is 64   19. 0.80, 1.80, 10,
1.10; 7 kilograms of $0.80 candy, 3 kilograms of the $1.80 candy

Exercises 3.7

1. 4, 12   3. 6, 10   5. 7 dimes, 13 quarters   7. 6 quarters, 8 dimes, 12 nickels   9. 7 cm, 21 cm   11. 13 m., 31 m.
13. 24 Holsteins, 16 Jerseys   15. $24,000   17. Jack, 13 yrs.; Sister, 9 yrs.   19. Father, 36 yrs.; Mother, 34 yrs.   21. 10, 5
23. 12 pennies, 9 nickels, 14 dimes   25. Father, 36; Son, 12
27. 158, 40   29. 26 quarters, 25 nickels, 14 dimes
31. Mother, 47; Felicia, 22   33. 150 feet, 1452 feet
35. 20 dimes, 30 nickels   37. 22 cm, 88 cm, 82 cm   39. 8 yrs.

Exercises 3.8

1. $7\frac{1}{5}$ min.   3. 1 hr., 12 min.   5. 1 hr., 12 min.   7. 600 g.
9. 7 lb.   11. 2.8 hrs.   13. 5 hrs., 70 miles   15. 55 km/hr., 66 km/hr.   17. 8%—$4000, 5%—$4500   19. 9%—$2940, 7%—4410, 6%—$1470   21. 36   23. 72   25. 36   27. 11 mph
29. $5000   31. 10:00 A.M.   33. 28.2 gal.   35. 2 qts.
37. 6 hrs.   39. 26.6 kg of $1.38 coffee, 43.4 kg of $2.38 coffee
41. $13\frac{1}{3}$ lbs.   43. $3750   45. 4 hrs.   47. $8000, $1500
49. 6.4 liters

Exercises 3.9

1. $d = \dfrac{C}{3.1416}$ 3. (a) $n = \dfrac{C}{p}$, (b) $p = \dfrac{C}{n}$ 5. (a) $l = \dfrac{V}{wh}$,

(b) $w = \dfrac{V}{lh}$, (c) $h = \dfrac{V}{lw}$,   7. (a) $W = EM$, (b) $E = \dfrac{W}{M}$ 9. $a = \dfrac{2S}{t^2}$

11. $p = \dfrac{C - 5}{5}$ 13. (a) $f = \dfrac{3T}{2w}$, (b) $w = \dfrac{3T}{2f}$ 15. (a) $w = \dfrac{2gK}{V^2}$,

(b) $g = \dfrac{wV^2}{2K}$ 17. (a) $E = IR + e$, (b) $R = \dfrac{E - e}{I}$

19. (a) $K = \dfrac{C(b - a)}{ab}$, (b) $a = \dfrac{Cb}{Kb + C}$ 21. (a) $V_t = 2V - V_0$,

(b) $V_0 = 2V - V_t$ 23. (a) $g = \dfrac{V^2 - V_0^2}{2h}$, (b) $h = \dfrac{V^2 - V_0^2}{2g}$

25. $F = \dfrac{9}{5} C + 32$ 27. (a) $A = P(1 + ni)$, (b) $n = \dfrac{A - P}{Pi}$

29. $l^2 = 6dR - 3d^2$ 31. (a) $n = Pod - 2$, (b) $d = \dfrac{n+2}{Po}$

33. (a) $E = Ir + \dfrac{IR}{n}$, (b) $r = \dfrac{E - \dfrac{IR}{n}}{I}$, (c) $R = \dfrac{n(E - Ir)}{I}$

Exercises 3.10

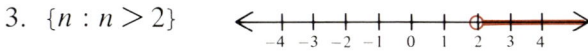

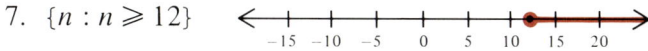

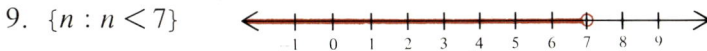

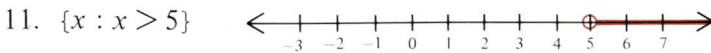

1. $\{x : x < 7\}$

3. $\{n : n > 2\}$

5. $\{y : y \leqslant 18\}$

7. $\{n : n \geqslant 12\}$

9. $\{n : n < 7\}$

11. $\{x : x > 5\}$

13. $\{n : n \geqslant 10\}$

15. $\{x : x < 20\}$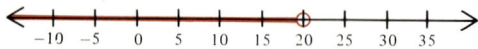

17. $\{n : n > -11\}$

19. $\{x : x \geqslant 6\}$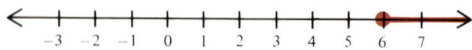

21. $\left\{n : n < \dfrac{1}{4}\right\}$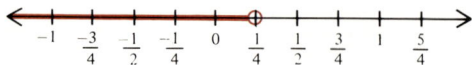

23. $\{x : x > 3\frac{2}{3}\}$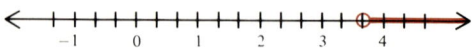

25. $\{x : x < 2\}$

27. $\{x : x < 7\}$

29. $\{x : x \geqslant -120\}$

31. $\{x : x > -34\frac{1}{8}\}$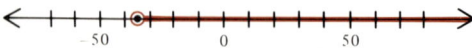

33. $\{x : x < -4\frac{8}{33}\}$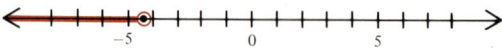

35. $\{x : x > -14\}$

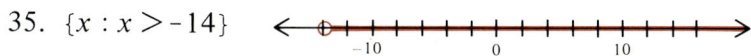

37. $\left\{y : y > -\dfrac{7}{10}\right\}$

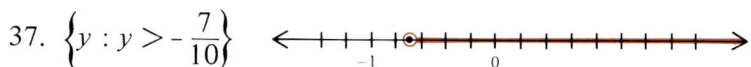

39. $\{x : x > 3\}$

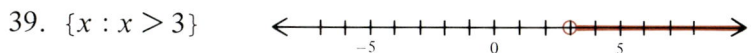

41. $\{x : x > 57.548148\}$ 43. $\{w : w < 26.231958\}$
45. $\{x : x \geqslant -5.096828\}$ 47. $\{x : x \leqslant 2.651712\}$

Review Exercises: Chapter 3

1. False, $>$ 3. True   5. True   7. False, $<$ 9. False, $<$
11. $\{3\}$ 13. $\{4, 5, 6, 7, 8, 9\}$ 15. $\varnothing$ 17. $\{8, 9\}$

19. $\{1, 2, 3, 4, 5, 6, 7, 8\}$ 21. $\{4\}$ 23. $\{6\}$ 25. $\left\{-\dfrac{1}{2}\right\}$

27. $\{-4\}$ 29. $\left\{\dfrac{1}{2}\right\}$ 31. $\{-10\}$ 33. $\left\{-\dfrac{5}{4}\right\}$ 35. $\{1\}$

37. $\{-8\}$ 39. $\{8\}$ 41. $\{4\}$ 43. $\{3\}$ 45. $\{10\}$ 47. $\{-21\}$

49. $\{2.6\}$ 51. $\dfrac{7}{9}$ 53. $\dfrac{4}{33}$ 55. $6\frac{2}{9}$ 57. $\dfrac{1706}{333}$ 59. $\dfrac{20,621}{4950}$

61. $x + 11$ 63. $3x$ 65. $x + y$ 67. $2x + 8$ 69. $\dfrac{2}{3}x + 5$

71. $10x$ 73. $\dfrac{N - 6}{3}$ 75. $52n$ 77. $175, 176, 177$

79. $40\frac{2}{3}$ ft., $89\frac{1}{3}$ ft.   81. 8 quarters, 44 dimes   83. Horse, $250; carriage, $175   85. 15 hrs., 45 min.   87. 4 hrs., 40 min.

89. $33\frac{1}{3}$ kg   91. $33\frac{1}{3}$ gal.   93. 40 min.   95. 36   97. $c = \dfrac{1}{2\pi f X}$

99. $R = \dfrac{wv}{Fg}$ 101. $h = \dfrac{3V}{\pi r^2}$ 103. $r = \dfrac{S - a}{S}$ 105. $M = \dfrac{E}{c^2}$

107. $t = \dfrac{4.18H}{Ei}$ 109. $b = \dfrac{Sa}{a - S}$ 111. $w = fk^2 + k$

113. $\left\{ x : x > -\dfrac{8}{5} \right\}$

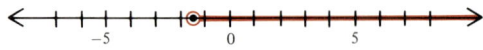

115. $\{ y : y \leqslant 63 \}$

117. $\{ x : x < 4 \}$

119. $\left\{ x : x > \dfrac{5}{7} \right\}$

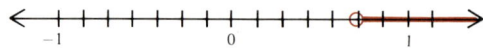

121. $\{ k : > -10 \}$

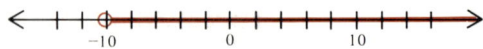

123. $\{ a : a \leqslant 1 \}$

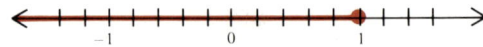

125. $\{ x : x < 1 \}$

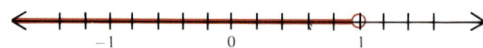

127. $\{ x : x < -1\frac{23}{32} \}$

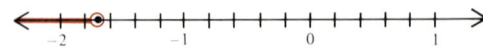

129. $\{ y : y > 6 \}$

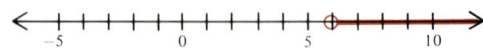

131. $\{ z : z \geqslant -2 \}$

Cumulative Review A

1. $-\dfrac{1}{24}$ 3. $\dfrac{41}{72}$ 5. $-\dfrac{46}{49}$ 7. $13x$ 9. $15ab^2$ 11. $3a + 3b$

13. $11a + 11b$ 15. $10x^2 y^2 + 8x^2 yz$ 17. Commutative property

of multiplication   19. 75   21. $\dfrac{14}{99}$ 23. $\{-4\}$ 25. $\{-1\}$

27. $\{2\}$ 29. $\varnothing$ 31. $\{-5\}$ 33. $\{0\}$

35. $\{x : x > -6\}$

37. $\{x : x \geqslant 2\}$

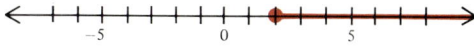

39. $\{x : x \leqslant 1\}$

41. $\{x : x > -3\}$

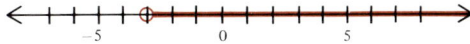

43. 21   45. 24 m, 6 m   47. Small cake, 5 people; Large cake,
10 people   49. $106\frac{2}{3}$ lb. of \$1.80 candy, $53\frac{1}{3}$ lb. of \$2.25 candy.

CHAPTER 4

Exercises 4.1

1. $\{(-2, 4), (-1, 3), (0, 2), (1, 1), (2, 0)\}$
3. $\{(-2, -4), (-1, -3), (0, -2), (1, -1), (2, 0)\}$
5. $\{(-2, -2), (-1, -1)\ (0, 0), (1, 1), (2, 2)\}$
7. $\{(-2, -3), (-1, -2), (0, -1), (1, 0), (2, 1)\}$
9. $\{(-2, -3), (-1, -2), (0, -1), (1, 0), (2, 1)\}$
11. $\{(-2, 7), (-1, 6), (0, 5), (1, 4), (2, 3)\}$
13. $\{(-2, -4), (-1, -2), (0, 0), (1, 2), (2, 4)\}$
15. $\{(-2, 0), (-1, 1), (0, 2), (1, 3), (2, 4)\}$
17. $\left\{(-5, 6), (-2, 3), (0, 1), \left(\dfrac{3}{2}, -\dfrac{1}{2}\right), (7, -6)\right\}$

19. $\left\{(-5, 5), (-2, 2), (0, 0), \left(\frac{3}{2}, -\frac{3}{2}\right), (7, -7)\right\}$

21. $\left\{(-5, -16), (-2, -10), (0, -6), \left(\frac{3}{2}, -3\right), (7, 8)\right\}$

23. $\left\{(-5, 15), (-2, 6), (0, 0), \left(\frac{3}{2}, -\frac{9}{2}\right), (7, -21)\right\}$

25. $\left\{\left(-5, -\frac{5}{3}\right), \left(-2, -\frac{2}{3}\right), (0, 0), \left(\frac{3}{2}, \frac{1}{2}\right), \left(7, \frac{7}{3}\right)\right\}$

27. $\left\{\left(-5, -\frac{5}{2}\right), (-2, -1), (0, 0), \left(\frac{3}{2}, \frac{3}{4}\right), \left(7, \frac{7}{2}\right)\right\}$

29. $\left\{(-5, -30), (-2, -12), (0, 0), \left(\frac{3}{2}, 9\right), (7, 42)\right\}$

31. Equivalent   33. Not equivalent, (1, 2)   35. Equivalent
37. Equivalent   39. Not equivalent, (1, 2)   41. Addition, 3
43. Multiplication, 2   45. Addition, $-y$ 47. Multiplication, 2
49. Multiplication, 12

Exercises 4.2

1. $(-9, 9)$ 3. $(-7, 5)$ 5. $(-3, -2)$ 7. $(-3, -7)$ 9. $(10, -10)$
11. $(0, -6)$ 13. $(8, 0)$ 15. $(10, 3)$ 17. $(6, 6)$ 19. $(0, 9)$

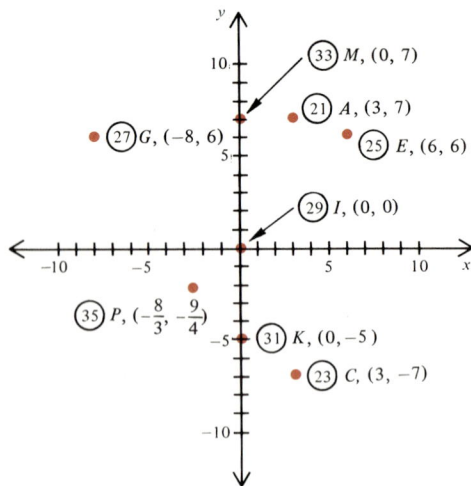

37. 2   39. 3   41. Positive, *x*   43. Negative, *x*   45. Negative, positive   47. Graphs will vary with each individual

Exercises 4.3

1.

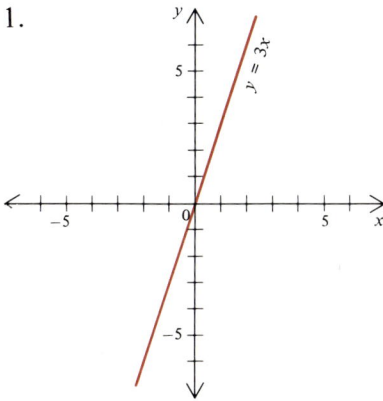

3.

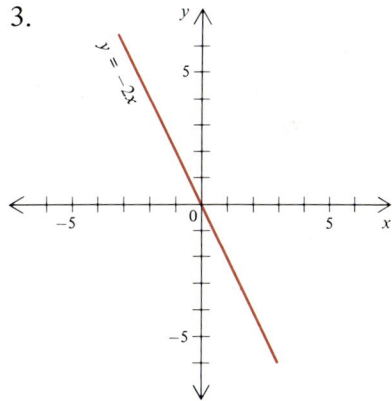

5.

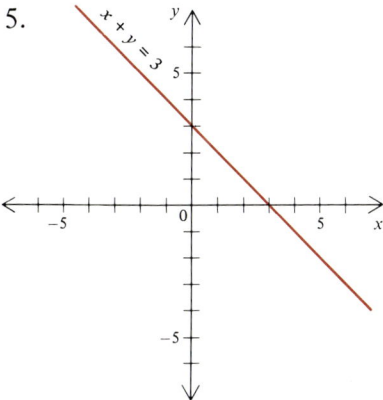

7.

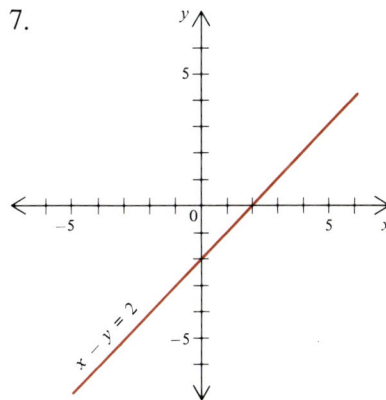

9.

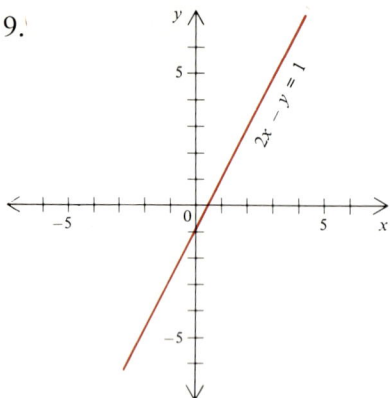

$2x - y = 1$

11.

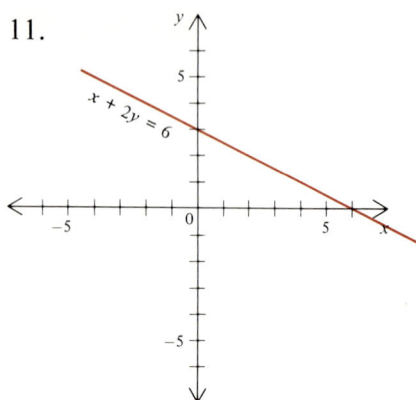

$x + 2y = 6$

13.

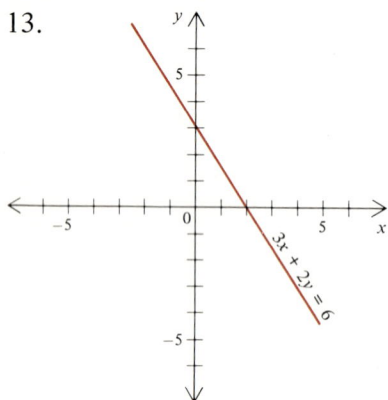

$3x + 2y = 6$

15.

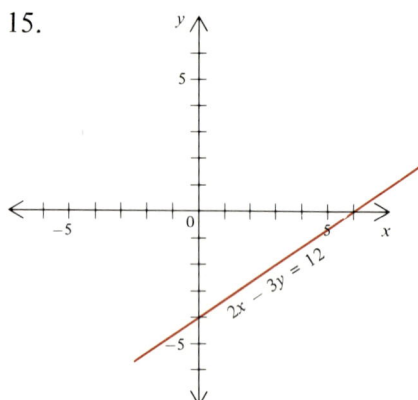

$2x - 3y = 12$

17.

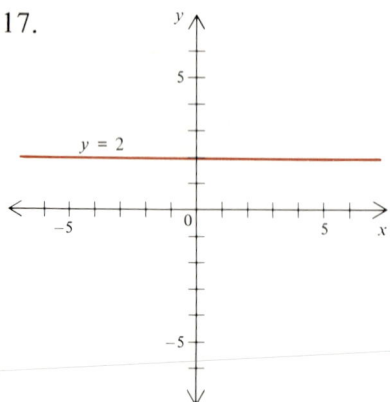

$y = 2$

19.

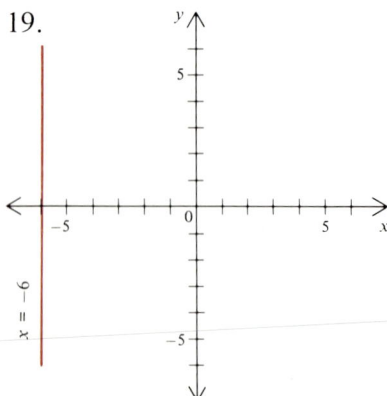

$x = -6$

21.

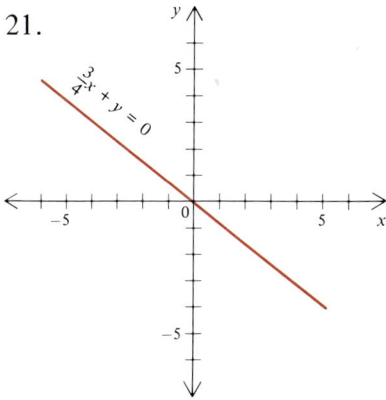

$\frac{3}{4}x + y = 0$

23.

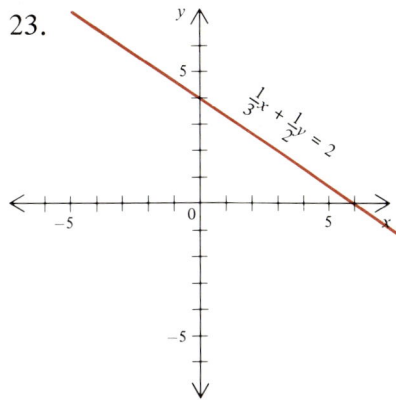

$\frac{1}{3}x + \frac{1}{2}y = 2$

25.

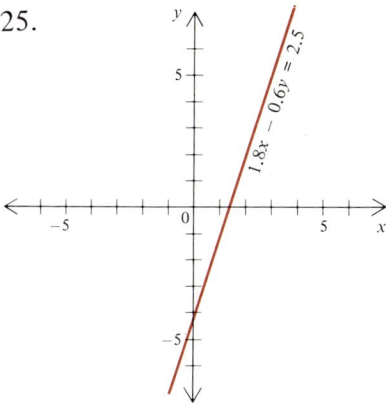

$1.8x - 0.6y = 2.5$

27.

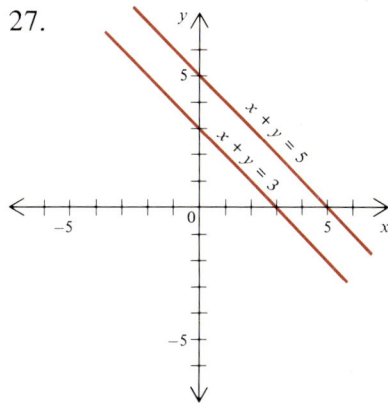

$x + y = 5$

$x + y = 3$

29.

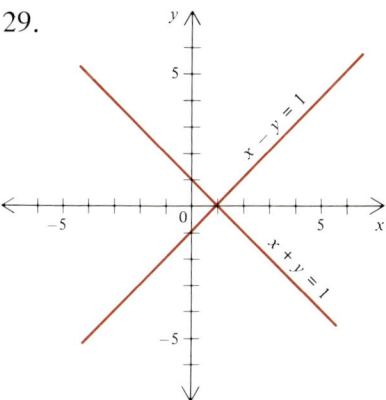

$x - y = 1$

$x + y = 1$

31.

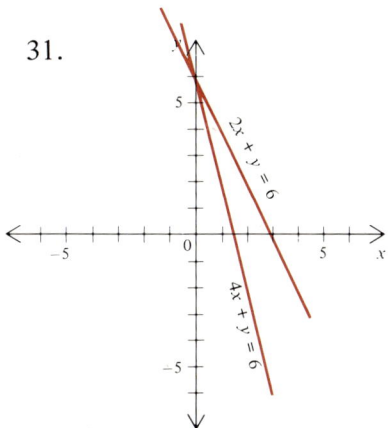

$2x + y = 6$

$4x + y = 6$

33.

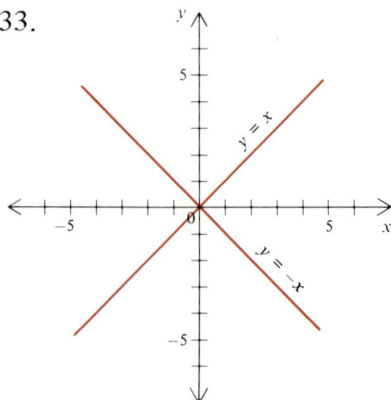

35. #26, $(-2, -6)$; #27, no common solution; #28, many common solutions; #29, $(1, 0)$; #30, $(1, 1)$; #31, $(0, 6)$; #32, $(-6, 3)$; #33, $(0, 0)$; #34, no common solution   37. #26, #29, #30, #31, #32, #33. The lines intersect at one point.   39. #27, #34. The lines are parallel.

41. (a)

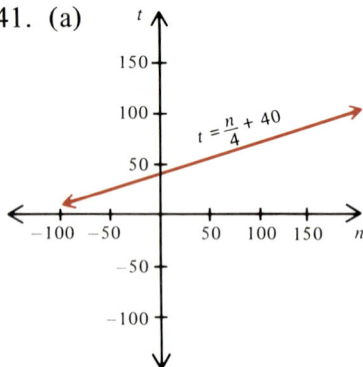

(b) When n is negative. A cricket cannot chirp a negative number of times. Also, at a very low or very high temperature, a cricket would cease to chirp.

Exercises 4.4

1.

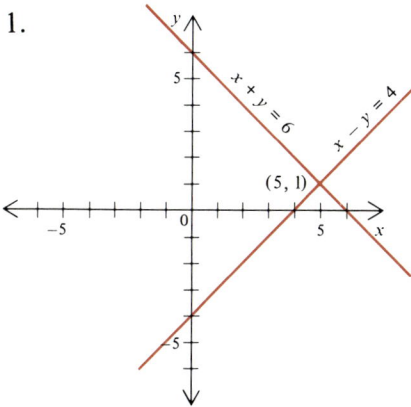

$x + y = 6$
$x - y = 4$
$(5, 1)$

3.

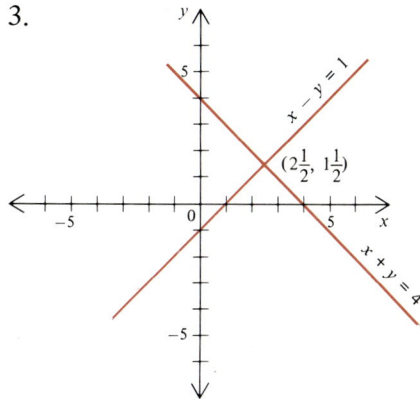

$x - y = 1$
$x + y = 4$
$(2\frac{1}{2}, 1\frac{1}{2})$

5.

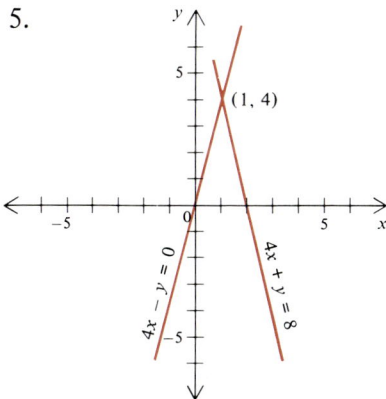

$(1, 4)$
$4x - y = 0$
$4x + y = 8$

7.

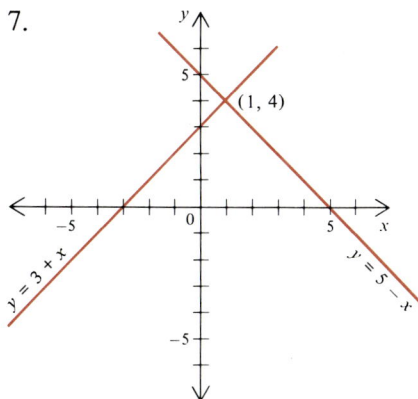

$(1, 4)$
$y = 3 + x$
$y = 5 - x$

9.

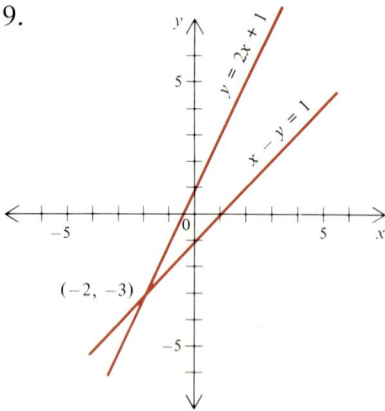

$y = 2x + 1$

$x - y = 1$

$(-2, -3)$

11.

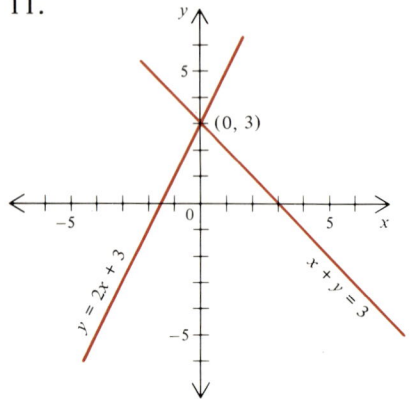

$(0, 3)$

$y = 2x + 3$

$x + y = 3$

13.

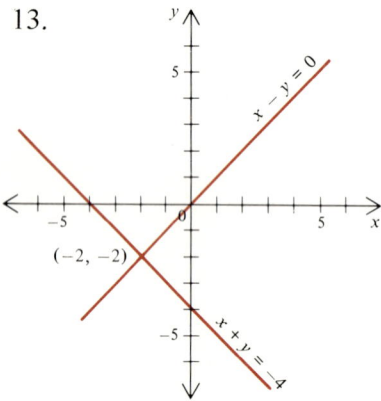

$x - y = 0$

$(-2, -2)$

$x + y = -4$

15.

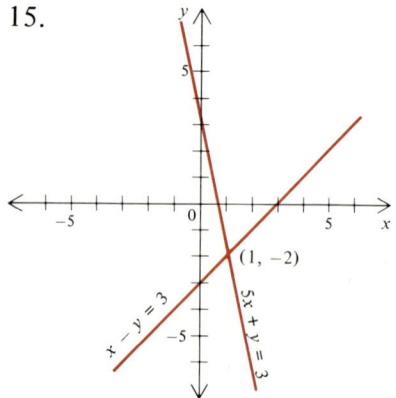

$(1, -2)$

$x - y = 3$

$5x + y = 3$

17.

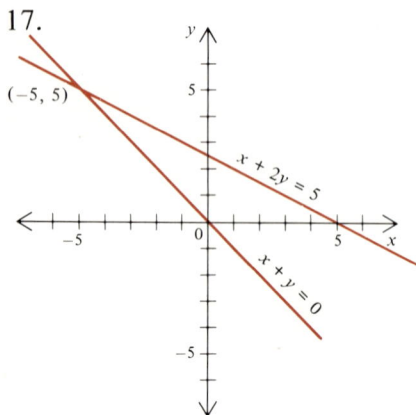

$(-5, 5)$

$x + 2y = 5$

$x + y = 0$

19.

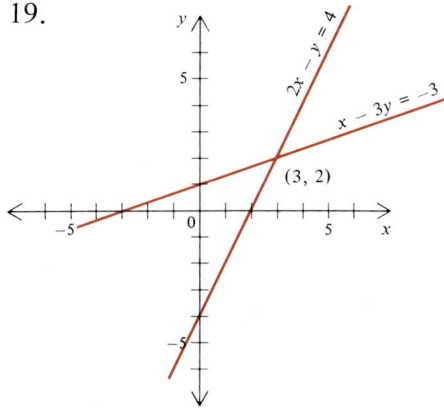

21.

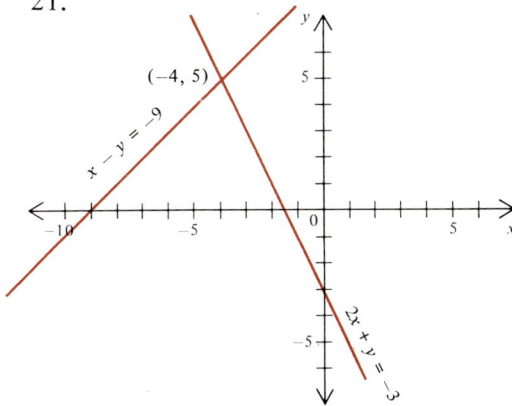

23.

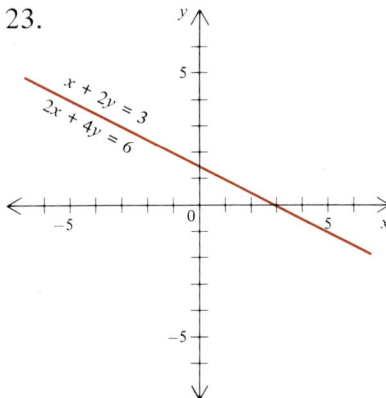

Any solution of one is a solution of the other.

25.

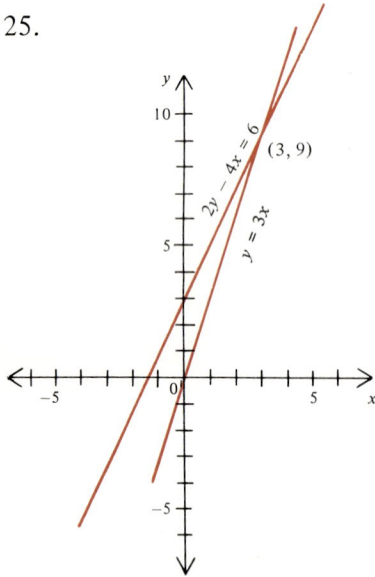

27.

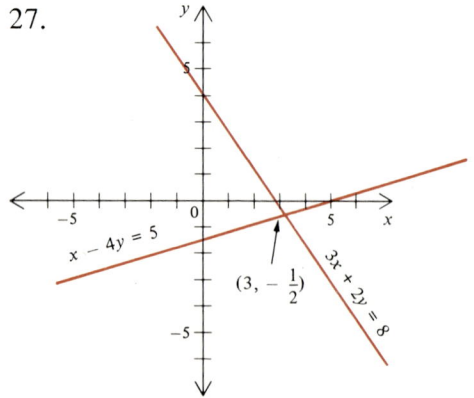

29.

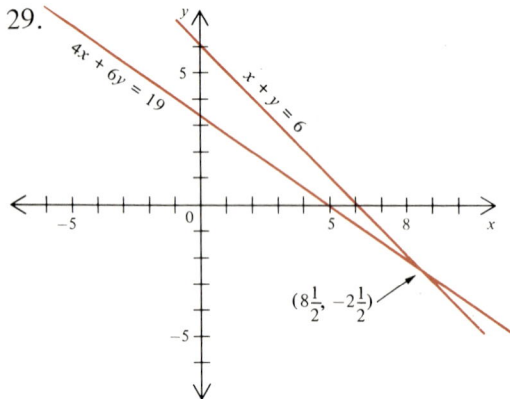

31.

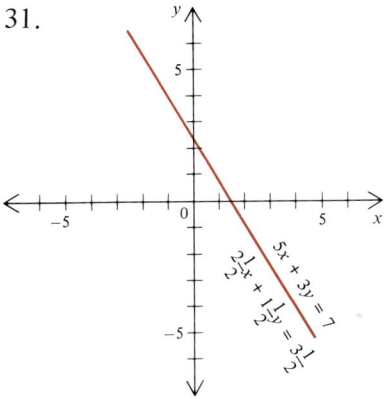

$5x + 3y = 7$
$2\frac{1}{2}x + 1\frac{1}{2}y = 3\frac{1}{2}$

Any solution of one is a solution of the other.

33.

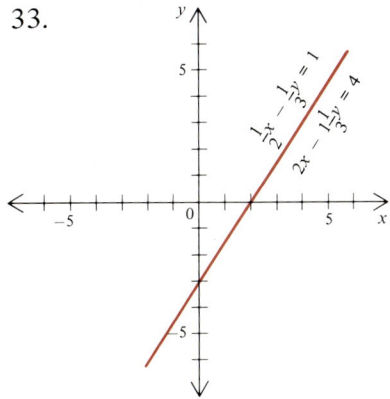

$\frac{1}{2}x - \frac{1}{3}y = 1$
$2x - 1\frac{1}{3}y = 4$

Any solution of one is a solution of the other.

35.

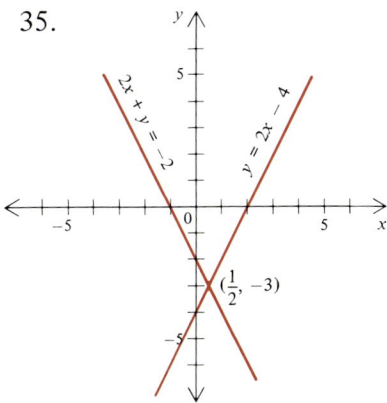

$2x + y = -2$
$y = 2x - 4$
$(\frac{1}{2}, -3)$

37.

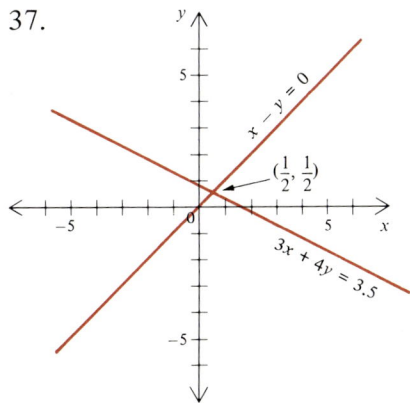

$x - y = 0$
$(\frac{1}{2}, \frac{1}{2})$
$3x + 4y = 3.5$

39.

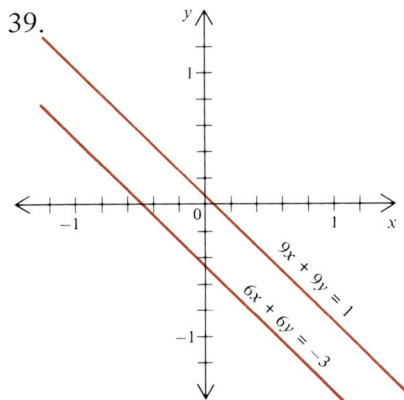

$9x + 9y = 1$
$6x + 6y = -3$

There is no common solution. Solution set $= \emptyset$

Exercises 4.5

1. $\{(5, 3)\}$ 3. $\{(2, 3)\}$ 5. $\{(17, 14)\}$ 7. Dependent equations
9. $\{(-2, 5)\}$ 11. $\{(3, 3)\}$ 13. $\{(98, 25)\}$ 15. $\{(-5, 2)\}$

17. $\{(-2, -3)\}$ 19. Inconsistent equations, $\varnothing$ 21. $\left\{\left(-3, -\dfrac{2}{3}\right)\right\}$

23. $\left\{\left(\dfrac{2}{3}, -\dfrac{7}{3}\right)\right\}$ 25. $\left\{\left(\dfrac{17}{2}, -\dfrac{5}{2}\right)\right\}$ 27. Inconsistent equations, $\varnothing$

29. $\left\{\left(-\dfrac{7}{10}, \dfrac{7}{2}\right)\right\}$ 31. $\{(4, -2)\}$ 33. $\{(-1, -5)\}$ 35. $\{(7, 2)\}$

37. $\{(4, -1)\}$ 39. $\{(6, 10)\}$ 41. Dependent equations

43. $\{(3, 2)\}$ 45. $\{(11, 6)\}$ 47. $\left\{\left(\dfrac{1}{2}, 0\right)\right\}$

49. $\{(1.3625954, -0.432061)\}$ 51. $\{(-120.76849, 25.140249)\}$
53. $\{(-1.6477416, 14.539927)\}$ 55. $\{(0.4602522, -1.5395702)\}$
57. $\{(-3.5893221, 22.439126)\}$

Exercises 4.6

1. 38, 12   3. 17, 19   5. 7 m, 18 m   7. $3975   9. Round,
$1.11 per lb.; sirloin, $1.48 per lb.   11. 7 quarters, 13 dimes
13. 1¢, 4¢   15. 867-9800   17. 75¢ coffee − 24 lbs., $1.20
coffee − 21 lbs.   19. $2500 at 10%, $1500 at 8%   21. $7000 at
4%, $14,000 at 7%   23. 564 mph, 364 mph   25. 11:00 A.M.
27. 15 n.m.   29. Mother, 10 hrs.; daughter, 15 hrs.   31. Ray,
15 hrs.; Jane, 22.5 hrs.   33. 36   35. 1¢, 4¢   37. 6 dimes,
10 nickels, 16 pennies   39. 12 km   41. For, 32; against, 22

43. $1\frac{1}{4}$ hrs.   45. $\dfrac{10}{9}$

Exercises 4.7

1.

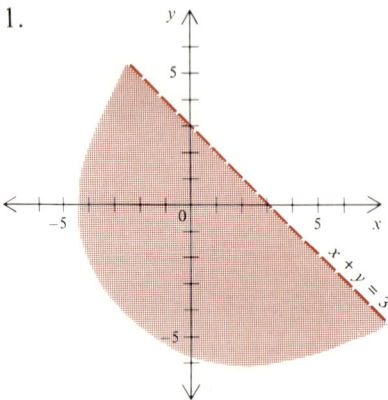

3.

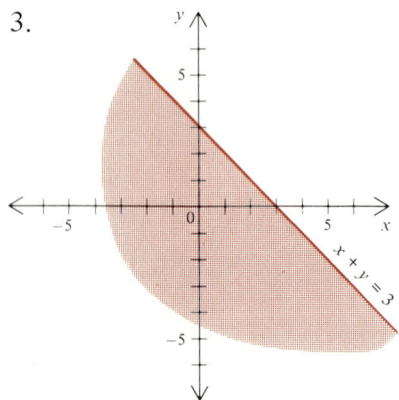

5.

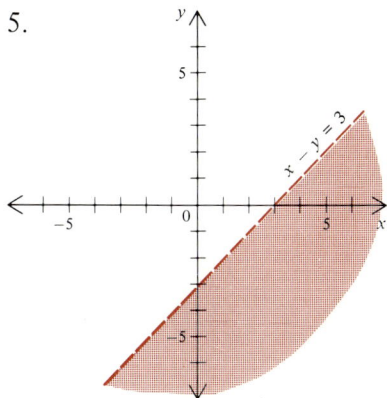

7.

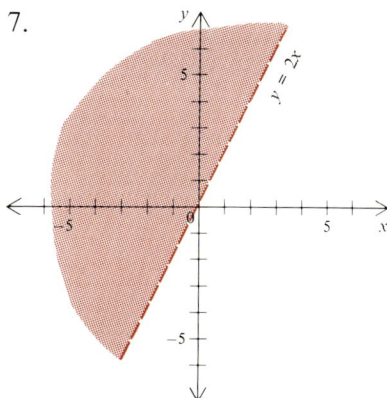

9.

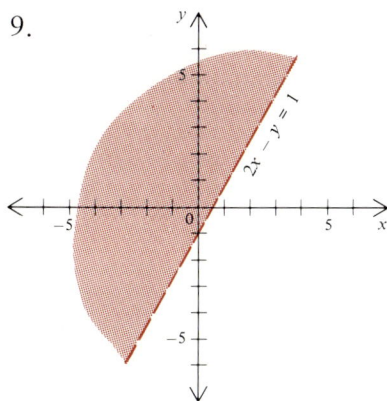

11.

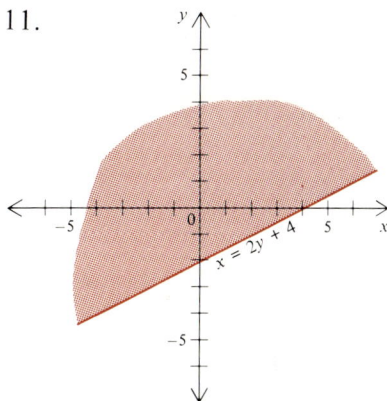

13.

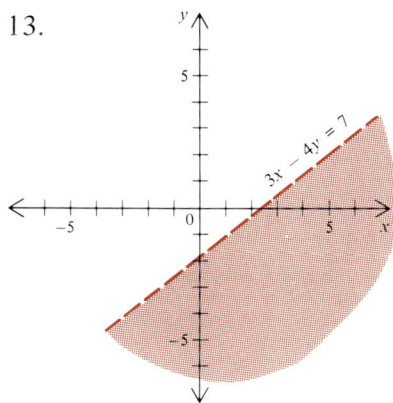

15.

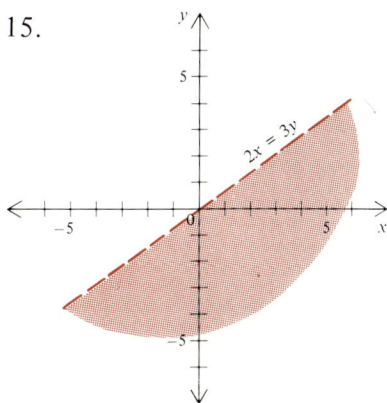

17.

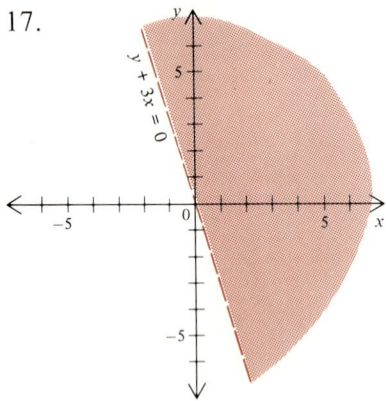

19.

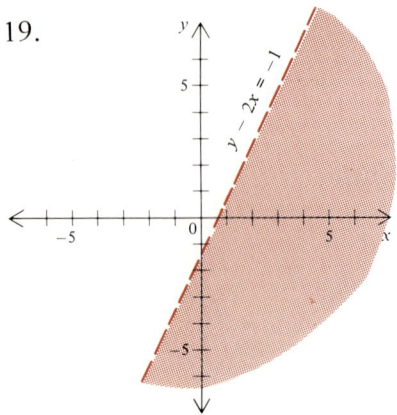

21.

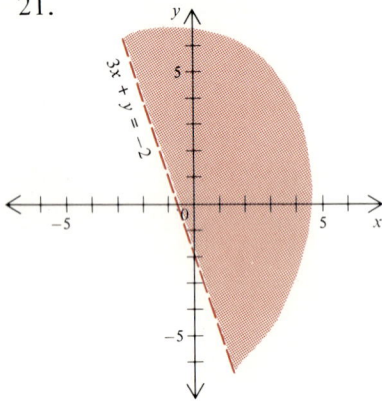

23.

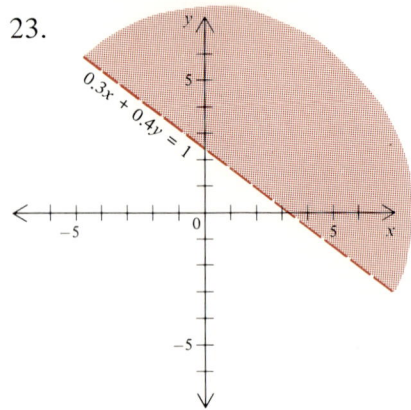

25.

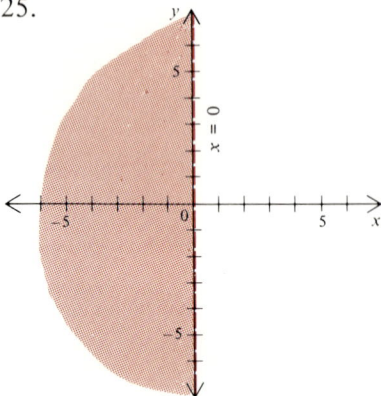

27.

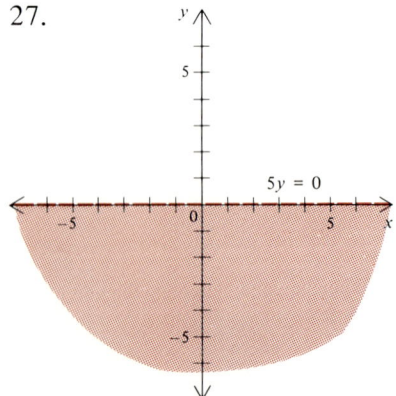

29.

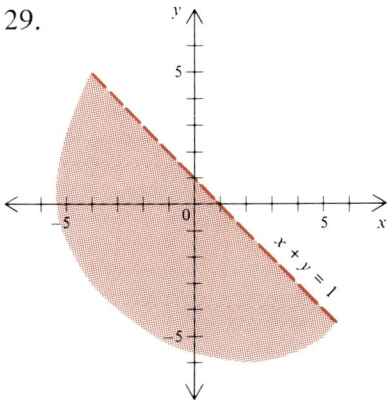

$x + y = 1$

31.

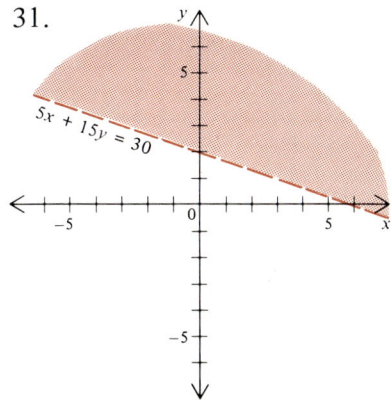

$5x + 15y = 30$

33.

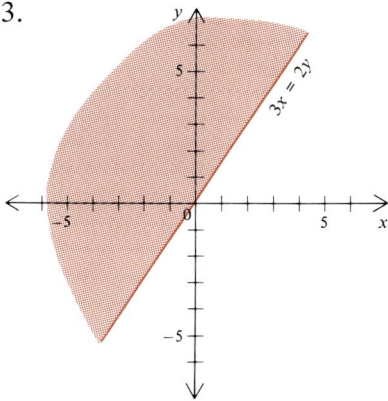

$3x = 2y$

35.

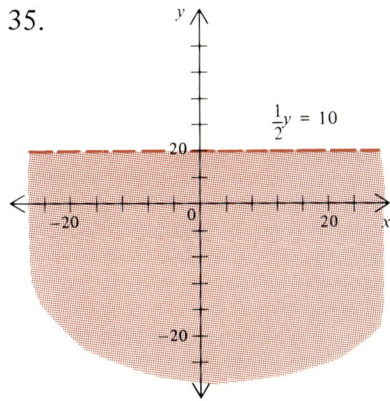

$\frac{1}{2}y = 10$

37.

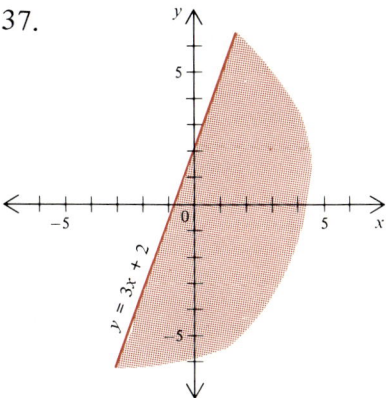

$y = 3x + 2$

39

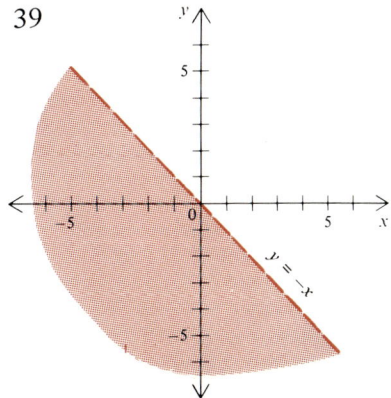

$y = -x$

Exercises 4.8

1. B, C, D, G 3. A, B, C 5. A 7. A, B 9. C

11. $x - 2y > 10$ 13. $\begin{cases} x - 2y < 10 \\ x + y > 2 \end{cases}$ 15. $\begin{cases} x - 2y > 10 \\ x + y < 2 \\ 10x + 3y > -30 \end{cases}$

17.

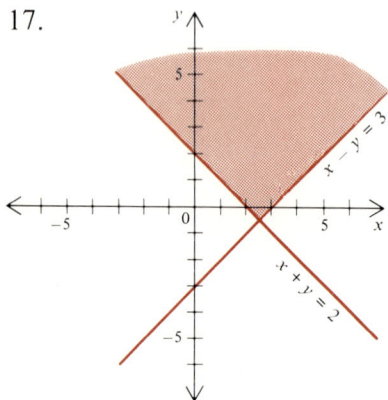

19.

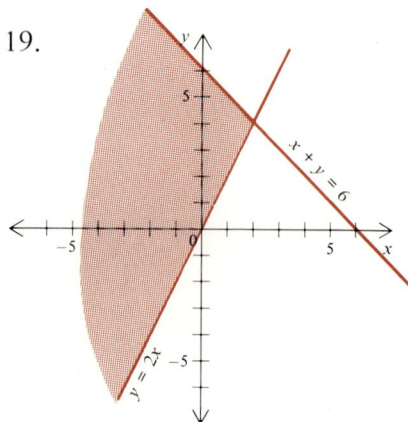

21.

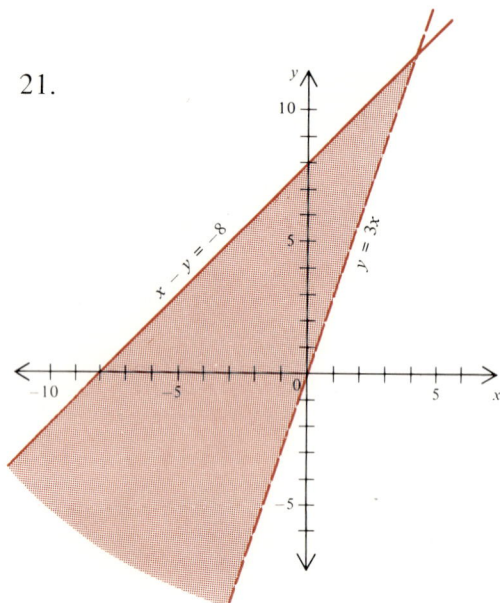

23.

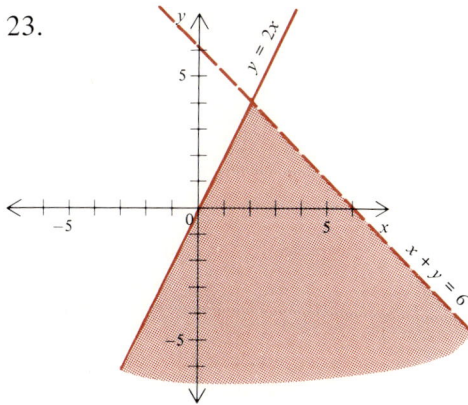

25.

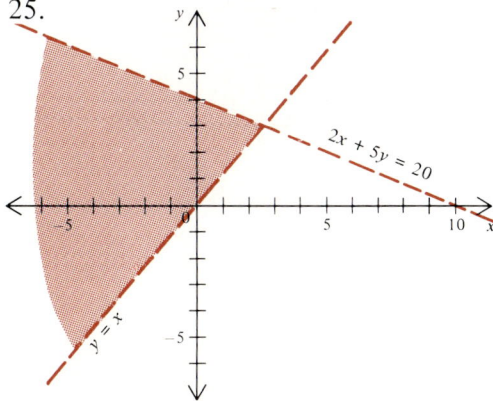

27.

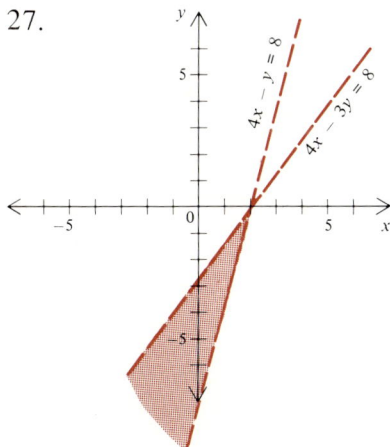

29.

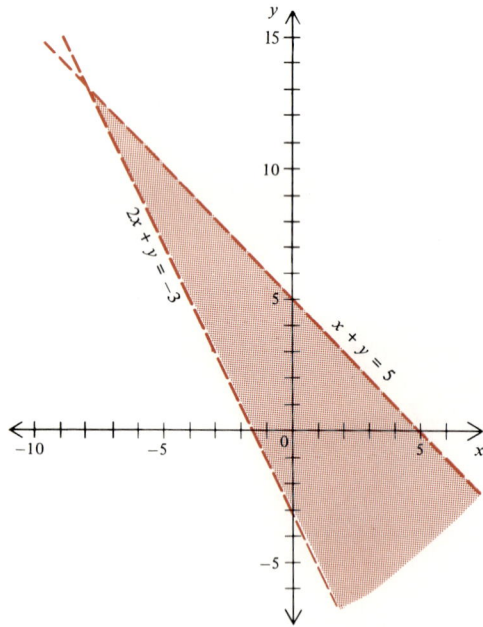

31.

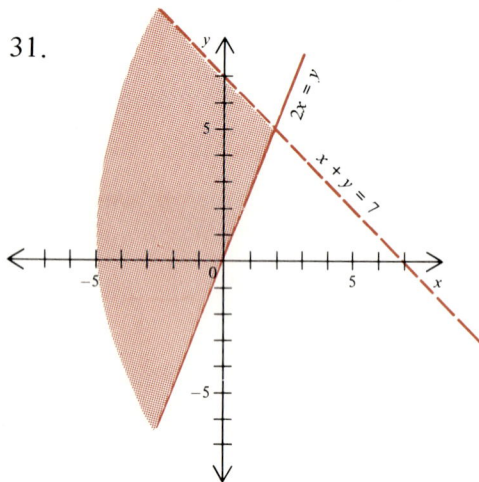

33.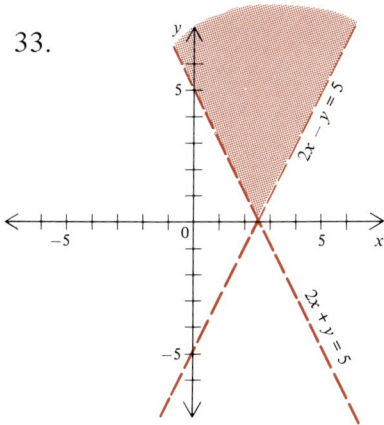

35.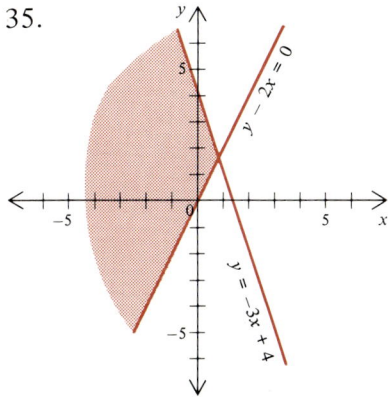

37.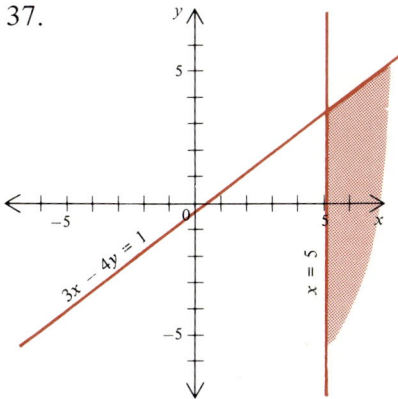

39.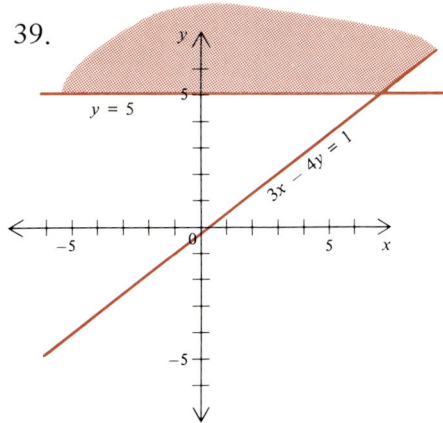

41.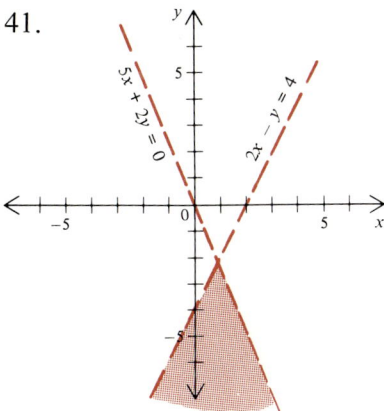

43.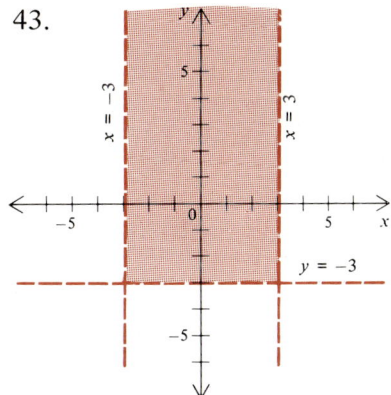

45.

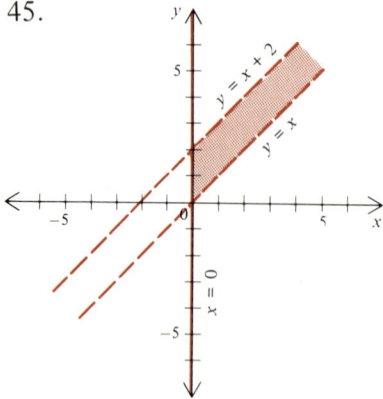

47.

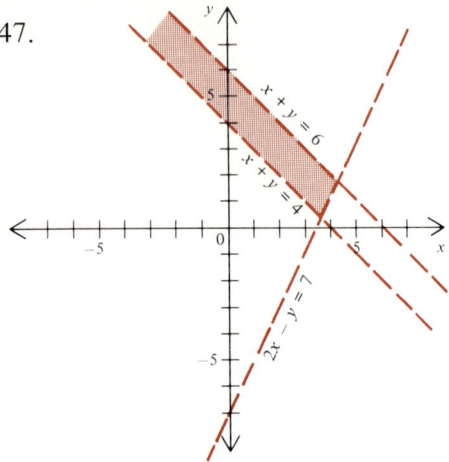

49.

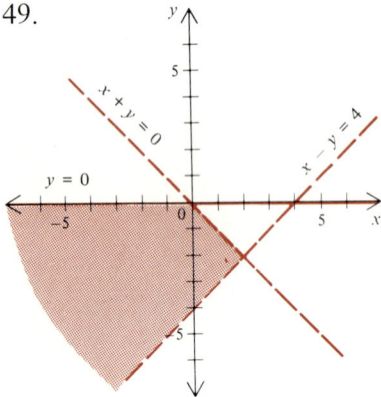

51.

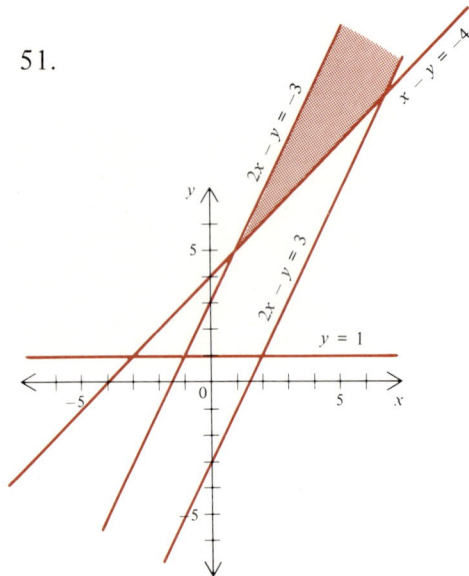

Review Exercises: Chapter 4

1. Equivalent   3. Equivalent   5. Equivalent   7. Equivalent
9. Equivalent   11. Equivalent   13. Equivalent   15. Equivalent
17. $(1, 1)$ 19. $(5, 4)$ 21. $(-3\frac{1}{2}, 2)$ 23. $(-7, 1)$ 25. $(-5, 0)$
27. $(-2, -2)$ 29. $(2, -2)$ 31. $(6, -4)$

33.

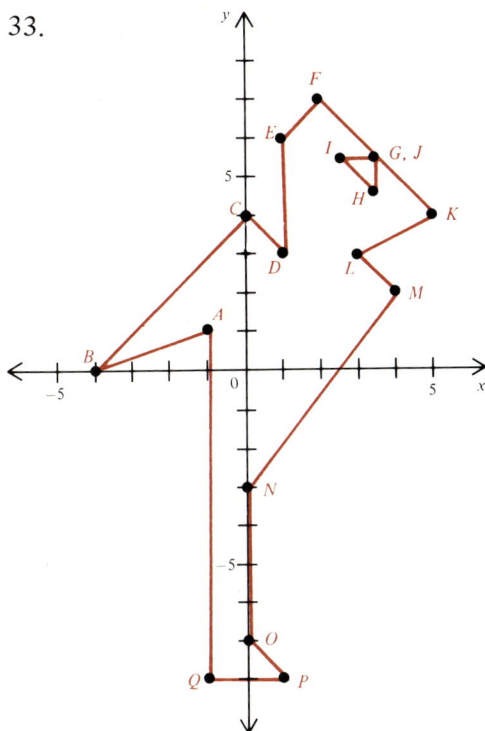

35. $y = 2x - 8$ 37. $y = \dfrac{2x}{3} - 4$ 39. $y = \dfrac{-3x}{4} + 6$ 41. $y = \dfrac{x}{3} - \dfrac{8}{3}$

43. $y = \dfrac{2x}{3}$ 45. $y = -\dfrac{x}{3} - \dfrac{4}{3}$ 47. $y = \dfrac{-x}{3} + 4$ 49. $y = \dfrac{-15x}{8} + 10$

51. $y = \dfrac{x}{16}$

53.

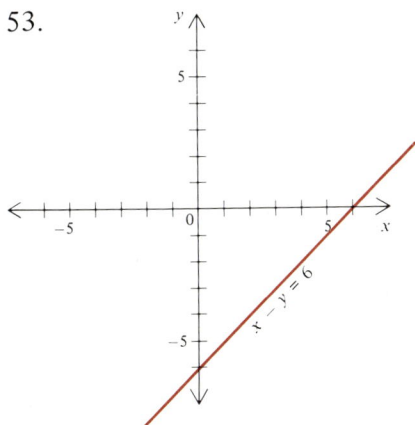

55.

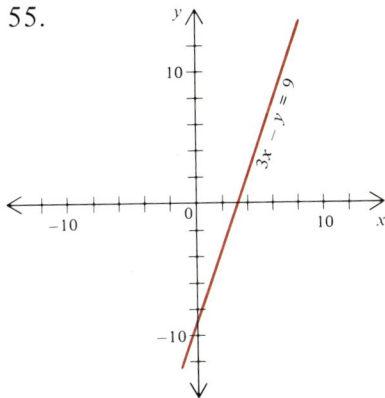

57.

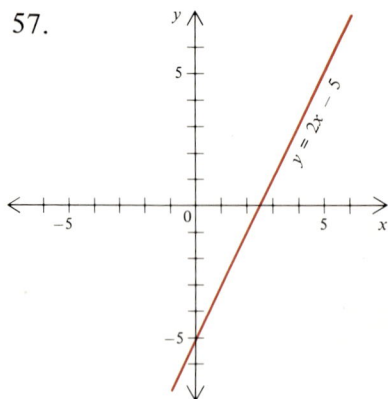

$y = 2x - 5$

59.

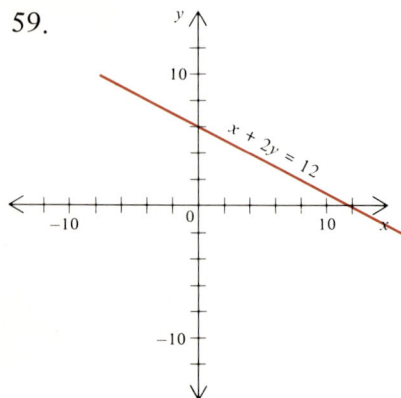

$x + 2y = 12$

61.

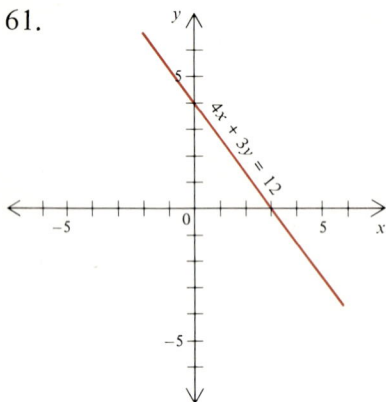

$4x + 3y = 12$

63.

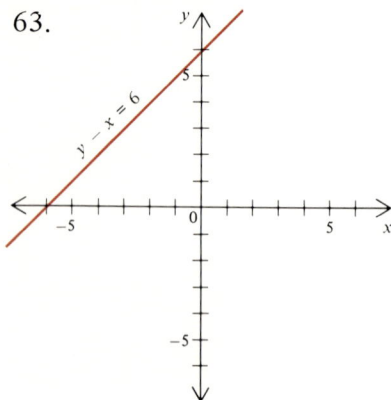

$y - x = 6$

65.

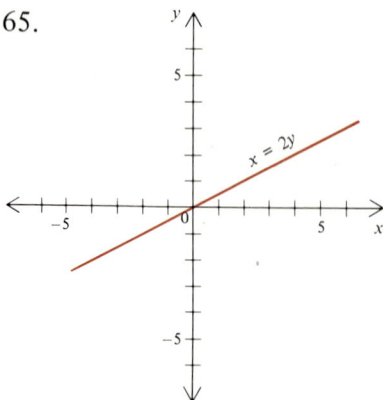

$x = 2y$

67.

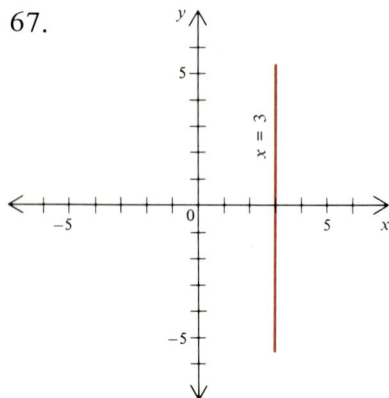

$x = 3$

69.

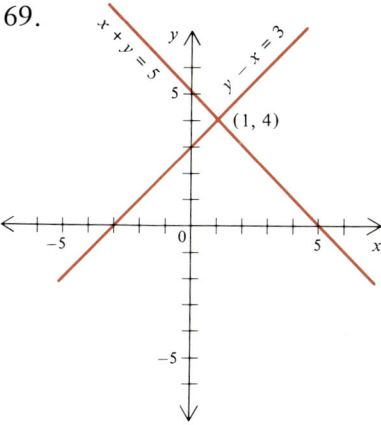

$x + y = 5$
$y - x = 3$
$(1, 4)$

71.

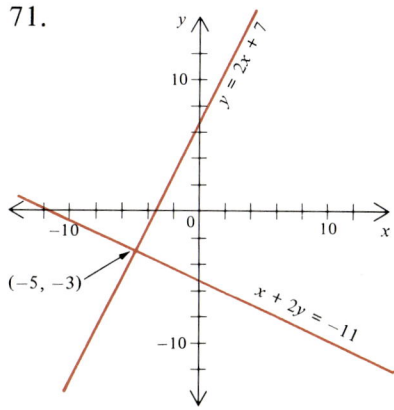

$y = 2x + 7$
$x + 2y = -11$
$(-5, -3)$

73.

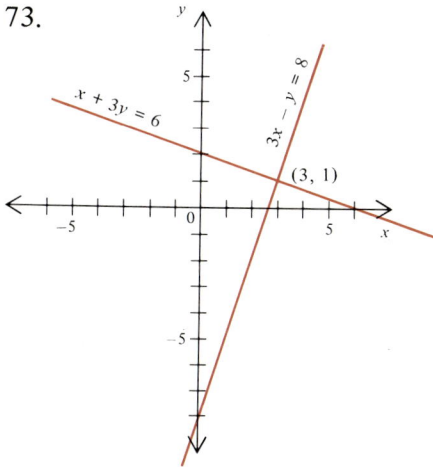

$x + 3y = 6$
$3x - y = 8$
$(3, 1)$

75.

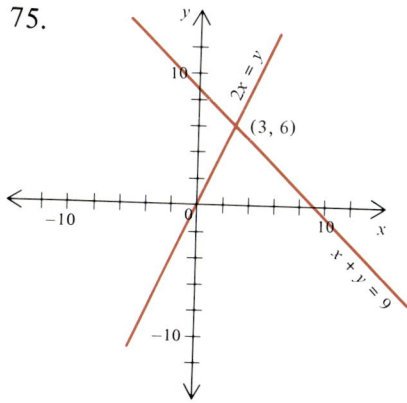

$2x = y$
$x + y = 9$
$(3, 6)$

77.

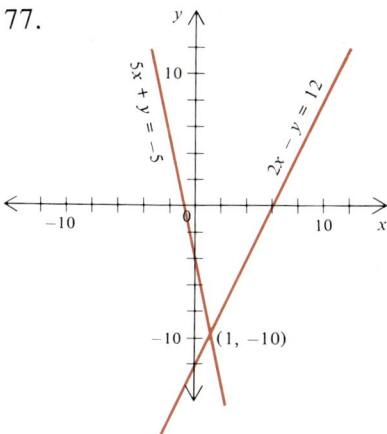

$5x + y = -5$
$2x - y = 12$
$(1, -10)$

79.

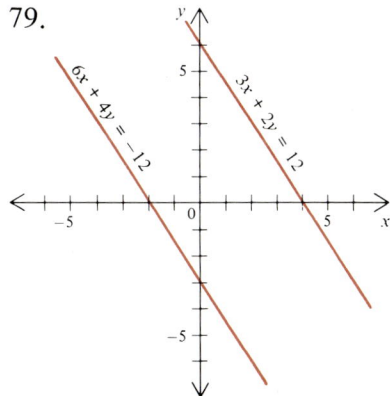

$6x + 4y = -12$
$3x + 2y = 12$

Equations are inconsistent.

81.

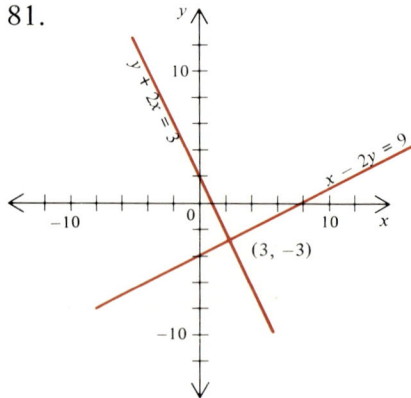

$y + 2x = 3$

$x - 2y = 9$

$(3, -3)$

83.

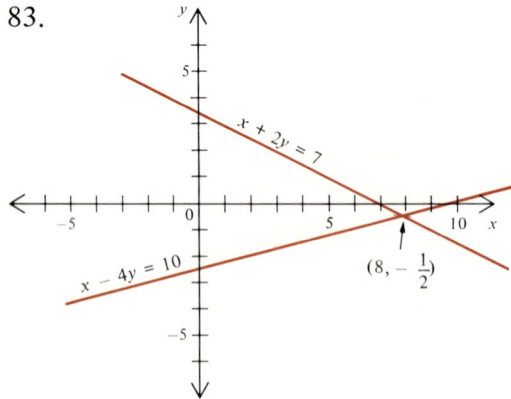

$x + 2y = 7$

$x - 4y = 10$

$(8, -\frac{1}{2})$

85.

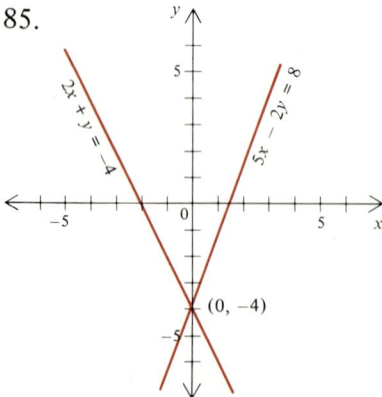

$2x + y = -4$

$5x - 2y = 8$

$(0, -4)$

87.

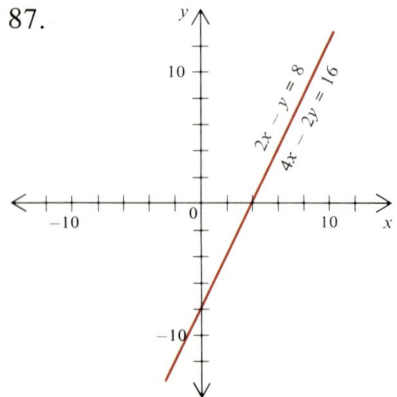

$2x - y = 8$

$4x - 2y = 16$

Equations are dependent.

89. $\{(10, -2)\}$ 91. $\{(-2, 3)\}$ 93. $\left\{\left(\frac{115}{7}, \frac{-92}{7}\right)\right\}$ 95. $\{(-5, 0)\}$

97. $\left\{\left(\frac{3}{4}, 6\right)\right\}$ 99. $\left\{\left(-\frac{1}{2}, \frac{3}{4}\right)\right\}$ 101. $\{(12, 8)\}$ 103. $\{(-20, 16)\}$

105. $\{(1, -8)\}$ 107. $\{(-2, -3)\}$ 109. $\{(-6, -1)\}$ 111. 88, 40
113. \$5600 at 10%, \$8400 at 11%   115. 4 liters   117. $46\frac{2}{3}$ km/hr.
$118\frac{1}{3}$ km/hr.   119. Tea, \$2.68; coffee, \$2.88   121. Still water,
5 km/hr.; current, $1\frac{2}{3}$ km/hr. 123. 8, 10   125. 54 dimes, 36 quarters
127. Bob, 30; Jim, 8   129. Guilder = 40¢, franc = 24¢
131. \$170,000, \$90,000   133. 12 hrs.   135. 152, 72

137.

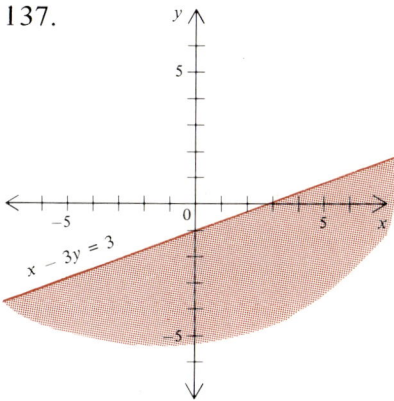

139.

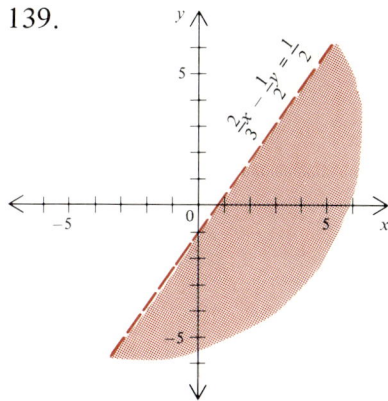

141.

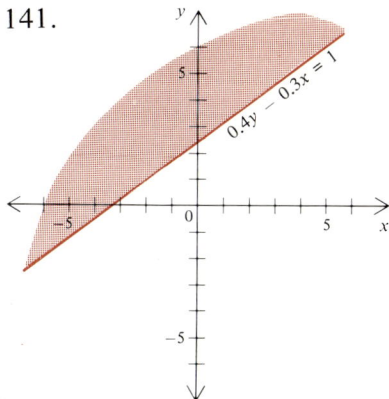

143.

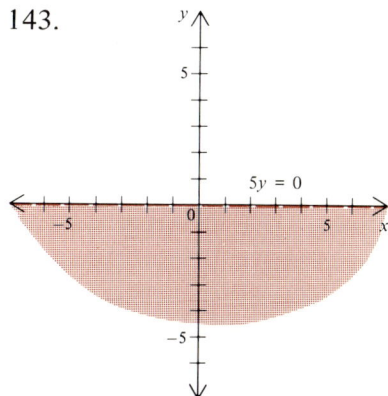

145.

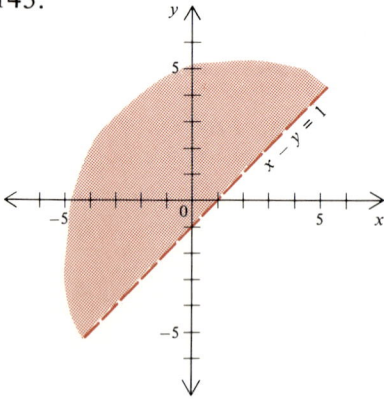

$x - y = 1$

147.

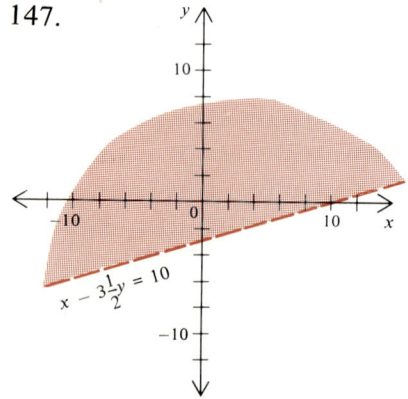

$x - 3\frac{1}{2}y = 10$

149.

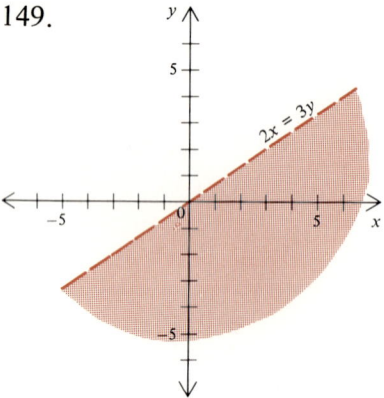

$2x = 3y$

151.

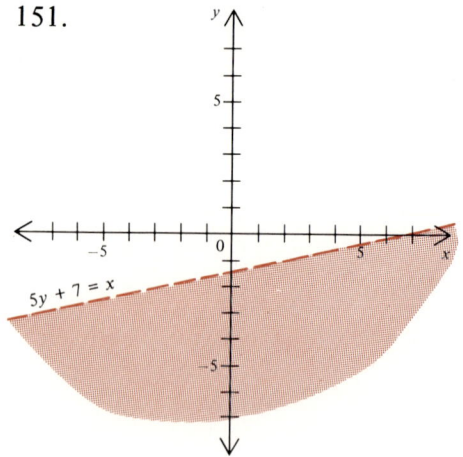

$5y + 7 = x$

153.

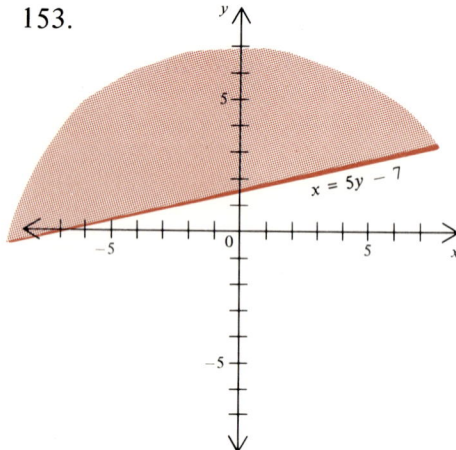

$x = 5y - 7$

155.

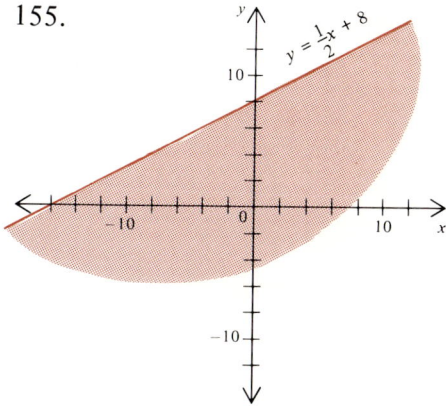

$y = \frac{1}{2}x + 8$

157.

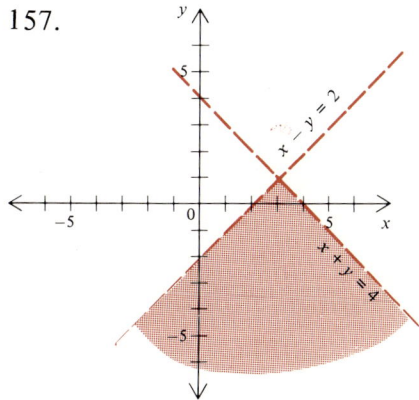

$x - y = 2$

$x + y = 4$

159.

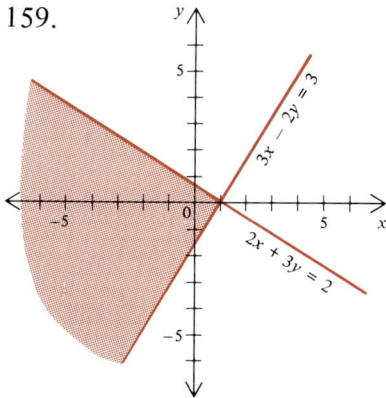

$3x - 2y = 3$

$2x + 3y = 2$

161.

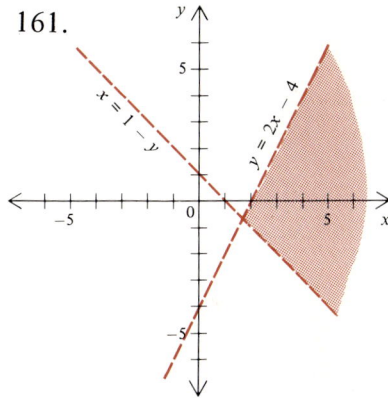

$x = 1 - y$

$y = 2x - 4$

163.

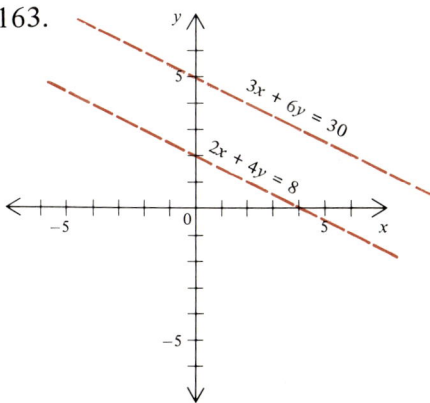

$3x + 6y = 30$

$2x + 4y = 8$

Solution set $= \emptyset$

165.

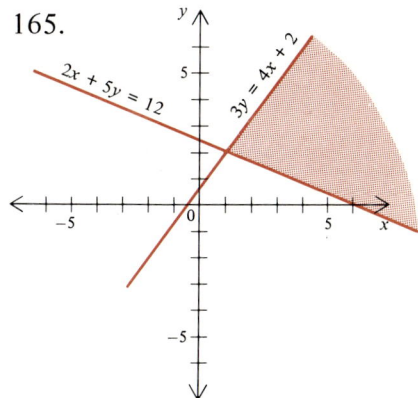

$2x + 5y = 12$

$3y = 4x + 2$

167.

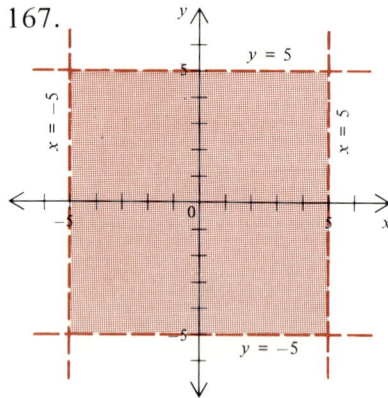

Cumulative Review B

1. $\dfrac{4}{21}$ 3. $3\frac{1}{16}$ 5. 164   7. $37ab$ 9. $0.41\overline{6}$ 11. $\varnothing$

13. $\{4\}$ 15. $\{0\}$

17. $\{x : x > -4\}$

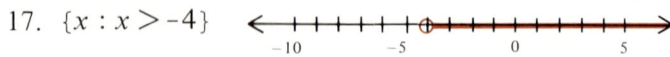

19. $\{x : x < 2\}$

21. $\{x : x < -3\}$

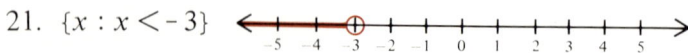

23. Solutions will vary.   25. $y = -\dfrac{1}{2}x + \dfrac{1}{2}$ 27. $k = \dfrac{Pr^4}{8lV}$

29.

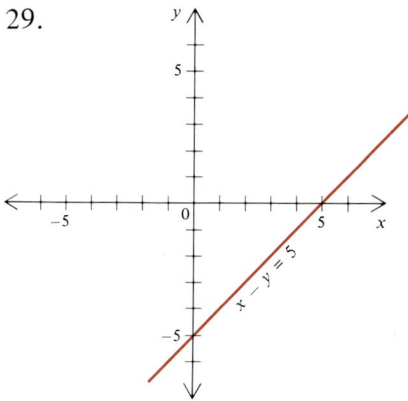

$x - y = 5$

31.

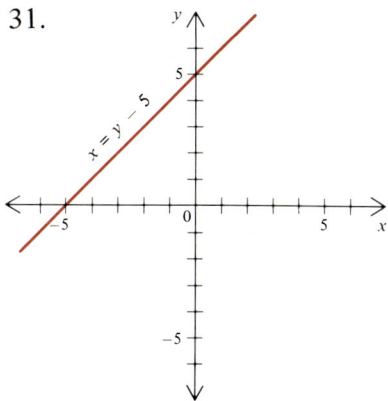

$x = y - 5$

33.

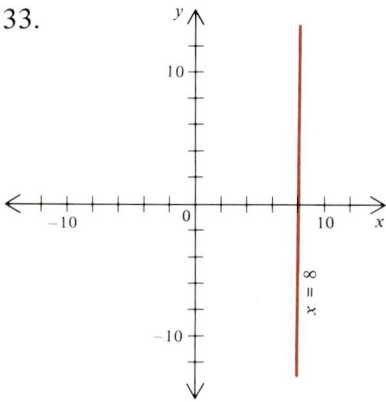

$x = 8$

35.

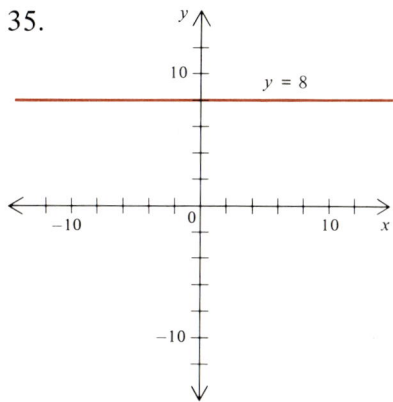

$y = 8$

37.

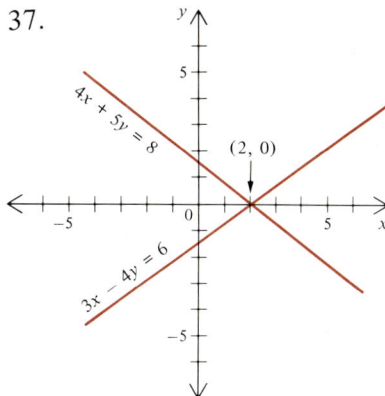

$4x + 5y = 8$

$(2, 0)$

$3x - 4y = 6$

39. $\{(8, 6)\}$ 41. $\{(3, 4)\}$ 43. $\left\{\left(-\frac{6}{5}, \frac{4}{5}\right)\right\}$ 45. 2 hrs.

47. $7000 at 8%, $21,000 at 4%

49.

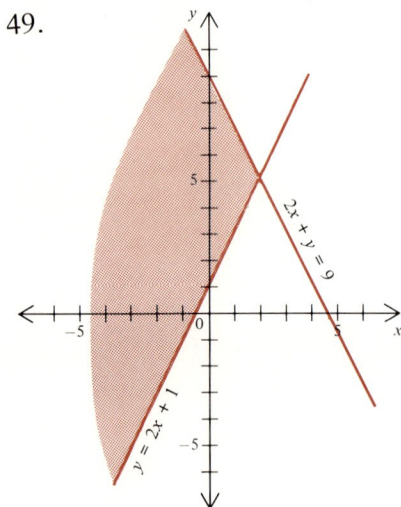

CHAPTER 5

Exercises 5.1

1. $2 \cdot x \cdot x \cdot y \cdot y \cdot y$ 3. $7 \cdot a \cdot a \cdot a \cdot a \cdot b \cdot b \cdot b$
5. $2 \cdot 5 \cdot a \cdot a \cdot a \cdot a \cdot b \cdot b \cdot b \cdot b \cdot b$
7. $4 \cdot 3 \cdot x \cdot x \cdot x \cdot y \cdot y \cdot y \cdot y \cdot y$
9. $2(x - y)(x - y)(x - y)(x + y)(x + y)$
11. $3 \cdot x \cdot x \cdot y \cdot y \cdot y \cdot (2x + y)(2x + y)$

13. $\dfrac{3 \cdot 3 \cdot x \cdot x \cdot x \cdot x \cdot x \cdot x}{(2y - 3)(2y - 3)(2y - 3)}$ 15. $\dfrac{2(x - y)(x - y)}{3 \cdot 3 \cdot (x + 2y)(x + 2y)}$

17. $16x^2 y^2$ 19. $-3x^2 y^3$ 21. $-50x^3 y^5$ 23. $12s^2 t^3$
25. $36x^2 y^2$ 27. $108xy^2 z^3$ 29. $12a^5 b^5 c^6$ 31. $-60x^{13} y^9$
33. $12a^8 b^5$ 35. $-864a^{10} b^8$ 37. $3a$ 39. $8x$ 41. $-5a^2 b^5 c$

43. $\dfrac{1}{3a^2 bc}$ 45. $\dfrac{8x^3 y^3}{5z^2}$ 47. $\dfrac{5b^3}{2a}$ 49. $-\dfrac{40a^3}{27b}$

Exercises 5.2

1. $\dfrac{y^2}{x}$ 3. $\dfrac{x^2}{y^2}$ 5. $\dfrac{1}{8a^6}$ 7. $\dfrac{1}{8a^3b^2}$ 9. $\dfrac{3}{b^3}$ 11. 16   13. $\dfrac{3z^3}{y^2}$

15. $12x^2$ 17. x^2y^3 19. $\dfrac{c^3}{d^7}$ 21. $\dfrac{ab^4y^3}{x^2}$ 23. $x^2 + x^8$

25. $1 + x^6$ 27. $\dfrac{1}{4}$ 29. 9   31. $\dfrac{1}{9}$ 33. -3 35. -1 37. 1

39. 1   41. $\dfrac{6}{x^3}$ 43. $\dfrac{9s^6}{r^4}$ 45. $\dfrac{a^4}{9b^2}$ 47. $\dfrac{1}{x^2}$ 49. $\dfrac{1}{9x}$

51. $\dfrac{1}{r^3s^5}$ 53. $\dfrac{x}{3y}$ 55. $\dfrac{r^3 + r^2s}{s^2}$ 57. $\dfrac{5a^3(a+b)^4}{3}$ 59. $\dfrac{1}{a^4x^5}$

Exercises 5.3

1. True   3. True   5. False   7. True   9. True   11. True
13. False   15. False   17. True   19. True   21. 9   23. 49
25. 78,125   27. -25 29. 625   31. 144   33. 0

35. 3.0976   37. 9   39. 13   41. $\dfrac{3}{4}$ 43. 5   45. $\dfrac{1}{9}$ 47. 0.6

49. 0.07   51. 3.46   53. 1.15   55. 1.22   57. 3.15   59. 0.32

Exercises 5.4

1. $\sqrt{21}$ 3. $4\sqrt{15}$ 5. $6\sqrt{35}$ 7. 12   9. 150   11. $2\sqrt{2}$
13. $2\sqrt{5}$ 15. $3\sqrt{5}$ 17. $2\sqrt{14}$ 19. $3\sqrt{7}$ 21. $9\sqrt{2}$
23. $11\sqrt{2}$ 25. 45   27. $7\sqrt{3}$ 29. $3\sqrt{6}$ 31. $7\sqrt{66}$

33. $30\sqrt{6}$ 35. 24   37. 3   39. $\dfrac{4}{5}\sqrt{2}$ 41. $\dfrac{1}{2}\sqrt{2}$ 43. $\dfrac{1}{5}\sqrt{15}$

45. $\dfrac{1}{2}\sqrt{10}$ 47. 6   49. $\sqrt{3}$ 51. 7   53. 4   55. $\dfrac{5}{4}\sqrt{6}$

57. $\dfrac{3}{4}$ 59. $\dfrac{3}{10}\sqrt{30}$ 61. $\dfrac{1}{3}\sqrt{105}$ 63. $\dfrac{1}{15}\sqrt{21}$ 65. $\dfrac{1}{2}\sqrt{33}$

Exercises 5.5

1. $2x$ 3. $3b\sqrt{b}$ 5. $2y\sqrt{6y}$ 7. $6x\sqrt{xy}$ 9. $3xy\sqrt{2xy}$
11. $3x\sqrt{2}$ 13. $36x\sqrt{xy}$ 15. $240a^2x\sqrt{ax}$ 17. $2xy^2\sqrt{3xy}$

19. $x^3y\sqrt{y}$ 21. $2a^3b\sqrt{2a}$ 23. $3x^5y^4\sqrt{6}$ 25. $5x^2y^3\sqrt{2x}$

27. $0.7b\sqrt{a}$ 29. $0.3xy\sqrt{3}$ 31. $\frac{2}{5}x$ 33. $\frac{a}{2}\sqrt{6}$ 35. $\frac{a^2}{b^2}\sqrt{3a}$

37. $\frac{x^2y}{3a}\sqrt{3axy}$ 39. $\frac{5y}{4x}\sqrt{10x}$ 41. $\frac{\sqrt{14x}}{7x^2}$ 43. $\frac{35x^3y^2\sqrt{6y}}{12}$

45. $\frac{3x^2\sqrt{2xy}}{5y^4}$ 47. $x^2y^2\sqrt{2y}$ 49. $\frac{2\sqrt{3x}}{3x}$ 51. $\frac{\sqrt{3x}}{3x}$

53. $(x-y)^2$

Exercises 5.6

1. $8\sqrt{2}$ 3. $6\sqrt{5}$ 5. $-\sqrt{5}$ 7. $-2\sqrt{3}$ 9. $7\sqrt{2}$ 11. $\sqrt{5}$
13. $7\sqrt{5}$ 15. $2\sqrt{3}$ 17. $\sqrt{2}$ 19. $18\sqrt{7}$ 21. $-5\sqrt{10}$
23. $-\sqrt{2}$ 25. $-2\sqrt{2}$ 27. $4\sqrt{2}+\sqrt{3}$ 29. $\sqrt{6}-17\sqrt{5}$
31. $-5\sqrt{5}-15\sqrt{2}$ 33. $25\sqrt{2}-15\sqrt{3}-10\sqrt{5}$
35. $67\sqrt{3}-42\sqrt{5}$ 37. $7x\sqrt{x}$ 39. $2\sqrt{2a}$
41. $(-2x^2+4x+1)\sqrt{x}$ 43. $(y-9y^2)\sqrt{2y}$

45. $(6x+4y)\sqrt{xy}+(3-2x)\sqrt{2xy}$ 47. $\frac{a}{4}\sqrt{5}$ 49. $\frac{\sqrt{xy}}{y}$

51. $\frac{10\sqrt{3}}{3}$ 53. $3\sqrt{2}$ 55. $\frac{11\sqrt{2}}{4}-\frac{1}{2}$ 57. $\frac{1}{3}\sqrt{6}$ 59. $\frac{7}{3}\sqrt{6}$

61. $\frac{5}{3}\sqrt{6}$ 63. $-\frac{3}{2}\sqrt{5}$ 65. $\sqrt{3}$ 67. $3\sqrt{3}$ 69. 0

Exercises 5.7

1. Index, 3; radicand, 5.   3. Index, 2; radicand, 3.   5. Index, 7; radicand, xy.   7. Index, 27; radicand, 1372.   9. Index, 4; radicand, 32.   11. Index, 3; radicand, $\frac{1}{8}$.   13. 2   15. 2

17. -3 19. 4   21. $-\frac{2}{3}$ 23. $-\frac{5}{4}$ 25. 4   27. $\frac{3}{5}$ 29. 5

31. 5   33. 9   35. 6   37. 15   39. $\frac{5}{6}$ 41. $-\frac{15}{2}$ 43. 1

Review Exercises: Chapter 5

1. P^{11} 3. $6x^5y^3$ 5. $75x^4y^3z^5$ 7. $169x^{28}$ 9. x^{32}

11. $-\dfrac{27}{16}x^6y^8$ 13. x^{4m} 15. xy^2z^4 17. $\dfrac{2}{3}xy^2$ 19. $6xyz^7$

21. 4 23. $\dfrac{17}{18}$ 25. 9 27. 1 29. $\dfrac{6x}{y}$ 31. $8y^6(y-2)^2$

33. x 35. $\dfrac{3}{2}x^7y^6$ 37. $\dfrac{x^4}{y^3}$ 39. $\dfrac{a^4}{b^4}$ 41. 6 43. 20

45. $\dfrac{5}{2}$ 47. $\dfrac{3}{2}$ 49. 0.02 51. 3.46 53. 1.34 55. 0.77

57. $2\sqrt{3}$ 59. $3\sqrt{3}$ 61. 48 63. $10\sqrt{3}$ 65. 60 67. $\dfrac{1}{3}\sqrt{3}$

69. $\dfrac{3}{2}\sqrt{2}$ 71. $2x\sqrt{3x}$ 73. $4x^2\sqrt{2}$ 75. $2x\sqrt{7y}$ 77. $3a^2\sqrt{2}$

79. $x^2y^2\sqrt{y}$ 81. $\dfrac{\sqrt{15y}}{6y}$ 83. $\dfrac{2c\sqrt{3c}}{3}$ 85. $\dfrac{\sqrt{6x}}{4x^2}$ 87. $-2\sqrt{2}$

89. $-\sqrt{6}$ 91. $8\sqrt{3}$ 93. $\dfrac{1}{4}\sqrt{2}$ 95. $\dfrac{11}{5}\sqrt{10}$ 97. $14x\sqrt{2}$

99. $4xy\sqrt{x}$ 101. $6x\sqrt{3}$ 103. $7x^2\sqrt{2}$ 105. $-3a\sqrt{2}$

107. $-24x$ 109. $\dfrac{70\sqrt{ab}}{a}$ 111. 5 113. -5 115. 3

117. $\dfrac{4}{5}$ 119. $-\dfrac{5}{6}$ 121. -6 123. 10 125. $\dfrac{22}{15}$ 127. $\dfrac{20}{7}$

129. 8 131. -49 133. $-\dfrac{3}{5}$ 135. $\dfrac{1}{2}$ 137. $\dfrac{5}{2}$ 139. -10

Cumulative Review C

1. 19 3. $-63a^2b$ 5. $-\dfrac{10}{7}$ 7. $0.\overline{428571}$ 9. $\{3\}$ 11. $\{7\}$

13. $\{-7\}$ 15. $\varnothing$ 17. $\{5\}$

19. $\{x : x > 3\}$

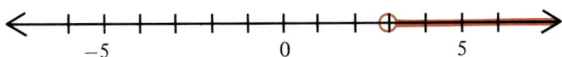

21. $\{x : x < -9\}$

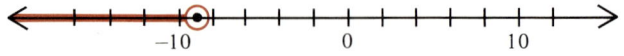

23.

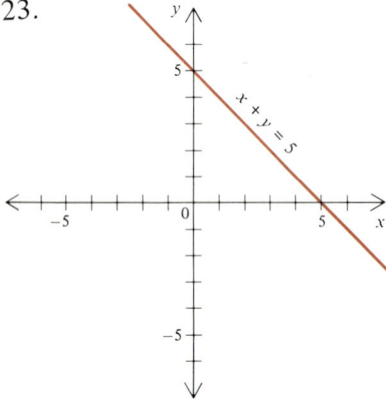

25.

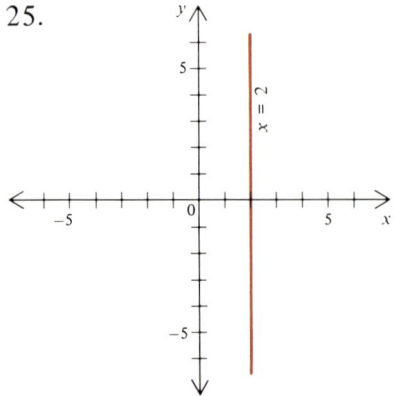

27.

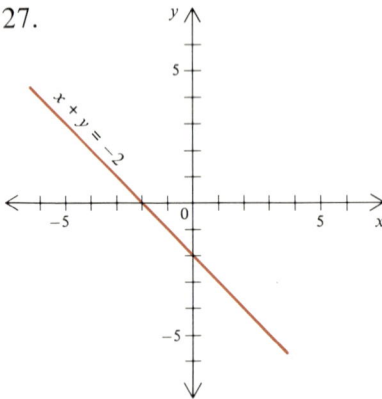

29.

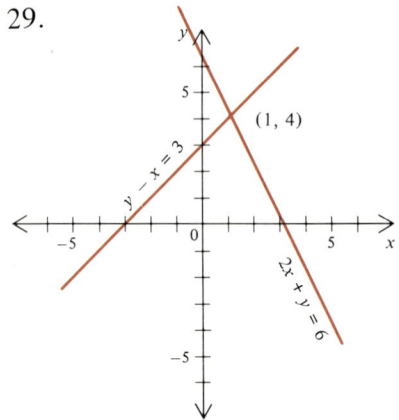

31. $\{(2, 5)\}$ 33. $\{(2, 6)\}$ 35. Large cake, 10; small cake, 5.

37. $2\frac{2}{3}$ days   39. 6 hrs.   41. 6   43. $\dfrac{3xy}{z^2}$ 45. $3\sqrt{3}$

47. $5\sqrt{3} + 8\sqrt{5}$ 49. 3

CHAPTER 6

Exercises 6.1

1. 2   3. 6   5. 1   7. 1   9. 13   11. 2   13. 15   15. 4
17. 4   19. 3   21. 3   23. 3   25. 2   27. 7   29. 0
31. Yes   33. No   35. No   37. Yes   39. Yes   41. Yes
43. No   45. Yes   47. No   49. No   51. *A, B, C*   53. *C*
55. *A, B, C*   57. *C*   59. *A, B, C*   61. *A, B, C*   63. *C*
65. *C*   67. *A, B, C*   69. *B, C*

Exercises 6.2

1. $12a$ 3. $236x^6$ 5. $-7a$ 7. $-6a^2$ 9. $16x^3$ 11. 0
13. $-17d$ 15. $11x$ 17. x^2y 19. $6abc$ 21. $17a^3bc^2$
23. a^2b 25. $2bc^2$ 27. $2xy + 6$ 29. $5x^2 + 5x$ 31. $ab + 10$
33. $4x + 15$ 35. $9x + y + 6w$ 37. $15 - 9a$ 39. $2ab - 10b^2$
41. $10y^2 + 5y$ 43. $-4\sqrt{5}xy + 38\sqrt{3}yz$ 45. $29a^2b - 13ab^2$

47. $\frac{3}{2}a^2 + \frac{1}{6}a$ 49. $-\frac{3}{4}m^2n^2 + 7\frac{3}{4}mn$ 51. $9x^2 - x - 2$

53. $9a^2 - 2ab - 10b^2$ 55. a^2 57. $-a - 8b + 12c$
59. $6x^2 + xy - 9y^2$ 61. $4\sqrt{2}x^2 + 9x + 5\sqrt{3}$
63. $2x^2y - 7xy + 7xy^2$ 65. $-4x^3 + x^2y - 7xy^2$
67. $m^2 - 3mn + 2n^2$ 69. $7a^2 + 2b^2 - 11c$ 71. $-8x + 8y + 18z$

73. $-x - 14$ 75. $27x^3 - 6x^2 + 10x + 10$ 77. $\left\{\frac{1}{2}\right\}$ 79. $\{5\}$

81. $\{b : b \geqslant 3\}$ 83. $\left\{\frac{6}{5}\right\}$ 85. $\{r : r > 3\}$ 87. $\left\{-\frac{2}{7}\right\}$

89. $\{x : x > 9\}$ 91. $\{0\}$ 93. $\{0\}$ 95. $\{x : x \leqslant 4\}$ 97. $\left\{-\frac{6}{5}\right\}$

99. $\left\{\frac{\sqrt{2}}{16}\right\}$

Exercises 6.3

1. $8x^5y$ 3. $-5y^2x$ 5. $20a^2b^3$ 7. $27s^5rw$ 9. $7x^4y^2$

11. $24x^2yz^4$ 13. $7x^5y^5z^5$ 15. $\frac{2}{5}t^5$ 17. $0.18a^{10}b^4$

19. $-0.0007m^4ny$ 21. $105x^2$ 23. $20x^3y^4$ 25. $-x^3$

27. $-30ab^2c^2d$ 29. $81x^4$ 31. $\frac{2}{3}x^5y^5$ 33. $\frac{8}{3}m^3n^3$

35. $-\frac{7}{2}a^4b^5$ 37. $6a^{3x}$ 39. $3x^{2b}y$ 41. $2x+10$

43. $-10a-12$ 45. $30y-70x$ 47. $3x^5+5x^3$ 49. $-4x^6+x^2$
51. $x^4+3x^3+x^2$ 53. $-x^4y^3+2x^3y^3-3x^3y^4$ 55. x^3+x^2
57. $6-4x+8x^2$ 59. $9y^4-6y^3$ 61. $x^{2a}-x^ay^a$
63. $x^{2+b}+x^2y^b$ 65. $a^{b+1}x^2+a^bbx^3$

Exercises 6.4

1. $x^2+7x+12$ 3. a^2+2a+1 5. $a^2+15a+36$
7. $y^2+7y+10$ 9. $s^2+9s+20$ 11. $x^2-7x+12$
13. a^2-2a+1 15. $a^2-15a+36$ 17. n^2-n-72
19. $a^2-9a-36$ 21. n^2+n-72 23. $a^2+9a-36$
25. $12c^2+23c+10$ 27. $10y^2-y-21$ 29. $4x^2-9$
31. $6a^2-4a-16$ 33. $4n^2-4n-35$ 35. $12c^2+7c-10$
37. $10y^2-y-21$ 39. $12c^2-23c+10$ 41. $10y^2-29y+21$
43. y^3-3y^2+3y-1 45. x^3+3x^2+3x+1
47. $y^3+10y^2+15y-2$ 49. $12x^4+30x^3y+14x^2y+33xy^2-5y^3$
51. $6x^3-23x^2+29x-12$ 53. $a^6+2a^5-6a^4+4a^3-5a^2+7a-3$
55. $4a^4+17a^3-a^2-22a-8$ 57. $9x^2-5x$ 59. $12x+33$
61. $9n^2-30n$ 63. $4x^2+10x+1$ 65. $-2a^2+9a-25$
67. $3t^2-4t+15$ 69. $11a^2+17ab-29b^2$ 71. $-2y^4+2x^2y^2$
73. $-x^3-x^2-x-1$ 75. $-9x^2y-3x^2-2xy^2+5y^2-3y-4x+9$

Exercises 6.5

1. $\{0\}$ 3. $\left\{a:a<\frac{3}{4}\right\}$ 5. $\{-2\}$ 7. $\{20\}$ 9. $\{n:n<-1\}$

11. $\{y:y\geqslant 10\}$ 13. $\left\{a:a>\frac{5}{2}\right\}$ 15. $\varnothing$

17. $\{n:n$ is a real number$\}$ 19. $\{y:y<2\}$ 21. $\{1\}$ 23. $\{-1\}$

25. $\left\{n:n<\frac{7}{25}\right\}$

Exercises 6.6

1. $3(a + b)$ 3. $5(a + 3)$ 5. $4x(y - 2z)$ 7. $3(a + 3b)$
9. $x(4x + 15y)$ 11. $9(y + 3x)$ 13. $t(3t + 1)$ 15. $6a(3b + 2c)$

17. $6x^2(2 + 3x)$ 19. $7a(a - 7b)$ 21. $2\pi r(h + r)$ 23. $\frac{1}{4}h(a + b)$

25. $5(x + y - 2z)$ 27. $rs(s - 2 + r)$ 29. $12a^2(3a + 2b - a^2 b)$
31. $xy(xz - y + z^2)$ 33. $7c^2(c^2 - 4c - 7)$ 35. $a^2 b^2(1 - 3ab - 2a^2 b^2)$
37. $6d^2(5n + d - 4p)$ 39. $0.2w(2 + 3w + w^2)$
41. $7a^3 b^2(2a^2 b^2 + 3ab - 7)$ 43. $12x^2 y^2(4x^2 - 12x + 9y^2)$
45. $xy^2(x^3 + x^2 y + xy^2 + y^3)$ 47. $3(3ax^2 - 4a^2 x^2 - 20x + a)$
49. $3x(-3x^2 + 12xy^2 - 2y^3)$ 51. $(r + n)(s + t)$ 53. $(2a - b)(x - y)$
55. $(2a + c)(x - 3y)$ 57. $(a - 3b)(2a - 3b)$
59. $(x^2 + y^2 + z^2)(a - b)$ 61. $(a + m)(b + n)$ 63. $(r + s)(x + y)$
65. $(x - t)(y - z)$ 67. $(a + c)(a + b)$ 69. $(x - y)(3m - n)$
71. $(x^2 - 1)(4x - 5)$ 73. $(3x^2 + 1)(x + 2)$

Exercises 6.7

1. $(a + b)(a - b)$ 3. $(x - 2)(x + 2)$ 5. $(ax + by)(ax - by)$
7. $(3ab + 5c)(3ab - 5c)$ 9. $(2a - 3x)(2a + 3x)$
11. $(x + y - 1)(x + y + 1)$ 13. $(9x - 8y)(9x + 8y)$
15. $(a^2 + b)(a^2 - b)$ 17. $(2 + x + y)(2 - x - y)$
19. $(x + y - a - b)(x + y + a + b)$ 21. $3(x - y)(x + y)$
23. $2a(3x - 2)(3x + 2)$ 25. $(x^2 + y^2)(x + y)(x - y)$
27. $6(a^2 + 2b^2)(a^2 - 2b^2)$ 29. $x^2(4x - 1)(4x + 1)$
31. $(5d - 3c)(5d + 3c)$ 33. $(3y + 2)(3y - 2)$
35. $(a + 0.1b)(a - 0.1b)$ 37. $4(4a - 3b)(4a + 4b)$
39. $t^2(1 + t)(1 - t)$ 41. $(0.2x - 0.3y)(0.2x + 0.3y)$

Exercises 6.8

1. $(x + 1)(x + 2)$ 3. $(y + 2)(y + 3)$ 5. $(x + 2)(x + 5)$
7. $(b + 3)(b + 9)$ 9. $(x + 6)(x + 9)$ 11. $(x - 1)(x - 3)$
13. $(a - 1)(a - 3)$ 15. $(x - 3)(x - 4)$ 17. $(a - 3)(a - 7)$
19. $(a - 8)(a - 9)$ 21. $(x - 3)(x + 5)$ 23. $(a - 1)(a + 2)$
25. $(x - 2)(x + 5)$ 27. $(w - 3)(w + 5)$ 29. $(x - 5)(x + 8)$
31. $(x + 3)(x - 5)$ 33. $(b - 7)(b + 6)$ 35. $(x + 6)(x - 8)$
37. $(s + 1)(s - 8)$ 39. $(x + 7)(x - 9)$ 41. $2(x - 2)(x + 6)$
43. $3(x - 2)(x + 5)$ 45. $(x^2 + 2)(x^2 - 14)$ 47. $(4 + x)(3 - x)$
49. $(8 + b^3)(9 - b^3)$ 51. $4, (x + 2)^2$ 53. $18, (a + 9)^2$

55. $9, (t + 3)^2$ 57. $14, (n + 7)^2$ 59. $49, (x - 7)^2$
61. $(y + 4x)(y - 2x)$ 63. $(x + 3y)(x + 6y)$ 65. $(x + y)^2$
67. $(c + 7d)(c - 4d)$ 69. $2(x + 2)^2$ 71. $y(yx + 10z)(yx - z)$
73. $3xy(3 - y)(4 + y)$ 75. Irreducible   77. $2(w - 18x)(w + 4x)$

Exercises 6.9

1. $(x + 1)(2x + 3)$ 3. $(x - 1)(2x - 3)$ 5. $(x - 1)(2x + 3)$
7. $(x + 1)(2x - 3)$ 9. Irreducible   11. $(a + 2)(2a + 1)$
13. $(3y - 5)(y + 1)$ 15. $2(x + 1)(2x + 1)$ 17. $(x + 5)(3x + 2)$
19. $(a + 7)(2a - 1)$ 21. $(t - 1)(3t - 2)$ 23. $(2x + 1)(3x + 4)$
25. $(2x + 3)(6x - 5)$ 27. $(x - 2y)(3x + 2y)$ 29. $(b + 7)(3b - 1)$
31. $(3t + 1)(9t - 2)$ 33. $(2s + t)(2s - 5t)$ 35. $(3y + 4)(2y - 1)$
37. $(t + 3)(2t + 5)$ 39. Irreducible   41. $2(1 + x)(1 - 4x)$
43. $(1 - 3a)(7 + 2a)$ 45. $(2t + 3)(6t - 5)$ 47. Irreducible
49. $(3a + 4b)(7a - 5b)$

Exercises 6.10

1. $8(a - b)(a + b)$ 3. $2(x + 6)(x - 6)$ 5. $ab(x + 2)(x + 4)$
7. $2(5 + y)(3 - y)$ 9. $m(m + 2)^2$ 11. $x^2(x - a)(x + a)$
13. $3(2t - 3)(t - 1)$ 15. $3b^2(8a^2 - ab - 12b^2)$
17. $7a(a - 3)(a + 3)$ 19. $a(x - 5)^2$ 21. $2a(2x + 5)(x + 7)$
23. $(x^2y^2 + 1)(xy + 1)(xy - 1)$ 25. $(x^4 + y^4)(x^2 + y^2)(x + y)(x - y)$
27. $2a(x - 2)(x - 6)$ 29. $8(a + 7)^2$ 31. $n(m + 3)^2$
33. $3(y^2 + 1)(y + 1)(y - 1)$ 35. $xyz(1 - x^3y^2z)(1 + x^3y^2z)$
37. $(m + 1)(m - 1)(n + 1)(n - 1)$ 39. $(xy - z)(a + bc)$
41. $(x + b)(y - z)(y + z)$ 43. $y(y^2 + 1)(3y - 2)$
45. $(a + b + c)(a + b - c)$ 47. $(2x + 1)(x^2 - 8x + 4)$
49. $(3x - 4)(x - 2)$

Review Exercises: Chapter 6

1. (a), (b), (c), (d), (e), (f), (g), (h), (i), (j), (m), (n)   3. (a), (f), (j)
5. (b), (d), (f), (h), (j)   7. (a) 1, (b) 2, (c) 13, (d) 0, (e) 1, (f) 2, (g) 3,
(h) none, (i) 0, (j) 2, (m) 1, (n) 0   9. xy 11. $5mn$

13. $-\dfrac{1}{4}x$ 15. $\dfrac{2}{15}xy$ 17. $-10abx$ 19. 0 21. $4x^2 - 2$

23. $-x - 11$ 25. $14x^2 + 3$ 27. $2a + 4b - 2c$
29. $x^2 + 2xy + y^2$ 31. $9x + 3y$ 33. $a^2 - ab - b^2$
35. $-x^2 + 4y^2$ 37. $7m^3 - 5n^2$ 39. $3a^2 + 2ab - 10b^2$

41. $r^3 + 4r^2 + 5r - 1$ 43. $-x^2 - 2x + 4$ 45. $9y^2 - 10y + 2$
47. $-2y^2$ 49. $-4b - 2a - 4c$ 51. x^5 53. $25b^{10}$ 55. $8a^5b$
57. $-28x^2yz^4$ 59. $27ab^4c^2$ 61. xy 63. $-16m^8$ 65. $21a^6b^6$
67. $-a^3 + 3a$ 69. $-6ab^2 + 15ab$ 71. $-6x^4y^2 + 6x^3y^4 + 18xy^6$
73. $x^2 + 2xy + y^2$ 75. $15x^2y^2 - 49xy + 24$ 77. $t^4 + 2t^2 - 15$
79. $-12x^2 - 12x + 45$ 81. $10a^2 - 2a - 36$ 83. $20a^2 - 3ab - 9b^2$
85. $2 + 17y + 6y^2 - y^3$ 87. $x^4 - 5x - 6$
89. $x^5 - 2x^4 + 3x^3 - 3x^2 + 2x - 1$ 91. $2x + 13$ 93. $5x - 19y$
95. $2y^2 + 2y + 8$ 97. $-7x^2 - 14x - 15$ 99. $x^2 - 32xy + 31y^2$

101. $\{10\}$ 103. $\{x : x > 3\}$ 105. $\{2\}$ 107. $\left\{\dfrac{4}{3}\right\}$

109. $\left\{x : x \leqslant \dfrac{8}{5}\right\}$ 111. $\{0\}$ 113. $\{-6\frac{2}{3}\}$ 115. $\left\{x : x \leqslant -\dfrac{5}{4}\right\}$

117. $\left\{-\dfrac{13}{10}\right\}$ 119. $\left\{\dfrac{400}{127}\right\}$ 121. $4(x - 2y)$ 123. $m(m - 2n)$

125. $3a(a - 4)$ 127. $xy(x + 3x^2y - 5y)$ 129. $2(2mn - 4m + 3n)$

131. $2xy(x^2 + 2xy + 8y^2)$ 133. $2(x^2 - 2xy + 6y^2)$

135. $\dfrac{1}{16}t(4t - 6tr + 7r^2)$ 137. $6a^3(4a^2 - 2a + 7)$ 139. $t\left(v + \dfrac{1}{2}gt\right)$

141. $(s + t)(r + n)$ 143. $(a^2 + b^2)(4 - x)$ 145. $(3z - 1)(a^2 + b^2)$
147. $(x + b)(x - a)$ 149. $2(x + y)$ 151. $(a + 2)(a - 2)$

153. $(2b - 1)(2b + 1)$ 155. $(7x - 4y)(7x + 4y)$ 157. $\left(\dfrac{1}{4} + m\right)\left(\dfrac{1}{4} - m\right)$

159. $(x^3 - 1)(x^3 + 1)$ 161. $2(a + 2b)(a - 2b)$
163. $a^3(ab - 1)(ab + 1)$ 165. $5(a - 2b^2)(a + 2b^2)$ 167. $(b - 6)^2$
169. $(x + 1)^2$ 171. $(m - 2)(m - 8)$ 173. $(a + 3x)^2$
175. $(x + 3y)(x - y)$ 177. $5(a + 3)(a - 1)$ 179. $(x + 9)(x - 4)$
181. $(2x + 3)(x - 3)$ 183. $(2m - n)(m + 5n)$

185. $(3a + b)(2a - 3b)$ 187. $\dfrac{1}{4}(4x + 1)(2x - 3)$

189. $2(3x - 1)(x + 5)$ 191. $(m + n)(m - n)$ 193. $3(x - 5)(x + 1)$
195. Irreducible   197. $(3ab + k)^2$ 199. $6(x + 2)(x - 2)$
201. $(x + y)(3x - a)$ 203. $3(x^2y^2 + 1)(xy + 1)(xy - 1)$
205. $a(b + 10)(b - 3)$ 207. $(y^2 + 7)(y - 6)$ 209. $(a^2 + 6)(a + 3)$
211. $x(4a^2 + 9a - 2)$ 213. $(x - a)(x - 3)$ 215. $4(3x + 7)^2$
217. Irreducible   219. $3y(x^2 + 5)(x^2 + 6)$

Cumulative Review D

1. -15 3. True   5. $-\dfrac{1}{14}$ 7. $\dfrac{8}{25}$ 9. $\{-14\}$ 11. $\{4\}$

13. $\{-1\}$ 15. $\{x : x < -2\}$ 17. $\{x : x < 2\}$ 19. $\{x : x > 0\}$

21.

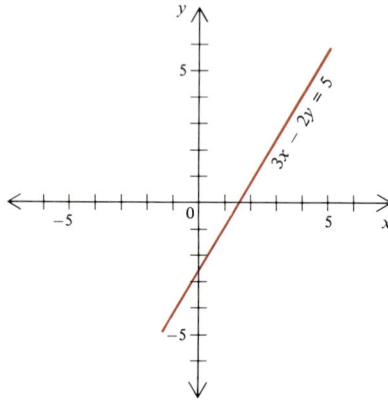

23. $\{(3, -2)\}$ 25. 12 kg   27. 4:35 P.M.   29. 2 hrs., 15 min.

31. $8\sqrt{2}$ 33. $12\sqrt{2}$ 35. $\dfrac{\sqrt{6}}{2}$ 37. $2p + q - 4r + 2$

39. $x^2 + 7x + 12$ 41. $x^3 + 8$ 43. $2x(7y^5 - 10z)$ 45. $(8 - x)^2$
47. $(x - b)(x + 5)$ 49. $(2a + 5b)^2$

CHAPTER 7

Exercises 7.1

1. $(-1) \cdot \dfrac{y}{(-1)(x - y)}$ 3. $\dfrac{(-1)(5)}{(-1)(3x - 2)}$ 5. $\dfrac{(-1)(a - b)}{(-1)(x - y)}$

7. $(-1) \cdot \dfrac{x - 1}{(-1)(2 - x)}$ 9. $\dfrac{(-1)(x^2 - xy - y^2)}{(-1)(a - b)}$

11. $\dfrac{(-1)(a - b)(a + b)}{(-1)(x - y)}$ 13. $(-1) \cdot \dfrac{(-1)(-6m^2 + 2m - 4)}{m + 2}$

15. $\dfrac{(x + y)(-1)(2x - y)}{(a + b)(-1)(2a - b)}$ 17. $\dfrac{4}{7}$ 19. $\dfrac{5}{9}$ 21. $\dfrac{5}{11}$ 23. $\dfrac{5a}{3}$

25. $\dfrac{x}{2y}$ 27. $\dfrac{4x}{7z}$ 29. $\dfrac{2}{3a}$ 31. $\dfrac{1}{7}$ 33. $\dfrac{x-1}{x+1}$

35. $\dfrac{2}{3a}$ 37. $\dfrac{4}{m+n}$ 39. $\dfrac{p+q}{p-q}$ 41. $\dfrac{2(2r-3s)}{4r^2+9s^2}$ 43. $\dfrac{2(a-b)}{3}$

45. $\dfrac{x-1}{x+1}$ 47. $\dfrac{c+3}{c-3}$ 49. $\dfrac{a+b}{a-2b}$ 51. $\dfrac{x-3}{x-1}$ 53. $\dfrac{a}{r}$

55. $\dfrac{2b+1}{b-1}$ 57. $\dfrac{x-1}{x+2}$ 59. $\dfrac{-2}{x+3}$ 61. $\dfrac{x-y}{x+y}$ 63. $\dfrac{3(1-t)}{4}$

Exercises 7.2

1. $\dfrac{10}{21}$ 3. $\dfrac{3}{10}$ 5. $\dfrac{2}{3}$ 7. $\dfrac{3}{4}$ 9. $\dfrac{3ac}{b^2}$ 11. $\dfrac{3}{4t}$ 13. $5y^3z$

15. $\dfrac{4}{3}$ 17. $\dfrac{2r(s-1)}{3s(r+2)}$ 19. $\dfrac{a^2-9}{16-b^2}$ 21. $\dfrac{a(a+b)}{2(a-b)}$ 23. 1

25. $\dfrac{rt}{5w}$ 27. $\dfrac{b}{2a}$ 29. $\dfrac{5}{12}$ 31. $\dfrac{7}{3}$ 33. $\dfrac{5}{6}$ 35. $\dfrac{a}{a-2}$ 37. $\dfrac{t-3}{6t}$

39. $\dfrac{2(x-2)}{7(x+2)}$ 41. $\dfrac{x+y}{x+5y}$ 43. $\dfrac{(a-7)(2a+1)}{(a+1)^2}$ 45. $\dfrac{1}{x}$

47. $\dfrac{2a}{5}$ 49. $\dfrac{x+2}{x-2}$ 51. $\dfrac{8}{15}$ 53. $\dfrac{3}{2}$ 55. $\dfrac{x^2}{y^2}$ 57. $\dfrac{3x(4y+1)}{4y(3x-1)}$

59. $\dfrac{x+1}{(x-1)^2}$ 61. $\dfrac{4mn^2}{3ac^2}$ 63. $\dfrac{8}{3}$ 65. $\dfrac{3}{8}$ 67. $\dfrac{ad}{bc}$

69. $\dfrac{m}{2(m+n)}$ 71. 2   73. $\dfrac{y-3}{y+7}$ 75. $\dfrac{x+3}{x-3}$ 77. $\dfrac{2(a+x)(a+2x)}{x(a-x)}$

79. $\dfrac{2x+1}{2x-1}$ 81. 3   83. $\dfrac{3}{2y}$ 85. $-x$ 87. $\dfrac{1}{x}$ 89. $\dfrac{-x}{6}$

Exercises 7.3

1. $\dfrac{4}{3}$ 3. $\dfrac{2}{3}$ 5. $\dfrac{4}{w}$ 7. $\dfrac{m-n}{r}$ 9. $\dfrac{25-4t}{x}$ 11. 1   13. $\dfrac{1}{r}$

15. $\dfrac{1}{x-2}$ 17. 2   19. $\dfrac{a-5}{a-3}$ 21. $12x$ 23. rst 25. $6r^2$

27. $6(x-y)$ 29. $2(x^2-16)$ 31. $6(a-3)(a+3)(a+2)$

33. $(x - 4)(x + 3)(x + 5)(x - 2)$ 35. $\dfrac{13}{8}$ 37. $\dfrac{23a}{10}$ 39. $\dfrac{11}{3y}$

41. $\dfrac{29y}{12x}$ 43. $\dfrac{23a}{20x}$ 45. $\dfrac{81x^2 + 40y^2}{144}$ 47. $\dfrac{a^2 - bax}{b^4}$

49. $\dfrac{x - 3y}{8}$ 51. $-\dfrac{2x + 11}{6}$ 53. $\dfrac{x^2 + 4x + 6}{2x}$

55. $-\dfrac{10x^2 + xy + 12y^2}{60x^2 y^2}$ 57. $\dfrac{5(2x^2 - y^2)}{12x^2 y^2}$

59. $\dfrac{-25a^2 + 45a^2 b + 6a - 5ab - 30ab^2 + 24b}{30a^3 b^3}$ 61. $\dfrac{11x - 31}{(x - 2)(x - 3)}$

63. $\dfrac{x(3x - 4)}{(x - 2)(x - 1)}$ 65. $\dfrac{x^2 + y^2}{x^2 - y^2}$ 67. $\dfrac{4y^2 - 8y + 25}{(y - 6)(y + 5)}$

69. $\dfrac{x^2 + 2x + 3}{(x - 1)(x + 2)}$ 71. $\dfrac{3 + 4a + 4b}{a + b}$ 73. $\dfrac{a^2 + 3ab + b^2}{a + b}$

75. $\dfrac{2b(3a + 4b)}{a + 3b}$ 77. $\dfrac{2(3x - 4y)}{3(x + y)}$ 79. $\dfrac{19}{6(x + 1)}$ 81. $\dfrac{2y^2 - x^2}{(x - y)^2}$

83. $\dfrac{3(8m^2 - 13mn - 5n^2)}{80(m + n)(m - n)}$ 85. $\dfrac{5a^2 + 12a + 6}{6(a + 3)(a - 3)(a + 2)}$

87. $\dfrac{2(x + 10)(x - 1)}{(x - 9)(x + 9)}$ 89. $\dfrac{13}{a + 5}$ 91. $\dfrac{y + 8}{y - 1}$ 93. $\dfrac{x + 1}{x - 1}$

95. $\dfrac{5a - 13}{3(a + 5)(a - 5)}$ 97. $\dfrac{a^3 - 4a^2 - a - 6}{(2a - 3)(a + 3)(3a - 2)(a - 3)}$

99. $\dfrac{3}{a - b}$ 101. $\dfrac{x}{2(x + 4)}$ 103. $\dfrac{-11a + 104}{6(a - 8)(a + 8)}$

Exercises 7.4

1. $x + 3$ 3. $b + 5$ 5. $d + 7$ 7. $x + 9$ 9. $y - 3$
11. $x + 3$ 13. $2x + 1$ 15. $2y^2 + 7$ 17. $5b + 4$
19. $x + 3yz$ 21. $a^2 + 2ab + b^2$ 23. $4x - 7y$

25. $2b^2 - 3bc + 4c^2$ 27. $\dfrac{1}{3}x + \dfrac{4}{9}$

29. $\dfrac{3}{4}x - \dfrac{1}{16}$ 31. $2x - 9 + \dfrac{55}{x + 6}$ 33. $a + \dfrac{b^2}{a + b}$

35. $6x + 13y + \dfrac{62y^2}{2x - 5y}$ 37. $x^2 - 3x + 13 + \dfrac{-33}{2x + 3}$

39. $-3x^2 - 7x + \dfrac{5}{4x + 5}$ 41. $x^2 - x + 1 + \dfrac{-2}{x + 1}$

43. $x^2 - xy + y^2 + \dfrac{-2y^3}{x + y}$ 45. $4x^2 + 12x + 15 + \dfrac{-6x + 12}{x^2 + x - 2}$

47. $3x - 2$ 49. $y^2 - y + 4$ 51. $3a + 4 + \dfrac{2}{a - 5}$

53. $2y - 3 + \dfrac{-2y}{4y^2 - 3y + 2}$ 55. $2x - 4 + \dfrac{8}{3x + 7}$

57. $x^2 + 2x + 4$ 59. $2y^2 - 3y - 4$ 61. $\dfrac{9}{2}x - 9 + \dfrac{-57}{2x^2 + 4x - 4}$

63. $\dfrac{5}{2}a + 11 + \dfrac{\frac{153}{2}a + 29}{2a^2 - 8a - 3}$ 65. $\dfrac{7}{2}x^2 - \dfrac{7}{4}x + \dfrac{7}{8} + \dfrac{-\frac{15}{8}}{2x + 1}$

Exercises 7.5

1. $\dfrac{8}{3}$ 3. 4   5. 29   7. 2   9. 1   11. 3   13. No solution

15. 3   17. $-\dfrac{1}{3}$ 19. 6   21. 4   23. $-\dfrac{7}{3}$ 25. 7   27. 6

29. $-\dfrac{24}{7}$ 31. 9   33. 3   35. 1   37. No solution   39. 3

41. 1   43. 1   45. -9

Exercises 7.6

1. $\dfrac{9}{14}$ 3. $\dfrac{22}{15}$ 5. 120, 176   7. 10, 15   9. 4 hrs.   11. 51

13. 87   15. 9 yrs.   17. 4 hrs., $21\dfrac{9}{11}$ min.   19. 5 hrs.,

$10\dfrac{10}{11}$ min.   21. $\dfrac{7}{13}$ 23. 27,107   25. 15   27. 12 hrs.

29. 120 eggs

Review Exercises: Chapter 7

1. $(-1) \cdot \dfrac{(-1)5}{a + b}$ 3. $(-1) \cdot \dfrac{(-1)(x - y)}{x + y}$ 5. $(-1) \cdot \dfrac{2a + 1}{(-1)(a)}$

7. $\dfrac{(x + y)(-1)(x - y)}{(-1)(x - 2)}$ 9. $\dfrac{(-1)(a - b - c)}{(-1)(a - b)}$ 11. $2x$

13. $\dfrac{4}{x - y}$ 15. $\dfrac{2x + 3y}{2x + y}$ 17. $\dfrac{a + 2}{a - 3}$ 19. $-\dfrac{2x + 3}{2x - 3}$ 21. $\dfrac{3x^2}{2y^2}$

23. $\dfrac{(m - n)^2}{(m + n)^2}$ 25. $6r$ 27. $\dfrac{4(x + 2)}{3}$ 29. $x + 2$

31. $2(x + y)$ 33. $\dfrac{(x + 2)(x + 4)}{x + 5}$ 35. $\dfrac{x + 2}{x - 2}$ 37. $\dfrac{1}{3}$

39. $\dfrac{x^2 + x}{y^2 + y}$ 41. $\dfrac{x + 1}{(x - 1)^2}$ 43. $\dfrac{a - b}{2a}$ 45. $\dfrac{1}{y(x + 2)}$

47. $x - 1$ 49. $\dfrac{x + y}{y - x}$ 51. $\dfrac{5}{3a}$ 53. $\dfrac{3x + 1}{w}$ 55. $\dfrac{8 - y - y^2}{y^2 - 16}$

57. $\dfrac{3a + 14}{15ab}$ 59. $\dfrac{5u - 2}{2u + 6}$ 61. $\dfrac{9 - 5y}{3y(y + 2)}$ 63. $\dfrac{1}{m + 1}$

65. $\dfrac{-x^2 + 8x - 18}{x(x + 3)(x - 3)}$ 67. $\dfrac{5a + b + 4}{a^2 - b^2}$ 69. $\dfrac{6x^2 - 2x + 2}{3x^2}$

71. $x + 2$ 73. $3x + 7 + \dfrac{18}{2x - 3}$ 75. $x - 6 + \dfrac{-26}{x - 7}$

77. $2x - 2 + \dfrac{-1}{x - 2}$ 79. $t - 8$ 81. $n^2 - \dfrac{1}{2}n + \dfrac{19}{4} + \dfrac{\frac{9}{4}}{2n - 3}$

83. $\dfrac{1}{2}x^2 - \dfrac{1}{4}x + \dfrac{3}{8} + \dfrac{-\frac{5}{8}}{2x - 1}$ 85. $\dfrac{1}{2}x^3 - \dfrac{1}{4}x^2 - \dfrac{1}{8}x + \dfrac{7}{16} + \dfrac{-\frac{57}{16}}{2x - 1}$

87. $\dfrac{3}{4}x + \dfrac{5}{16} + \dfrac{\frac{11}{16}}{4x + 1}$ 89. 2 91. 5 93. 1 95. 5

97. No solution   99. 3 101. 2 103. 1 105. 3

107. $\dfrac{1}{4}$ 109. 3 111. $18, 30$ 113. Train, 720 miles;

boat, 576 miles; bus, 240 miles   115. $\dfrac{17}{31}$ 117. 64

119. $15\frac{3}{4}$ hrs.   121. 4 hrs., $5\frac{5}{11}$ min.   123. $7\frac{1}{5}$ min.

125. $166\frac{2}{3}$ days

Cumulative Review E

1. -11 3. $14x^4$ 5. $-18a - 15b - 12$ 7. $30\sqrt{15}$

9. $\frac{2}{3}\sqrt{3}$ 11. $x^6 y$ 13. $\{x : x < -3\}$ 15. $\{x : x < -8\}$

17. $\left\{x : x < \frac{5}{3}\right\}$ 19. $\left\{x : x < \frac{2}{15}\right\}$ 21. $\left\{-\frac{14}{3}\right\}$ 23. $\{3\}$

25. $\left\{\frac{19}{3}\right\}$

27.

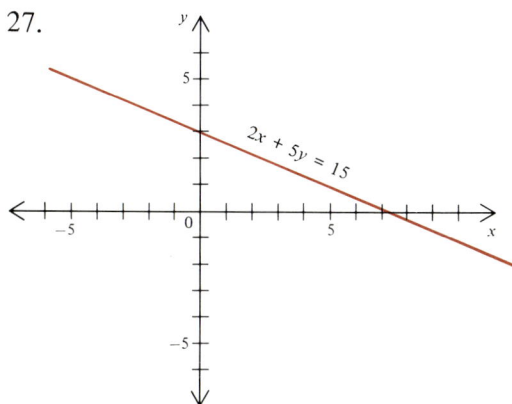

29. 2 km/hr.   31. $144   33. 80   35. $x^2 + 5xy + 6y^2$
37. $x^4 - x^2 y^2 + 4xy^3 - 4y^4$ 39. $4(x + 5)(x - 2)$

41. $3(y + 7)(y - 6)$ 43. $(x - 6)(x - 5)$ 45. $\frac{3}{4}$

47. $\frac{11x - 4}{15}$ 49. $(-3, 2)$

CHAPTER 8

Exercises 8.1

1. $47, 16, -5, -8, -22$ 3. $-40, -31, -7, \dfrac{1}{2}, 15$ 5. $-5\frac{3}{4}, -1\frac{3}{4},$

$6\frac{1}{4}, 14\frac{1}{4}, 46\frac{1}{4}$ 7. $1, 2, 3, 0, 5$ 9. $-5, -3, \dfrac{3}{4}, \dfrac{9}{5}, 10$

11. $\{-2, -6, 2, -3, -10\}$ 13. $\{1, 9, 25, 0, 49\}$ 15. $\{0, 1\}$
17. $\{-1, -3, -5, 0, -7\}$ 19. $\{2, 10, 26, 1, 50\}$
21. $\{-6\frac{1}{2}, \frac{1}{2}, 1\frac{1}{2}, 2\frac{1}{2}, 9\frac{1}{2}\}$ 23. $\{-2, -1, 0, 1, 2\}$
25. $\{-21, 0, 3, 6, 27\}$ 27. $\{-27, -6, -3, 0, 21\}$

29. $\left\{\dfrac{51}{4}, -\dfrac{9}{4}, -\dfrac{3}{4}, \dfrac{3}{4}, \dfrac{45}{4}\right\}$ 31. $y = x - 1$ 33. $b = a^2$

35. $t = s - 2$ 37. $B = \dfrac{A + 2}{2}$ 39. If u is an odd number, v is 1,
and if u is an even number v is 2.   41. $A = S^2, D = R =$ positive real numbers   43. $A = \pi r^2, D = R =$ positive real numbers
45. $y = |x|, D =$ all real numbers, $R =$ nonnegative real numbers
47. $C = \dfrac{I}{2.54}, D = R =$ nonnegative real numbers

49.

Date in June	1	2	3	4	5	6	7	8	9	10
Temp. in °C	17	22	26	20	14	23	17	27	24	21

$D = \{1, 2, 3, 4, 5, 6, 7, 8, 9, 10\}$,
$R = \{14, 17, 20, 21, 22, 23, 24, 26, 27\}$

Exercises 8.2

1. (a) $f(1) = 4, f(10) = 13, f(15) = 18, -f(2) = -5$
   (b) Domain $= \{1, 2, 3, 4, 5, 6, 10, 15, 20\}$
   (c) Range $= \{4, 5, 6, 7, 8, 9, 13, 18, 23\}$
   (d) $f(x) = x + 3$
3. (a) $f(1) = -2, f(10) = -20, f(15) = -30, -f(2) = 4$
   (b) Domain $= \{-10, -5, 0, 1, 2, 3, 4, 5, 10, 15\}$
   (c) Range $= \{20, 10, 0, -2, -4, -6, -8, -10, -20, -30\}$
   (d) $f(x) = -2x$
5. (a) $f(1) = 3\frac{1}{2}, f(10) = 8, f(15) = 10\frac{1}{2}, -f(2) = -4$
   (b) Domain $= \{-2, -1, 0, 1, 2, 3, 4, 5, 10, 15, 20\}$

(c) Range = $\{2, 2\frac{1}{2}, 3, 3\frac{1}{2}, 4, 4\frac{1}{2}, 5, 5\frac{1}{2}, 8, 10\frac{1}{2}, 13\}$

(d) $f(x) = \dfrac{x+6}{2}$

7. $2, 1\frac{5}{8}, \frac{1}{2}, \dfrac{4-\sqrt{3}}{2}, \dfrac{4-a}{2}$ 9. $-2, 8, 4x+2, -3, 20$ 11. $9, 19,$

$-149, -48$ 13. $3, \dfrac{10}{3}, 2, 5, 3s-1$ 15. $9, \dfrac{8}{5}, -\dfrac{1}{2}, 5, \dfrac{91}{4}$

17. All rational numbers that can be written with a denominator of 2.
19. The set of integers greater than or equal to 0 which are perfect

squares.   21. $\left\{\dfrac{1}{2}, 2, \dfrac{3}{2}, 4, \dfrac{5}{2}\right\}$ 23. 0   25. $\sqrt{a^2 - 2a - 15}$

27. All nonnegative real numbers.   29. No, the domain of f is the set of all rational numbers, while the domain of g is the set of all non-zero rational numbers.   31. $g(x) + 2$ 33. $|g(x) + 2| + 2$
35. $(g(x))^2 + 4g(x) + 2$

Exercises 8.3

1. $f(x) = 7 - x$ 3. $f(x) = \dfrac{1}{2}x + 3$ 5. $f(x) = \dfrac{1}{2}$ 7. $f(x) = \dfrac{3x-6}{5}$

9. $f(x) = \dfrac{10-x}{7}$ 11. $f(x) = \dfrac{7x+2}{3}$ 13. $f(x) = -x$ 15. $f(x) = 0$

17. $f(x) = \dfrac{1}{x}, x \neq 0$ 19. $f(x) = x^3 + 3x^2 + 3x - 1$

21. $f(x) = -\dfrac{1}{x-1}, x \neq 1$ 23. $f(x) = \dfrac{5}{2x-3}, x \neq \dfrac{3}{2}$

25. $f(x) = \dfrac{5x^2+1}{2x}, x \neq 0$

27.

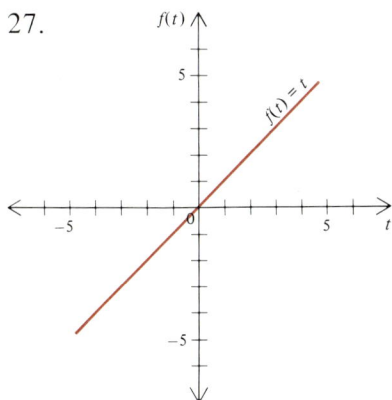

29.

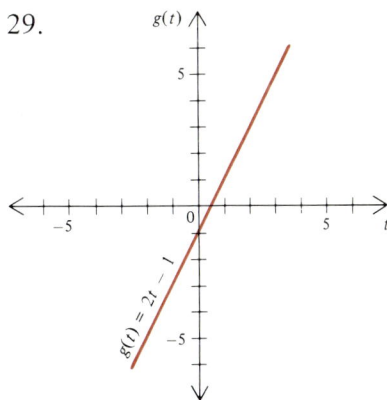

31.

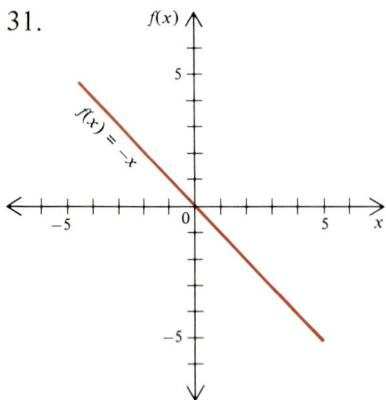

33.

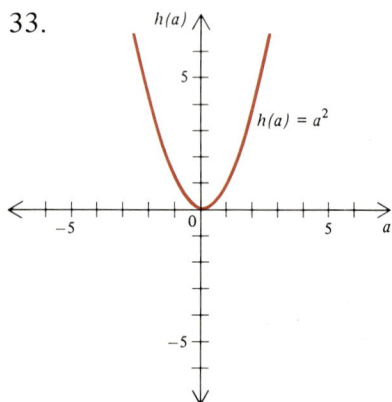

35.

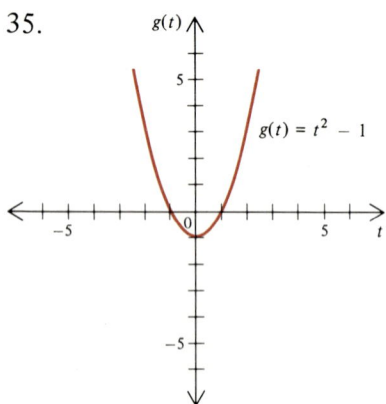

37.

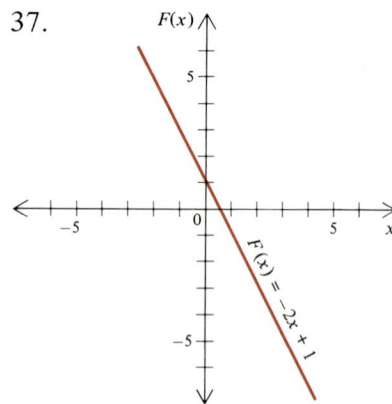

39.

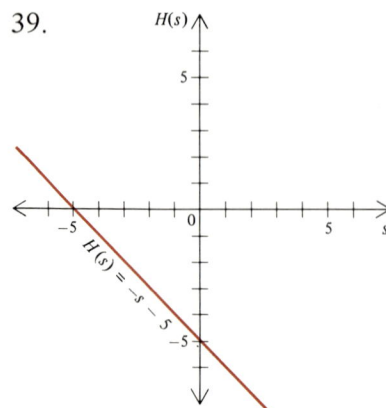

41.

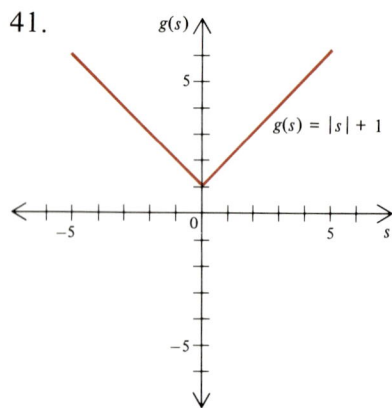

43.

$g(x) = |x + 1|$

45.

$g(y) = 2|y|$

47.

$f(x) = -2|x|$

49.

$f(x) = \frac{1}{2}x^2$

51.

$f(x) = |x| + 2$

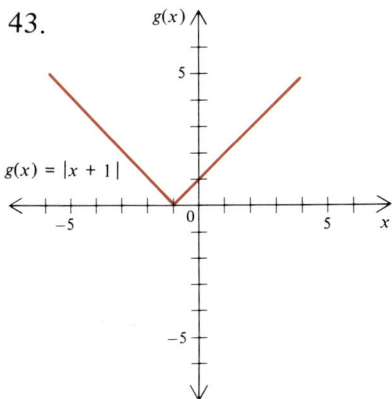

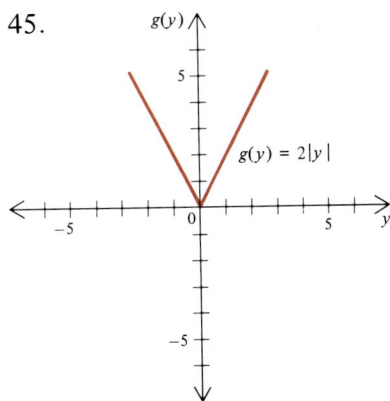

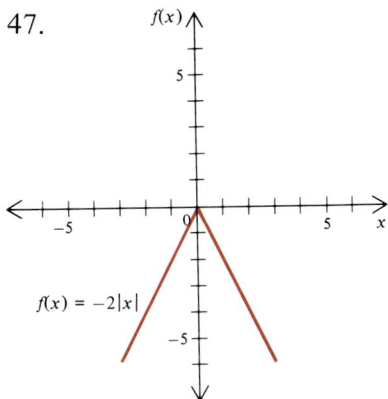

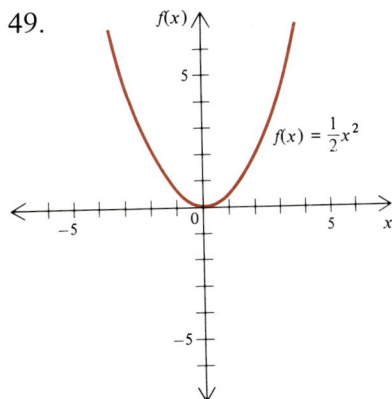

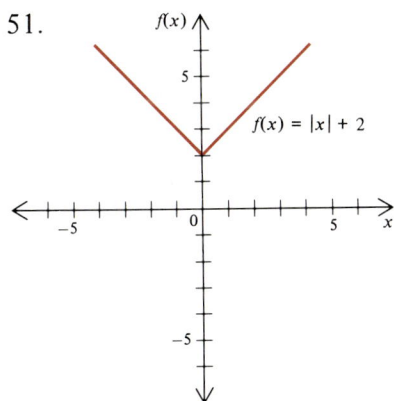

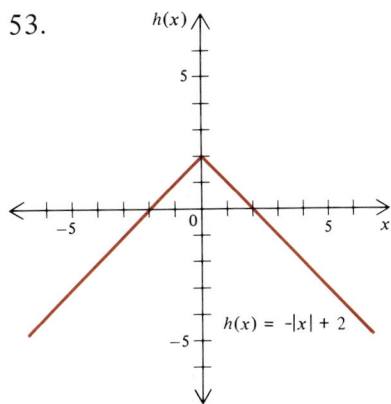

Range: All real numbers ≥ 2

53.

$h(x) = -|x| + 2$

Range: All real numbers ≤ 2

55.

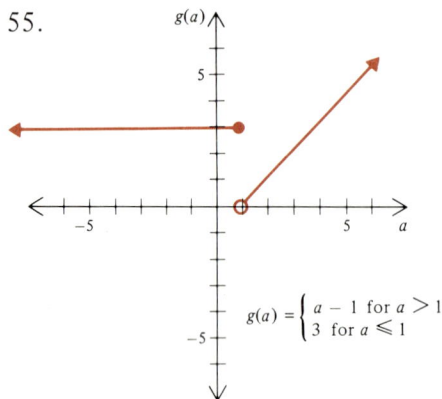

$$g(a) = \begin{cases} a - 1 & \text{for } a > 1 \\ 3 & \text{for } a \leqslant 1 \end{cases}$$

Range: All real numbers > 0

57.

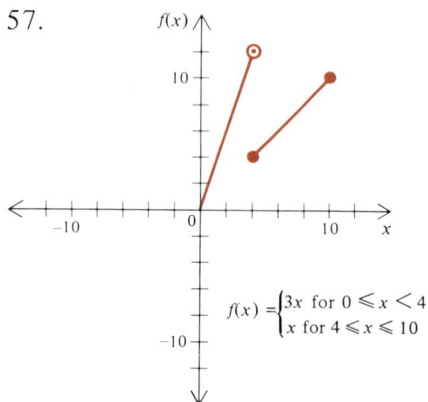

$$f(x) = \begin{cases} 3x & \text{for } 0 \leqslant x < 4 \\ x & \text{for } 4 \leqslant x \leqslant 10 \end{cases}$$

Range: All real numbers $\geqslant 0$ and $\leqslant 10$

59.

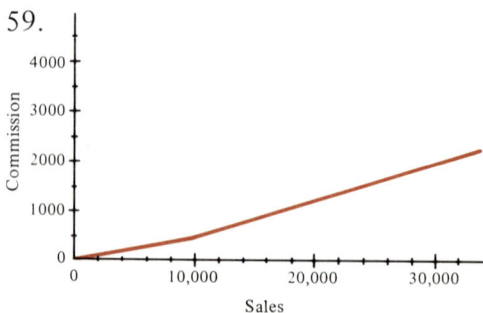

Review Exercises: Chapter 8

1. $7, 8, 9, 10, 11$ 3. $-23, -8, 13, 22, 37$ 5. $\frac{1}{4}, \frac{1}{3}, \frac{3}{8}, \frac{2}{5}, \frac{5}{12}$
7. $-13\frac{1}{2}, -8, -6\frac{1}{4}, 2, 3\frac{1}{8}$ 9. $5,000, 50,000, 500,000, 5,000,000,$
$50,000,000$ 11. $7, 16.5, 25, 38, 49$ 13. $2, 5, 12, 17, 107$
15. $5, 10, 12, 25, 101$ 17. $-4, -2, 0, 2, 4$ 19. $-3, -2, -1, 0, 1, 2$
21. $y = 3x$ 23. $y = x + 3$ 25. $y = 3x + 2$ 27. $y = 3x - 1$
29. $y = \frac{1}{2}x + 3$ 31. 7 33. $2p + 1$ 35. 4 37. $a + 1$
39. 3 41. 0 43. $a + 1$ 45. 22 47. $a^2 + 4a + 1$

49. 15   51. $f(x) = -2x - 3$ 53. $f(x) = \frac{3}{2}x$ 55. $f(x) = \frac{5}{2}x - 11$

57. $f(x) = -\dfrac{3x + 21}{2}$ 59. $f(x) = \dfrac{-10x + 3}{2}$

61.

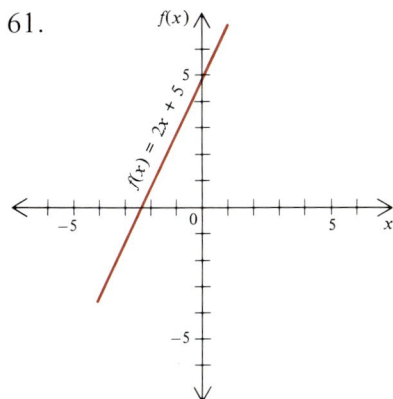

$f(x) = 2x + 5$

63.

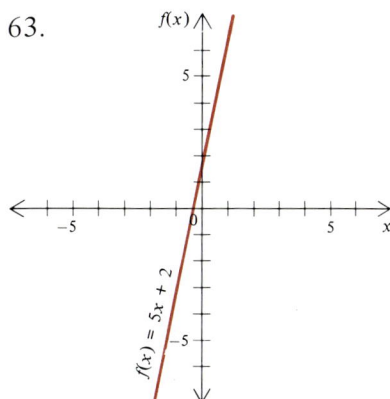

$f(x) = 5x + 2$

65.

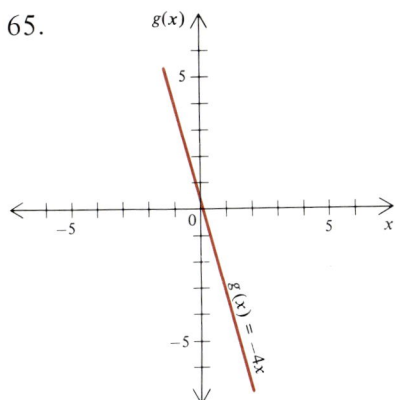

$g(x) = -4x$

67.

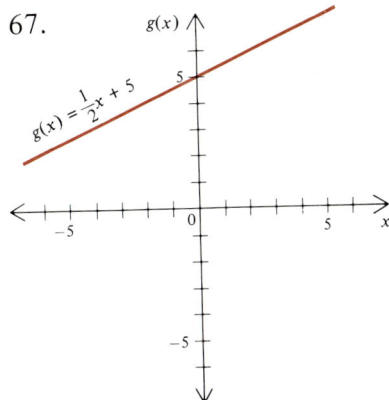

$g(x) = \frac{1}{2}x + 5$

69.

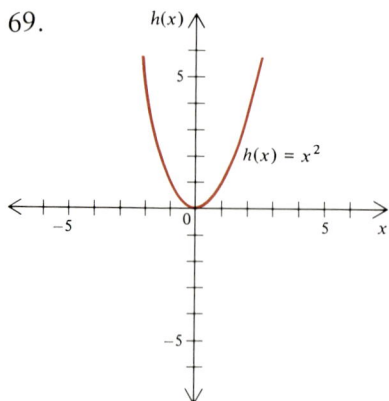

71.

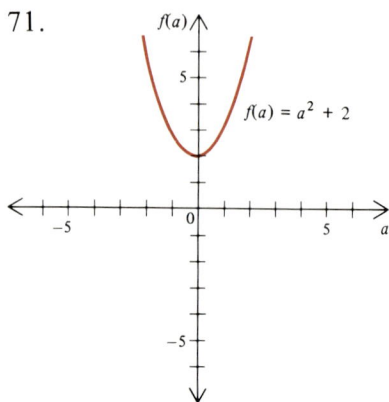

73.

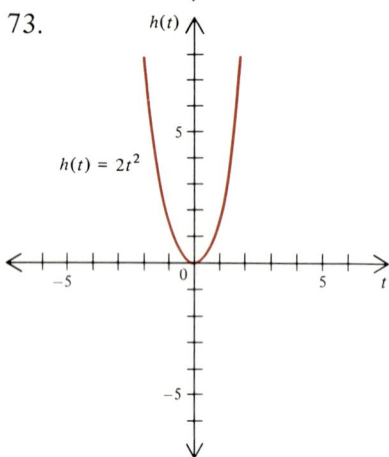

75.

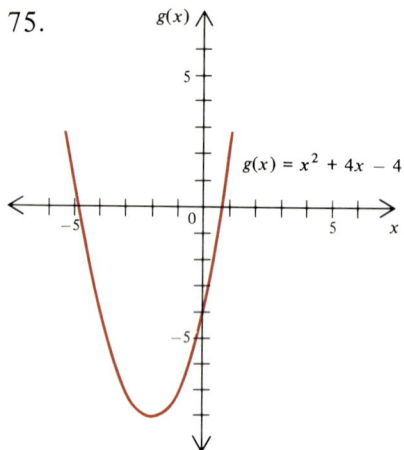

77.

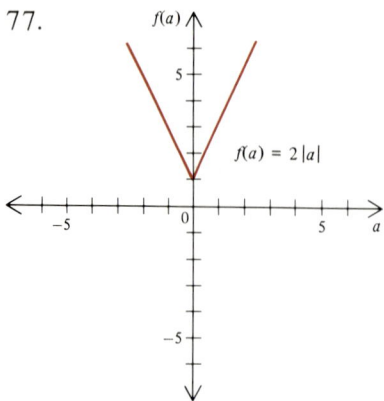

79.

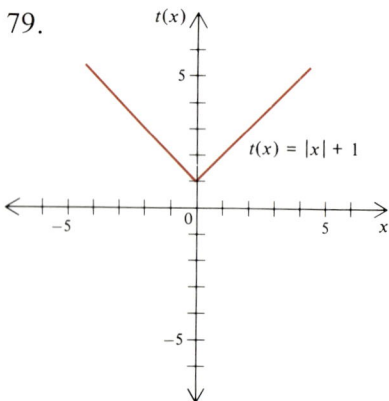

Cumulative Review F

1. $36ab + 8b$ 3. $6a^6b$ 5. $x^2 + 1$ 7. $-\dfrac{4}{9}x^3y^4z$

9. $(x - 7)(x + 8)$ 11. $(1 - x)(1 + x)(1 + x^2)(1 + x^4)$

13. $(2x - 5)(3x - 2)$ 15. $\dfrac{a}{2b}$ 17. $5x + 2y$ 19. $\dfrac{2(x + 4)}{3(x - 3)}$

21. $\dfrac{x - 5}{xy}$ 23. $\dfrac{3(a - 2)}{7(a + 3)}$ 25. $8\sqrt{3}$ 27. 15   29. $\dfrac{\sqrt{6x}}{3}$

31. $\{6\}$ 33. $\left\{\dfrac{15}{7}\right\}$ 35. $\{x : x \text{ is a real number}\}$ 37. $p = \dfrac{a}{1 + rt}$

39. 6

41.

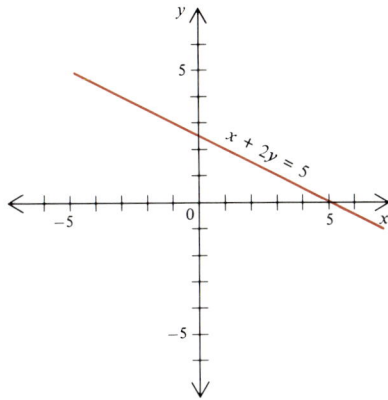

43. $(2, 3)$ 45. $\left(\dfrac{35}{17}, \dfrac{42}{17}\right)$ 47. 6 hrs.   49. 100 cubic centimeters

CHAPTER 9

Exercises 9.1

1. $\{-6, 6\}$ 3. $\{3\sqrt{2}, -3\sqrt{2}\}$ 5. $\left\{\dfrac{7}{2}, -\dfrac{7}{2}\right\}$ 7. $\{-9, 9\}$

9. $\{2\sqrt{2}, -2\sqrt{2}\}$ 11. $\{3, -3\}$ 13. $\{5, -5\}$ 15. $\{7, -1\}$

17. $\{-7, -1\}$ 19. $\{8, 2\}$ 21. $\{-7, 1\}$ 23. $\{9, -7\}$

25. $\{-5 + 2\sqrt{2}, -5 - 2\sqrt{2}\}$ 27. $\{3 + 3\sqrt{3}, 3 - 3\sqrt{3}\}$

29. $\left\{\dfrac{\sqrt{3}}{2}, -\dfrac{\sqrt{3}}{2}\right\}$ 31. $\{-3, 3\}$ 33. $\{\sqrt{2}, -\sqrt{2}\}$ 35. $\{\sqrt{3}, -\sqrt{3}\}$

37. 4   39. 49   41. $\dfrac{25}{4}$ 43. $\dfrac{1}{25}$ 45. $\dfrac{4}{49}$ 47. $\{1, 5\}$

49. $\{-2, -6\}$　　51. $\{-3, -5\}$　　53. $\{4, -1\}$　　55. $\{1, 4\}$

57. $\left\{3, -\dfrac{1}{2}\right\}$　　59. $\{1, -5\}$　　61. $\left\{\dfrac{5}{3}, 1\right\}$　　63. $\left\{-\dfrac{2}{3}\right\}$

65. $\{-2 + 2\sqrt{3}, -2 - 2\sqrt{3}\}$　　67. $\left\{\dfrac{1}{2} + \dfrac{\sqrt{5}}{2}, \dfrac{1}{2} - \dfrac{\sqrt{5}}{2}\right\}$

69. $\left\{\dfrac{2 + 2\sqrt{3}}{3}, \dfrac{2 - 2\sqrt{3}}{3}\right\}$　　71. $\left\{\dfrac{7}{12} + \dfrac{\sqrt{13}}{12}, \dfrac{7}{12} - \dfrac{\sqrt{13}}{12}\right\}$

73. $\left\{\dfrac{9}{2}, -\dfrac{3}{4}\right\}$　　75. $\left\{-\dfrac{1}{2}, -\dfrac{2}{3}\right\}$　　77. $\{3 + 2\sqrt{5}, 3 - 2\sqrt{5}\}$

79. $\{0, -5\}$

Exercises 9.2

1. $\{-1, -5\}$　　3. $\{-5, -2\}$　　5. $\{-8, 5\}$　　7. $\{-5, -7\}$　　9. $\{-11, -5\}$

11. $\{3\}$　　13. $\{3, -2\}$　　15. $\left\{-3, \dfrac{5}{2}\right\}$　　17. $\left\{-4, \dfrac{5}{2}\right\}$

19. $\left\{\dfrac{2}{3}, -\dfrac{1}{3}\right\}$　　21. $\left\{\dfrac{3}{2}, \dfrac{4}{3}\right\}$　　23. $\left\{\dfrac{1}{3}, -\dfrac{1}{4}\right\}$　　25. $\left\{-\dfrac{1}{2}, \dfrac{1}{2}\right\}$

27. $\left\{\dfrac{3}{4}, -\dfrac{3}{4}\right\}$　　29. $\left\{\dfrac{7}{8}\right\}$　　31. $\{\pm\sqrt{13}\}$　　33. $\left\{\dfrac{-7 \pm 3\sqrt{5}}{2}\right\}$

35. $\left\{\dfrac{-11 \pm \sqrt{117}}{2}\right\}$　　37. $\left\{\dfrac{-3 \pm 3\sqrt{5}}{2}\right\}$　　39. $\left\{\dfrac{-1 \pm \sqrt{13}}{4}\right\}$　　41. $\varnothing$

43. $\{-4, 2\}$　　45. $\left\{\dfrac{-5 \pm \sqrt{37}}{2}\right\}$　　47. $\left\{-2, \dfrac{9}{5}\right\}$　　49. $\left\{-\dfrac{11}{5}, \dfrac{6}{5}\right\}$

51. $\{8.541705, -0.491705\}$　　53. $\{224.09673, -24.096735\}$
55. $\{13.645311, -7.145314\}$　　57. $\{118.43908, 81.560915\}$
59. $\{2.9396602, -24.890878\}$

Exercises 9.3

1. $\{2, 3\}$　　3. $\{0, 4\}$　　5. $\left\{-\dfrac{1}{4}, \dfrac{2}{5}\right\}$　　7. $\{3, 5, -4\}$

9. $\left\{0, \dfrac{1}{3}, -\dfrac{5}{2}, \dfrac{5}{3}\right\}$　　11. $\{7, 1\}$　　13. $\{4, 1\}$　　15. $\{-4, -5\}$

17. $\{5, 10\}$　　19. $\{7, 11\}$　　21. $\left\{\dfrac{1}{2}, -3\right\}$　　23. $\left\{\dfrac{4}{3}, -1\right\}$

25. $\left\{-\dfrac{2}{5}, -2\right\}$　　27. $\left\{\dfrac{5}{3}, -1\right\}$　　29. $\left\{-\dfrac{2}{3}, \dfrac{5}{2}\right\}$　　31. $\left\{0, \dfrac{1}{6}\right\}$

33. $\{8, -8\}$　　35. $\left\{\dfrac{5}{2}, -\dfrac{5}{2}\right\}$　　37. $\left\{\dfrac{3}{2}\right\}$　　39. $\{-13\}$

41. $\left\{-\dfrac{1}{2}, -1\right\}$ 43. $\left\{-\dfrac{11}{2}, 4\right\}$ 45. $\left\{\dfrac{3}{4}, 8\right\}$ 47. $\{-4, 2\}$

49. $\varnothing$ 51. $\left\{\dfrac{3}{2}, -2\right\}$ 53. $\left\{\dfrac{1}{7}\right\}$ 55. $\left\{-\dfrac{3}{2}, -2\right\}$

Exercises 9.4

1. -9 or 4 3. 5 and 13, or -13 and -5 5. 15 and 3
7. 12 and 13, or -13 and -12 9. $3, 4$ 11. 11 and 13, or -13
and -11 13. 5 yds., 17 yds.   15. 8 in., 3 in.   17. 18 ft., 14 ft.
19. 3 in. or 2 in.   21. 12 shares   23. 16 yrs.   25. 36

Exercises 9.5

1. $-3, 3$ 39. $4, -4$ 5. $\pm \dfrac{2}{15}\sqrt{15}$ 7. $-2, 1$ 9. -1

11. $-8, 3$ 13. $-3, 2$ 15. $24, -1$ 17. $25, -7$ 19. $\dfrac{3}{4}, 5$

21. $\dfrac{3}{7}, 4$ 23. $-\dfrac{1}{3}, 2$ 25. $\pm\dfrac{\sqrt{2}}{4}$ 27. $1, -\dfrac{3}{8}$ 29. $7, \dfrac{20}{31}$

31. -2 33. $\dfrac{7 \pm \sqrt{53}}{2}$ 35. 1 37. $-2, -1$ 39. $\dfrac{45 \pm 5\sqrt{57}}{6}$

41. No real solution.   43. 40 km/hr.   45. 600 mph
47. Betty, 10 days; Jim, 15 days.   49. Large pipe, 6 hrs.; Small pipe,
10 hrs.   51. $17, -7$ 53. 5 and 7, or $\dfrac{1}{5}$ and $\dfrac{11}{5}$ 55. 21 sheep
57. 23 poles/mile   59. 4 and 5, or 5 and 6 61. -8 or 2

Exercises 9.6

1. $\{28\}$ 3. $\{12\}$ 5. $\{22\}$ 7. $\{-2\}$ 9. $\varnothing$ 11. $\{3, 4\}$

13. $\{7\}$ 15. $\varnothing$ 17. $\{9\}$ 19. $\left\{-\dfrac{3}{4}\right\}$ 21. $\{10\}$ 23. $\{0\}$

25. $\{0\}$ 27. $\{2\}$ 29. $\left\{\dfrac{5}{3}\right\}$ 31. $\left\{\dfrac{25}{36}\right\}$ 33. $\{2\}$

35. $\{-2, 6\}$ 37. $\{-10, 4\}$ 39. $\{-8, -2\}$ 41. $\varnothing$
43. $\left\{-\dfrac{1}{3}, 3\right\}$ 45. $\{7\}$ 47. $\left\{\dfrac{7}{3}, 7\right\}$ 49. $\left\{-\dfrac{11}{2}, -\dfrac{5}{2}\right\}$ 51. 2

53. 5 or 1 55. $\dfrac{9}{2}$ 57. 2 59. 20 cm

Exercises 9.7

1. $\{x : -3 < x < 3\}$

3. $\{x : x \geqslant 4 \text{ or } x \leqslant -4\}$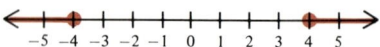

5. $\{x : -4 < x < 4\}$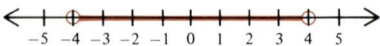

7. $\{x : x < -5 \text{ or } x > 2\}$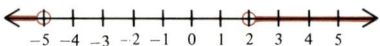

9. $\{x : x < -3 \text{ or } x > -2\}$

11. $\{x : x < -2 \text{ or } x > 7\}$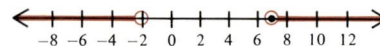

13. $\{x : -6 \leqslant x \leqslant 3\}$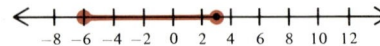

15. $\{x : -4 < x < 4\}$

17. $\{x : -2\sqrt{3} < x < 2\sqrt{3}\}$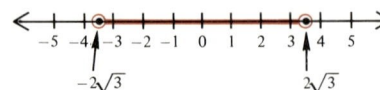

19. $\left\{x : x < -\dfrac{3}{2} \text{ or } x > \dfrac{4}{3}\right\}$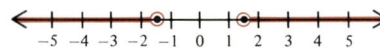

21. $\{x : 1 < x < 4\}$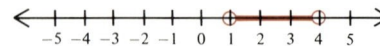

23. $\left\{x : x < -\dfrac{1}{2} \text{ or } x > 3\right\}$

25. $\left\{x : -\dfrac{2}{3} < x < 2\right\}$

27. $\left\{x : x \leqslant -\frac{2}{3} \text{ or } x \geqslant -\frac{1}{2}\right\}$

29. $\left\{x : -\frac{4}{3} < x < \frac{3}{4}\right\}$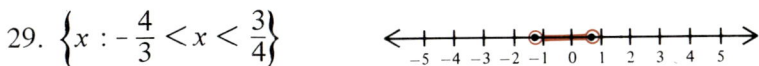

31. $\varnothing$

33. $\{x : x \text{ is a real number}\}$

35. $\left\{x : -\frac{2}{3} < x < \frac{5}{4}\right\}$

37. $\left\{x : \frac{3}{2} \leqslant x \leqslant \frac{9}{4}\right\}$

39. $\varnothing$

Exercises 9.8

1. $\{x : x > 2 \text{ or } x < -2\}$

3. $\{x : x \text{ is a real number}\}$

5. $\{x : -3 \leqslant x \leqslant 3\}$

7. $\{x : x \leqslant -2 \text{ or } x \geqslant 8\}$

9. $\{x : x < 1 \text{ or } x > 9\}$

11. $\{x : x < -3 \text{ or } x > 2\}$

13. $\{x : x \leqslant -51 \text{ or } x \geqslant 31\}$

15. $\{x : x \leqslant -3 \text{ or } x \geqslant 3\}$

Review Exercises: Chapter 9

1. $\{3, -3\}$ 3. $\{2, -2\}$ 5. $\{2\sqrt{5}, -2\sqrt{5}\}$ 7. $\varnothing$ 9. $\{2, -2\}$
11. $\{3\sqrt{2}, -3\sqrt{2}\}$ 13. $\{4, -4\}$ 15. $\{\sqrt{3}, -\sqrt{3}\}$ 17. $\{8, -4\}$
19. $\{\sqrt{11}, -\sqrt{11}\}$ 21. $\{3, 9\}$ 23. $\{4, -6\}$ 25. $\{-6, -8\}$

27. $\{6, -8\}$ 29. $\{0, -8\}$ 31. $\left\{\frac{2}{3}, 1\right\}$ 33. $\left\{\frac{5+\sqrt{13}}{6}, \frac{5-\sqrt{13}}{6}\right\}$

35. $\{7 - \sqrt{41}, 7 + \sqrt{41}\}$ 37. $\left\{\frac{5-\sqrt{13}}{6}, \frac{5+\sqrt{13}}{6}\right\}$ 39. $\{3, 12\}$

41. $\{4, 1\}$ 43. $\left\{-\frac{5}{3}, -1\right\}$ 45. $\left\{\frac{2\pm\sqrt{17}}{2}\right\}$ 47. $\left\{\frac{-1\pm\sqrt{5}}{2}\right\}$

49. $\{-11, 5\}$ 51. $\left\{\frac{3\pm\sqrt{17}}{2}\right\}$ 53. $\left\{\frac{-5\pm\sqrt{41}}{4}\right\}$ 55. $\left\{\frac{-3\pm\sqrt{17}}{4}\right\}$

57. $\varnothing$ 59. $\{-3\}$ 61. $\left\{\frac{-2\pm\sqrt{22}}{3}\right\}$ 63. $\{2\pm\sqrt{5}\}$ 65. $\varnothing$

67. $\{-7, -1\}$ 69. $\left\{-\frac{1}{3}, 2\right\}$ 71. $\{-8, 3\}$ 73. $\{5, -3\}$

75. $\left\{\frac{2}{3}, -\frac{3}{5}\right\}$ 77. $\{4, -4\}$ 79. $\{-8, -4\}$ 81. $\{0, 4\}$

83. $\{10, -8\}$ 85. $\{-9, -2\}$ 87. $\left\{-\frac{7}{2}, \frac{7}{2}\right\}$ 89. $\{1, 9\}$

91. $5, -5$ 93. $0, -\frac{7}{2}$ 95. $7, -4$ 97. $\frac{2}{3}, -\frac{2}{3}$ 99. $4, -\frac{5}{2}$

101. $-7, 1$ 103. $-\frac{1}{3}$ 105. $1\pm\sqrt{3}$ 107. No real solution.

109. -5.5 111. $-\frac{1}{5}, \frac{7}{2}$ 113. No real solution.   115. $-3\pm\sqrt{6}$

117. $2 \pm \sqrt{5}$ 119. $\dfrac{-2 \pm \sqrt{3}}{2}$ 121. 7, 9   123. 5   125. 10 feet

127. Base, 21 cm; altitude, 10 cm   129. 4 m   131. 1 m
133. Small square, 7 cm; large square, 8 cm   135. $2\sqrt{5}$ sec. ≈ 4.5
sec.   137. 25 sec.   139. 5 sec.   141. $\{-4, 2\}$ 143. $\{3, -1\}$

145. $\left\{4, \dfrac{13}{10}\right\}$ 147. $\{5, 2\}$ 149. $\left\{7, \dfrac{-10}{7}\right\}$ 151. $\{4, -4\}$

153. $\left\{\dfrac{1}{3}\sqrt{6}, -\dfrac{1}{3}\sqrt{6}\right\}$ 155. $\left\{\dfrac{11 \pm 4\sqrt{10}}{3}\right\}$ 157. $\left\{\dfrac{-3 \pm \sqrt{89}}{2}\right\}$

159. $\left\{\pm \dfrac{3}{23}\sqrt{161}\right\}$ 161. 30 km/hr., 45 km/hr.

163. Going, 35 mph; coming, 45 mph.   165. 56 mph and 70 mph
167. 30 km/hr.   169. Maria, 8 days; Carla, 12 days.   171. 9 days
173. 8   175. 3 m, 4 m   177. $\{12\}$ 179. $\{1\}$ 181. $\{6\}$
183. $\{5, -5\}$ 185. $\{9\}$

187. $\{5\}$ 189. ϕ 191. $\left\{7\dfrac{1}{4}\right\}$ 193. $\{7, -11\}$ 195. $\left\{\dfrac{5}{4}, -\dfrac{13}{4}\right\}$

197. $\left\{4, \dfrac{4}{3}\right\}$ 199. 3 or -1 201. 1 or 3   203. $\dfrac{1}{3}$ or -1

205. 2   207. 19

209. $\{x : x < -4 \text{ or } x > -1\}$

211. $\left\{a : -\dfrac{5}{2} \le a \le -1\right\}$

213. $\{3\}$

215. $\{x : x \le -1 \text{ or } x \ge 3\}$

217. $\left\{a : a \le -\dfrac{1}{2} \text{ or } a \ge 1\right\}$

219. $\{x : x \text{ is a real number}\}$

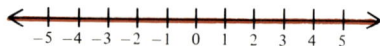

221. $\{x : 0 \leqslant x \leqslant 1\}$

223. $\{x : -2 \leqslant x \leqslant 2\}$

225. $\{x : x \geqslant 16 \text{ or } x \leqslant -10\}$

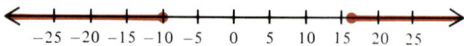

227. $\{x : x < 0\}$

Cumulative Review G

1. $\frac{5}{8}x^2 + \frac{11}{12}x + \frac{5}{2}$ 3. $a^2 - \frac{1}{4}$ 5. $9abc - 20a + bc$

7. $(3 + b)(2a + 3)$ 9. $\frac{1}{2}(a + 3)(a + 4)$ 11. $(x - 2)(x + 2)(x^2 + 4)$

13. $\frac{x - 8}{x + 2}$ 15. $\frac{1}{x + 3}$ 17. $\{6\}$ 19. $\left\{-\frac{100}{11}\right\}$ 21. $\{x : x < 1\}$

23. $\left\{-\frac{5}{13}\right\}$ 25. $\{-6, 1\}$ 27. $\{-7, 7\}$ 29. $\{x : 2 < x < 4\}$

31. $\varnothing$ 33. $S_1 = \frac{RS_2}{TS_2 - R}$ 35. $\{(4, -5)\}$ 37. $5y\sqrt{3y}$

39. $\frac{2\sqrt{3}}{3}$ 41. $12\sqrt{15}$

43.

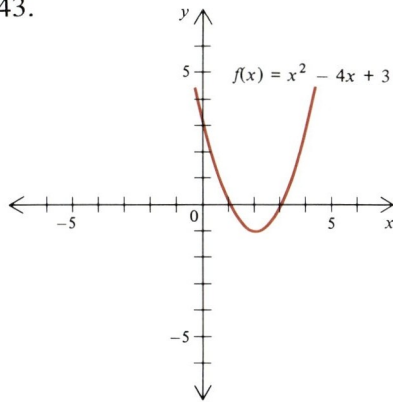

$f(x) = x^2 - 4x + 3$

45. 15 hrs., 20 hrs.   47. 20   49. 50 lbs., \$1.28; 100 lbs., 80¢

INDEX